MANUFACTURING TECHNOLOGY—I

MANUFACTURING TECHNOLOGY—I

By

C. ELANCHEZHIAN
Senior Faculty
Department of Mechanical Engineering
Sri Sairam Engineering College
Chennai-600044
Tamil Nadu.

B. VIJAYA RAMNATH
Senior Faculty
Department of Mechanical Engineering
Sri Sairam Engineering College
Chennai-600044
Tamil Nadu.

UNIVERSITY SCIENCE PRESS
(An Imprint of Laxmi Publications Pvt. Ltd.)
BANGALORE • CHENNAI • COCHIN • GUWAHATI • HYDERABAD
JALANDHAR • KOLKATA • LUCKNOW • MUMBAI • RANCHI
NEW DELHI • BOSTON, USA

Published by :
UNIVERSITY SCIENCE PRESS
(*An Imprint of Laxmi Publications Pvt. Ltd.*)
113, Golden House, Daryaganj,
New Delhi-110002

Phone : 011-43 53 25 00
Fax : 011-43 53 25 28

www.laxmipublications.com
info@laxmipublications.com

Price : **Rs. 175.00** *Only.*

First Edition : 2005
Second Edition : 2006, *Reprint* : 2008, 2009

OFFICES

India

© **Bangalore**	080-26 61 15 61
© **Chennai**	044-24 34 47 26
© **Cochin**	0484-239 70 04
© **Guwahati**	0361-254 36 69, 251 38 81
© **Hyderabad**	040-24 65 23 33
© **Jalandhar**	0181-222 12 72
© **Kolkata**	033-22 27 43 84
© **Lucknow**	0522-220 95 78
© **Mumbai**	022-24 91 54 15, 24 92 78 69
© **Ranchi**	0651-221 47 64

USA

Boston
11, Leavitt Street, Hingham,
MA 02043, USA

UMT-9342-175-MANUFACTURING TECHNOLOGY-I
Typeset at : Jay Emm Communications, Chennai-600030

C. 1412 /010/04
Printed at : Akashdeep Printers, Delhi.

PREFACE TO THE SECOND EDITION

This edition of **"Manufacturing Technology—I"** based on the latest syllabus for B.E./B.Tech., Mechanical, Production, Industrial Engineering and Manufacturing Engineering for Anna University as well as other universities. This book will also be useful to students appearing in various professional and competitive examinations.

This book is an attempt to provide all necessary information about Metal Casting Processes, Fabrication Processes, Bulk Deformation Processes, Sheet Metal Forming Processes and Forming and Shaping of Plastics. The subject matter has been presented in a simple and systematic way with understanding of the topics. To apprise the student about expected type of questions which can asked in the examination important Summary and Review questions have been provided at the end of each chapter.

We are grateful to our management who gave their continuous support to bring out this book successfully.

We also thank the publishers who have come forward to publish this book enabling us to release it in time.

Suggestions for the improvement of this book are eagerly welcome and would be incorporated in the next edition to make the book more useful.

—Authors

CONTENTS

Chapter 4 SHEET METAL FORMING PROCESS 195–244

Dedicated

to

Our Beloved Parents

and

Future Engineers

CHAPTER - 1

METAL CASTING PROCESSES

1.1 INTRODUCTION

Foundry or casting is one of the metal forming processes in which molten metal is poured into a pre-shaped mould cavity and allowed to cool. On cooling the metal solidifies and acquires the configuration of the mould. The solidified object is called "Casting".

The term founding refers both the solidified object as well as the process. Foundry is a section of a factory where the foundry or casting process is carried out.

1.2 SAND CASTING / FOUNDRY PROCESS

Foundry process consists of pattern making, making moulds, melting the metal and pouring the metal into the moulds. Thus making castings, founding or casting is the process of forming metallic products by melting the metals pouring it into a cavity of the required shape, known as the "Mould" and allowing it to solidify. In this way the molten metal takes the shape of the mould.

The product of the foundry is the casting which may vary from a fraction of a kilogram to several tonnes. Practically all ferrous and non-ferrous metals and alloys can be cast.

A shop where moulding, casting and their related processes are conducted is called "Foundry Shop".

1.3 SAND MOULD - TYPES OF MOULDING

1.3.1 Green sand moulding:

The step by step procedure of making a green sand mould is shown in the figure 1.1.

A two piece pattern is used. One half of the pattern in placed on a moulding board. The drag box is placed around the pattern.

20mm layer of facing sand is first placed around the pattern.

Then the drag is filled up with green sand.

Ramming is done by rammer.

Sand is filled upto the top of box.

Ramming is completed. Excess sand is removed by strike off bar.

Vent holes are made.

Then the drag is tilted upside down.

Parting sand is placed on the upper side to prevent two pieces of Pattern from sticking to each other.

Top half of the pattern is placed correctly in position.

Cope box is placed correctly in position on the drag.

Sprue pin and riser pin are placed in position.

Then the cope is filled up with moulding sand.

Ramming is completed in the cope.

Top surface of sand is smoothened. Vent holes are made. Sprue pin and riser pin are removed. Pouring basin is cut. Cope and drag are separated.

The top surfaces of cope and drag are smooth finished with trowel.

Draw spike is driven into pattern pieces and shaked slowly. Pattern pieces are withdrawn slowly.

Gate is cut to the mould.

Repair work is carried out in the mould if necessary.

Surfaces of the mould are coated with graphite to give smooth surface to the casting.

Core is set in position.

The two moulds are assembled in correct position.

A pouring weight is placed.

Now the mould is ready for pouring.

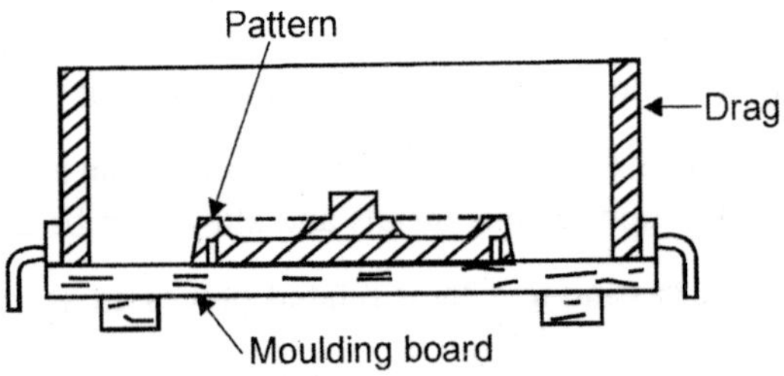

(i)

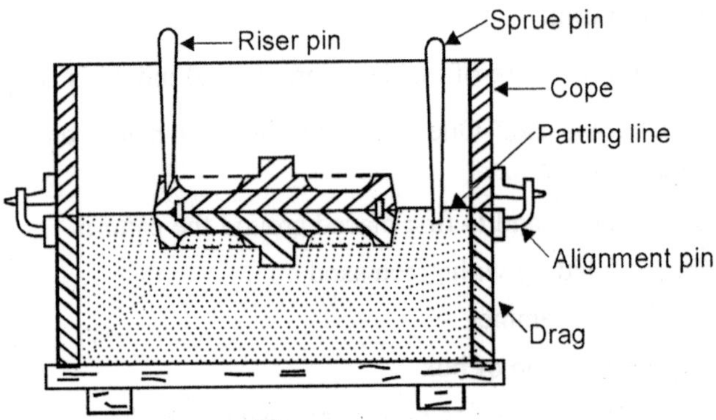

(ii)

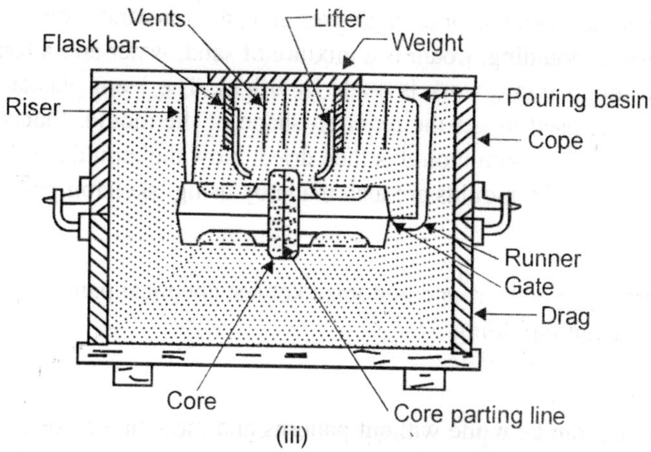

(iii)

Figure 1.1 Making a Green-sand Mould

1.3.2 Dry sand moulding:

The step by step procedure of making dry sand mould is the same as that of green sand moulding. The following points are to be added.

After the mould is made, the mould is (dried) heated. The heating is done by oxy acetylene flame for large moulds. Small moulds are heated in ovens.

Application:

Dry sand moulds are used for large castings such as engine cylinders, engine blocks and mill rolls.

Advantages:

1. Dry sand mould is stronger than green sand mould. Therefore it is not damaged while handling.
2. Dimensional accuracy of dry sand mould is better than green sand mould.
3. Dry sand moulds are coated with a wash. Therefore surface finish of coating is good.

1.3.3 Loam moulding:

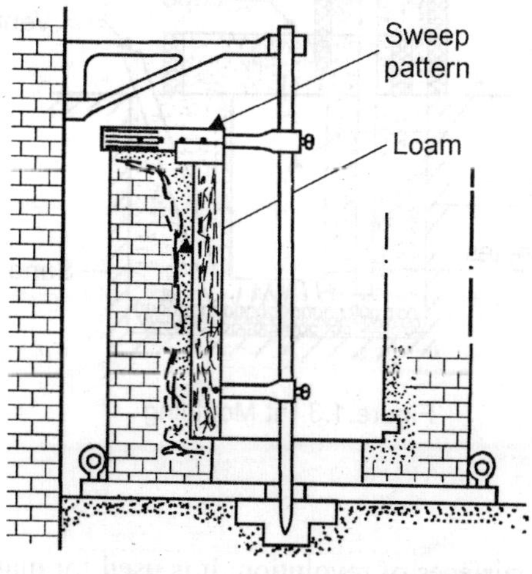

Figure.1.2 Loam Moulding

For some very large castings, moulding boxes and patterns cannot be used. Such large castings are made by the use of loam moulding. Loam is a mixture of sand, water and a large amount of clay. It is in a paste form. First a rough frame work is made by bricks. The loam plaster is applied over the brick work. A sweep pattern is used to get the required shape of the mould. Plates and bars can be used to reinforce the brick work. This loam mould is then dried. Now the mould is ready for pouring. Cores can be made by this process. Loam moulds are also made by using skeleton patterns.

Application:

Large cylinders, large bells, round bottomed kettles, chemical pans, large gears and other machine parts are produced by loam moulding.

Advantages:

1. Large castings can be made without patterns and moulding boxes.
2. Castings will have very good surface finish.
3. Accurate castings can be produced.

1.3.4 Floor moulding:

When castings are larger in size, it will be difficult to handle the mould. In this case, the moulding is done on the floor level itself. Medium size and large size castings are produced in this method. As the moulds can not be shifted, the metal is poured into the mould at the floor directly.

1.3.5 Pit moulding:

Very large and heavy castings like machine bodies are moulded in a pit. The pit is formed in the foundry floor itself. Pit acts as the drag part of the mould. A separate cope is used above the pit at the floor level. The sides of the pit are usually brick lined with refractory bricks. Vent pipes are provided from the bed to allow gases to escape. Sweep patterns are used for making the mould cavity.

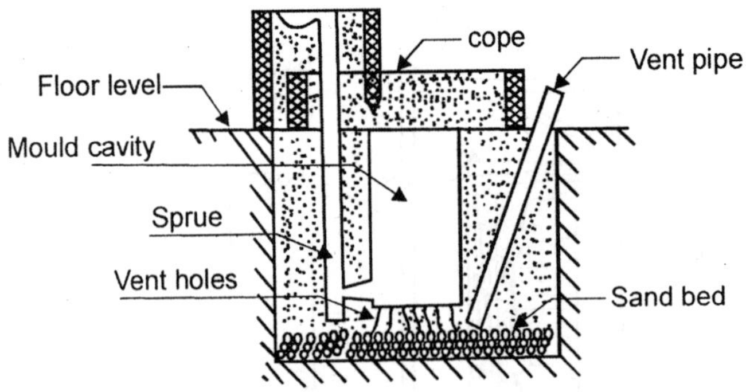

Figure.1.3 Pit Moulding

1.3.6 Sweep moulding:

It is used for object having surfaces of revolution. It is used for making moulds for objects whose axis are symmetrical.

1.3.7 Bench moulding:

Small moulds can be easily lifted by a single worker. These moulds are normally prepared on a moulding bench. The moulding bench is conveniently designed for storing moulding sand and keeping moulding tools, cores etc. The moulds prepared by bench moulding are shifted to a separate place for pouring the metal.

1.4 PATTERN

1.4.1 Types of Patterns:

The type of pattern selected for a particular casting depends upon the following conditions.

1. The shape and size of casting,
2. The number of castings required, and
3. The method of moulding employed.

The common types of patterns are discussed in detail in the following pages:

(i) Solid or Single Piece Pattern:

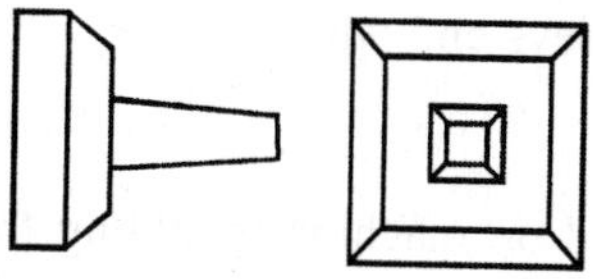

Figure.1.4 Solid Pattern

A solid or single piece pattern, as shown in Figure.1.4, has a compact form. It has no joints, partings or loose pieces in its construction. It has usually one broad surface that serves as a parting surface in the mould. This type of pattern is used for a limited number of castings because most of the moulding operations like parting surface formation, cutting of gating system, providing runners and risers, withdrawal of pattern etc. is done by hand.

(ii) Split Pattern:

The most of the patterns are not made in a single piece because of the difficulties encountered in moulding them. In order to eliminate this difficulty, some patterns are made into two or more pieces. A pattern consisting of two pieces is called a two piece split pattern, as shown in Figure 1.5. One-half of the pattern rests in the lower part of the moulding box known as drag and the other half in the upper part of the moulding box known as a cope. The line of separation of the parts is called parting line or parting surface.

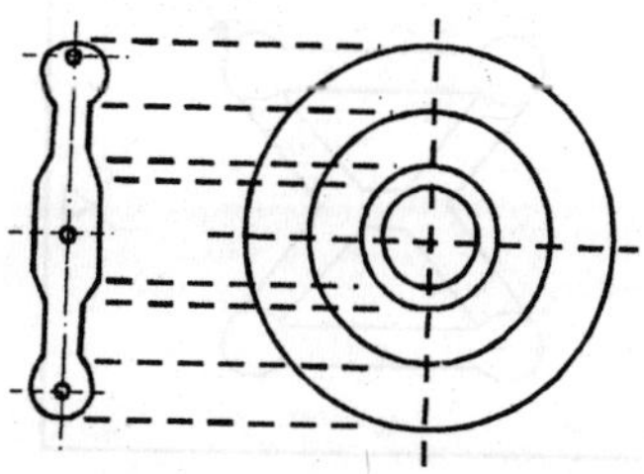

Figure 1.5 Split Pattern

Sometimes a pattern for complex castings is made in three parts as shown in Figure 1.6. Such a pattern is called a multi-piece pattern. A three piece pattern requires a moulding box with three parts, the middle one being called as cheek.

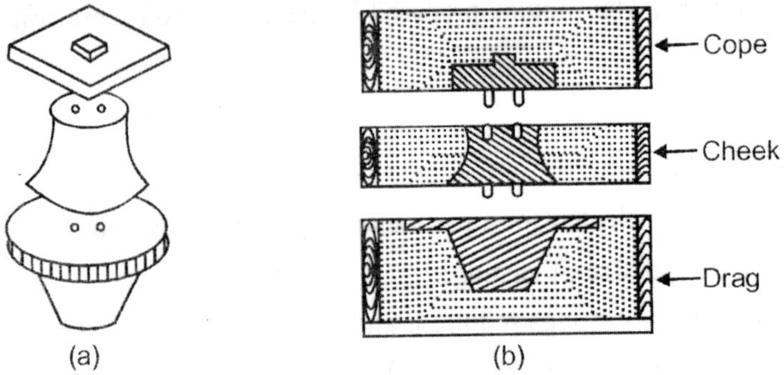

Figure 1.6 Multipiece Pattern

The split patterns are commonly used for casting of spindle, cylinders, steam valve bodies, water stop-cocks and taps, bearings, small pulleys and wheels.

(iii) Match Plate Pattern:

The match plate patterns are used on machine moulding for quantity production of castings. A single pattern or a number of patterns may be mounted on a match plate. When the cope and drag portions of the split pattern are mounted on opposite sides of the wooden or metal plate (usually aluminium plate), the pattern is called a match plate pattern. The gates and runners are permanently fastened to the drag side of the plate in their correct positions in order to form a complete match plate. When the match plate is lifted off the mould, all patterns are withdrawn and the gates and risers are completed in one operation.

Figure 1.7 shows a match plate pattern upon which the patterns of two small dumb bells are mounted. Though the cost of construction of the match plate patterns is quite high, yet it is easily compensated by the high rate of production, greater dimensional accuracy and minimum machining requirement of casting.

Application:

★ Piston rings of I.C. Engines are produced by this process.

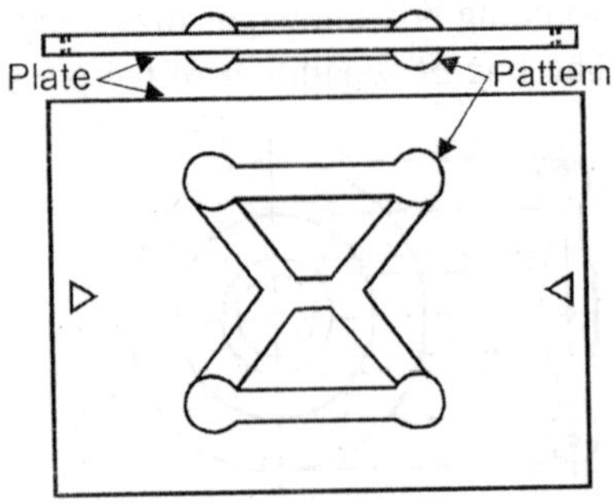

Figure 1.7 Match Plate Pattern

(iv) Cope and Drag Pattern:

When very large castings are to be made, the complete mould becomes too heavy to be handled by a single operator. In order to ease this problem, the cope and drag pattern is used. It is nothing but a two-piece pattern, split on a convenient joint line. One part is moulded in a cope and the other part in a drag of the moulding box.

(v) Loose Piece Pattern:

Sometimes a pattern has to be made with projections or overhanging parts, as shown in Figure.1.8. These projections make the removal of the pattern difficult. Therefore, such projections are made in loose pieces and are fastened loosely to the main pattern by means of wooden or wire dowel pins. These pins are taken out during the moulding operation. After moulding, the main pattern is withdrawn first and then the loose piece is removed with the help of a lifter.

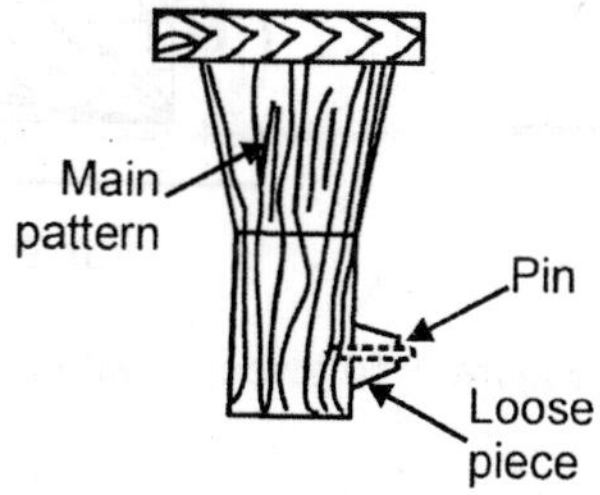

Figure 1.8 Loose Piece Pattern

(vi) Gated Pattern

The gated pattern, as shown in Figure.1.9 is used for mass production of small castings. When a number of small patterns are placed in a single mould, then each pattern may be provided with a gate with it. It consists of pieces of wood or metal fixed to the patterns to form the runner and rising channels in the mould. In this way, full supply of the molten metal flows into every part of the mould.

The gated pattern eliminates hand cutting of gates and thus makes the moulding easy, If a group of patterns is to be placed in one mould, the gate pattern has a further function of holding the patterns in the proper position with respect to each other.

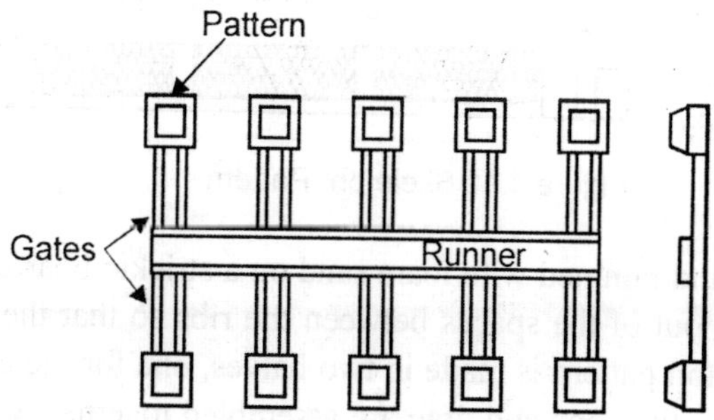

Figure 1.9 Gated Pattern

(vii) Sweep Pattern:

The moulds of large size and symmetrical in shape particularly of circular section can be easily prepared by using a sweep instead of a full pattern. It is not considered as a true pattern when compared with others. A sweep is a template of wood or other material which has the shape corresponding to the shape and size of casting. It is arranged to rotate about a central axis by mounting it on a spindle. The sand is rammed in place with a cavity of approximately similar shape and size to that of the required mould. Now by revolving the sweep, the desired shape of the mould is generated. The sweep and the spindle are then removed and the hole in the centre is filled up.

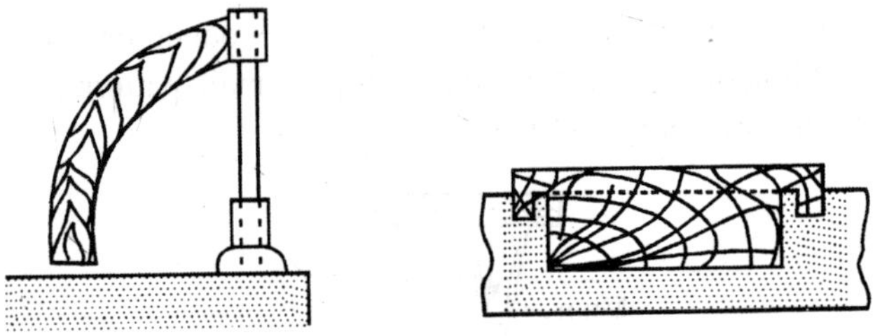

(a) Curved sweep (b) Straight sweep

Figure 1.10 sweep pattern

Figure.1.10(a) shows a curved sweep which may be used to form part of the mould for a large cast-iron kettle. Figure.1.10(b) shows a straight sweep for any type of groove or ridge. The principal advantage of this pattern is that it eliminates expensive pattern construction.

(viii) Skeleton Pattern

When a few and large-sized castings are required, it is not advisable to use a large solid pattern of that size, as it will require a lot of wood and time to make a full pattern. In such cases, a skeleton pattern in the hollow form, consisting of a wooden frame and strips, as shown in Figure.1.11 is used.

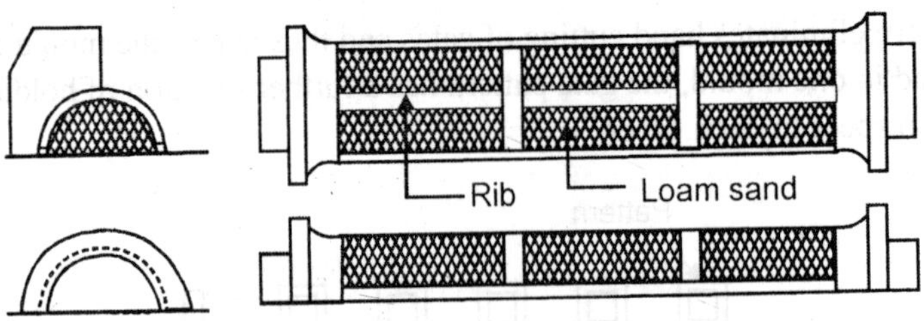

Figure 1.11 Skeleton Pattern

The frame work is filled and rammed with loam sand on a strickle board, as shown in Figure.1.11 is used to scrap the excess sand out of the spaces between the ribs so that the surface is even with the outside of the pattern. The skeleton pattern is made in two halves, one for the cope and the other for the drag. After taking the impression, the cope and drag are assembled together with the core in position to form the complete mould. The skeleton patterns are used for castings of hollow cast iron pipe (as shown in Figure 1.11), bends, valve bodies and boxes.

(ix) Shell Pattern

The shell pattern, as shown in Figure.1.12, is used largely for drainage fittings and pipe work. This type of pattern is usually made of metal mounted on a plate and parted along the centre line, the two sections being accurately doweled together. The short bends are usually moulded and cast in pairs. The shell pattern is a hollow construction like shell. The outside shape is used as a pattern to make the mould while the inside is used as a core box for making cores.

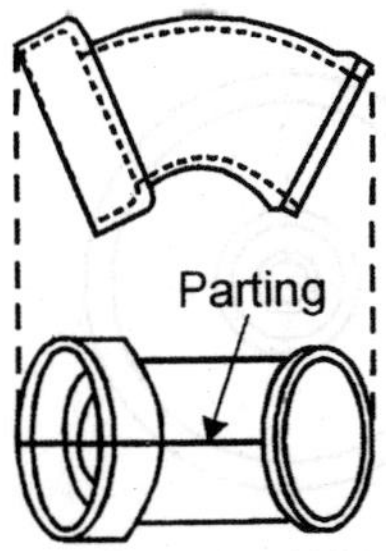

Figure 1.12 Shell Pattern

(x) Segmental pattern

The segmental patterns, as shown in Figure.1.13, are also known as part patterns. These patterns are sections of a pattern so arranged as to form a complete mould by being moved to form each section of the mould. These patterns are usually applied to circular work, such as rings, wheel rims, gears etc.

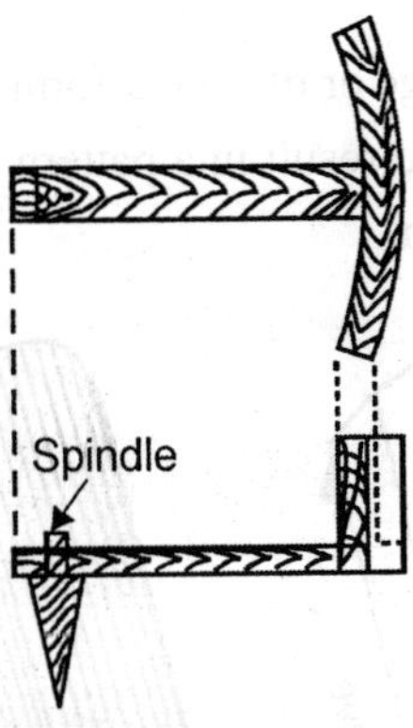

Figure 1.13 Segmental Pattern

When a mould is to be made using a segmental pattern, a vertical spindle is firmly fixed in the centre of a drag flask. The mould bottom is rammed, and swept well. Now the segmental pattern is fastened to the spindle. The moulding sand is rammed between the outside of the pattern and the flask and on the inside except at the ends of the pattern. After ramming one segment, the next segment is rammed, until entire perimeter of the mould is completed.

(xi) Follow Board Pattern:

A follow board pattern as shown in Figure.1.14, is used for solid patterns having an irregular parting line. It may be used with either single or multiple gated patterns. The patterns requiring follow boards are usually somewhat difficult to make as a split pattern. The board is routed out so that the patterns rests in it up to the parting line and this board then acts as a moulding board for the first moulding operation.

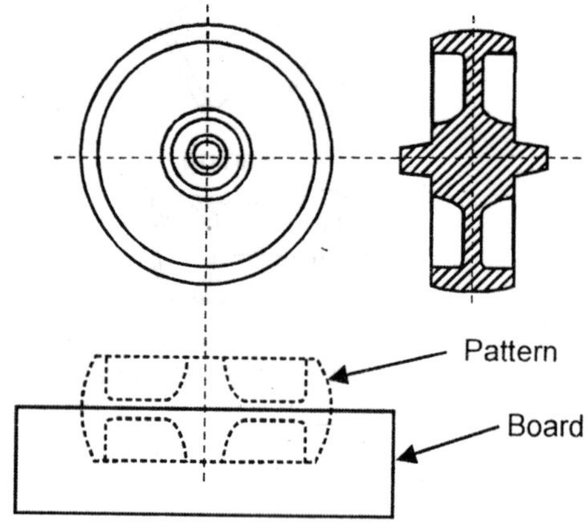

Figure 1.14 Follow Board Pattern

(xii) Lagged-up-Pattern:

When a pattern or core box is so large or of such a form that it cannot be made economically from a solid piece or when such a method would result in a pattern of little strength or excessive weight, it is necessary to use a lagged or staved pattern.

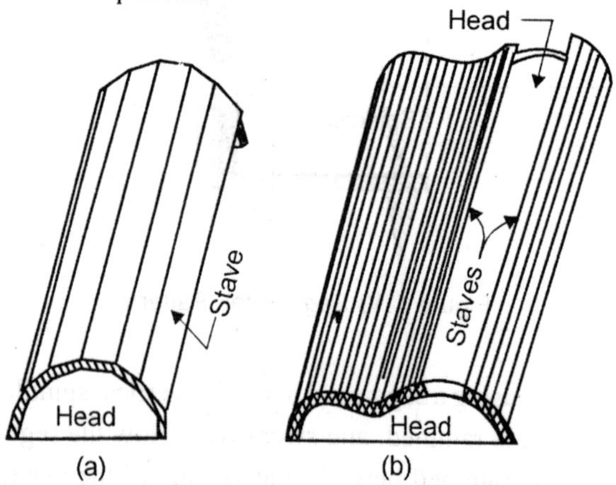

Figure 1.15 Lagged Up Pattern

The cylindrical patterns e.g., barrels, pipes or columns are built up with lag or stave construction to ensure proper shape. The lags are longitudinal strips of wood bevelled on each side for making the joint tight outside. These lags are glued and nailed or screwed to the wooden end pieces called heads, as shown in Figure.1.15(a). Such a construction gives the maximum amount of strength and permits building close to the finished outline of the pattern so that there is comparatively little excess stock to be removed to bring it to the required form. In building pieces that are not cylindrical, the heads are frequently cut to follow a line parallel to the finished outline of the pattern, so that the staves, when fastened in place, will closely approximate this outline, as shown in Figure.1.15(b).

(xiii) Left and Right Hand Pattern:

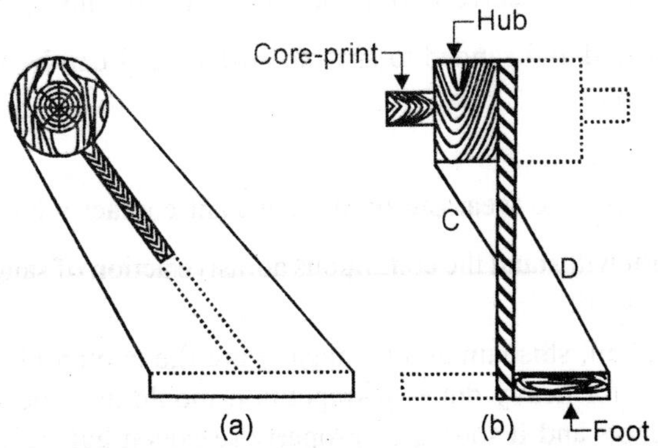

Figure 1.16 Left and Right Hand Pattern

Some patterns are required to be in pairs, and when their form is such that they cannot be reversed and have the centres of hubs, bosses, etc., opposite and in line, then they must be made right and left hand separately. A few examples where a pair of left and right hand patterns is required are legs for wood turning lathe, J-hangers for overhead shafting, legs for garden bench, legs for paddle type sewing machine, brackets for luggage racks in the railway carriages etc.

A bracket, as shown in Figure.1.16, is an example of left and right hand pattern. The hub and the foot flangers are fastened with screws and are moved from side to side to make the pattern right and left hand, as shown by dotted lines in the figure. The ribs are also loose. The rib C is reversed, but right and left hand ribs D are required because of the angle at the lower edge.

1.5 PATTERN MATERIALS

There are several materials commonly used in the construction of patterns. The modern pattern making employs the use of materials that can be easily shaped and are durable. The type of pattern material chosen depends upon the following factors:

(a) The design of casting,

(b) The number of castings to be produced,

(c) The type of casting and moulding process used in foundry, and

(d) The degree of accuracy and surface finish required.

Following are the commonly used materials for pattern making

i) *Wood:*

It is widely used material for patterns. It is used when small number of castings upto 500 are to be produced. The wood as a pattern material has the following advantages and disadvantages.

Advantages:

(a) It is cheap and light.

(b) It can be easily worked and shaped as desired.

(c) It can be cut and fabricated into numerous forms by gluing, bending and carving.

(d) It is easily planed and sanded to smooth surface and can be preserved fairly for a long time with shellac.

Disadvantages:

(a) It has the tendency to wear out by the constant contact with damp sand.

(b) Since it can not withstand the continuous abrasive action of sand, therefore it is unsuitable for repetition work.

The pine, deodar, teak, kail, shisham and mahogany are the commonly used woods for patterns, but mahogany is more durable for heavy duties of repetition moulding. The wood selected for pattern making should be free from knots and it should be properly seasoned before use.

2. *Metal:*

When large number of castings are required, the pattern is made of a metal. The metal patterns are more durable, have longer life and produces moulds to a close dimensional accuracy.

Following materials are commonly used for pattern making.

(a) *Cast iron:* It is cheap, easy to file and fit. It is strong, gives a good smooth mould surface with sharp edges and is resistant to abrasive action of the sand. But the cast iron patterns are heavy and are easily broken. Though the cast iron can be cast into fairly thin sections, but there is a danger that the edges of thin sections, which cool very quickly, may be too hard to machine. The patterns for small and light work are not ordinarily made of cast iron because of their tendency to rust.

(b) *Brass:* It is strong, tough, rust proof and takes a better surface finish than cast iron. It has the ability to withstand wear of the moulding sand. It can be used successfully for casting very thin sections without the danger of encountering unmachineable edges. The small patterns can be easily rectified, built up or fitted by soldering. Since the brass patterns are heavier and costlier than cast iron, therefore it is used only for patterns or sections of patterns and not for the plates on which the patterns are mounted.

(c) *Aluminium and its alloy:* The aluminium and its alloy having 80 percent aluminium, 14 percent zinc, 3 percent copper, 1.5 percent iron and 1.5 percent silicon, are widely used for production moulding due to its light weight, strength, corrosion resistance and machining characteristics. Since the aluminium and its alloys are softer than brass or cast iron, the patterns are liable to be damaged by accidental contact with the steel ramming tools. Though they are fairly resistant to sand abrasion, yet it is advisable to protect critical locations with steel or iron insertions.

(d) *White metal:* It is not used to any great extent of patterns, but is the best material used for lining stripping plates, because it can be cast into the narrow cavities between the master pattern and frame of the stripping plate. It is an alloy of lead, copper and antimony and has a low melting point about 260° C. The white metal patterns are soft and easily worn away by the moulding sand. Since it is commonly used for die casting, it may be called as die casting alloy.

Advantages of Metals :

★ More durable and accurate in size than wooden patterns.

★ Have a smooth surface.

★ Can with stand rough handling.

★ Donot deform in storage.

Limitations :

★ Expensive as compared to wood.

★ Not easily repared.

★ Heavier than wood.

★ Ferrous patterns can get rusted.

3. Plaster:

The plaster of paris or gypsum cement is successfully used as a pattern material because it can be easily casted into intricate shapes and can be easily worked. It has a high compressive strength (upto 30 MPa) and controlled expansion. It is used for making small patterns and core boxes of intricate shapes.

4. Plastics:

The thermo-setting resins, usually phenolic resin plastics, are now gaining popularity as a plastic material for patterns. The plastics are light in weight, have high strength, high wear resistance, high corrosion resistance, low solid shrinkage and have very smooth surface finish. In order to make plastic patterns, first of all a master pattern from wood is made and then a plaster of paris mould is prepared from this master pattern. The plastic resin is now poured in the mould and heated to some specific temperature. After solidification, it produces a plastic pattern. In large scale production, numerous plastic patterns of similar shape and size can be mounted on match plates so as to reduce the unit cost of casting.

Advantages of Plastics :

★ No moisture absorption.

★ Smooth surface of patterns.

★ Moulding sand stick less to plastics than to wood.

★ Economical in cost and labour.

5. Wax:

It is used for investment casting process. It helps in imparting a high degree of surface finish and dimensional accuracy to castings. The wax pattern is prepared by pouring heated wax into the split moulds or a pair of dies while the latter are kept water cooled. The dies after having been cooled down are parted off. The wax pattern is now taken out and used for moulding.

NOTE: Sometimes old castings are used as a pattern in case of emergency. The casting to be used as patterns must be well cleaned and made smooth so that they can be drawn out of the mould easily. In order to increase the thickness and other dimensions so as to provide for finish, machining and shrinkage etc., the surfaces of old casting are lagged with thin strips of wood or leather.

1.5.1 Moulding Tools / Foundry Tools:

Moulding Box or Flask:

It is a box like structure made of rectangular walls without any bottom or top cover. It may be made from wood but metallic (cast iron) boxes are more common in foundry shops.

It is made in two parts held in alignment by dowel pins. The upper part is called 'cope' and the lower part 'drag'.

Moulding Board:

It is smooth wooden board suitable for the moulding box. It is used to rest the moulding box and the pattern box while starting to make the mould.

Shovel:

It has a broad iron part fitted with a long wooden handle. It is used for mixing and tampering moulding sand. It is also used for carrying the moulding sand from the sand pit to the moulding box.

Riddle:

It is a metal screen used for removing foreign materials such as stones, nails, etc., from the moulding sand. It has a circular or square wooden frame with a wire mesh at the bottom. The riddle is specified by the diameter of the frame and mesh number.

Rammer:

It is used for pressing or ramming the moulding sand uniformly around the pattern. It has got two ends. One end is wedge shaped and is known as "peen". The other end is cylindrical in shape known as "butt end".

Peen end is used for ramming the sand near the corners. When the box is full and surface of the sand is to be rammed, the butt end is used.

Trowels:

Various sizes and shapes of trowels are available and each of them does different work. It consists of a metal blade fitted with a wooden handle. Square trowel is used for shaping and smoothing the square corners. Finishing trowel has round corner and is used for shaping and smoothening larger surfaces. It is also used to repair the damaged portions of the mould.

Slick:

It is a spoon shaped double ended trowel. It is used for repairing or smoothening the mould surface.

Sprue pin:

It is a tapered wooden piece. It is used for making a sprue hole in the mould. The size of the sprue pin depends upon the size of the mould. Molten metal is poured into the mould through the sprue hole.

Bellow:

Bellow is used for blowing out dust and sand particles from the mould.

Vent wire:

As the molten metal is poured into the mould, the moisture in the sand gets evaporated due to the heat of the metal. The steam and gas try to go out. If proper vent hole is not made in the moulding sand, then a part of it would pass through the mould also causing "blow holes" and other defects in the casting.

To safeguard against this, vents are made through the sand with the help of a steel wire whose one end is pointed sharp. This wire is called "vent wire".

Draw spike:

The pattern cannot be picked up with hand. Draw spike is a pointed steel rod with a ring at one end. It is used to remove the pattern from the mould.

Mallet:

The mallet is similar to wooden mallet used in carpentary shop. It is used for driving the draw spike into the pattern and then lifting it from the mould.

Lifter:

It is used for removing loose sand which has fallen to the bottom of the mould after the withdrawal of the pattern. It is also used to repair broken surfaces of the mould.

Strike-off bar:

It is made of metal or wood. It has a straight edge which is perfectly smooth. It is used to strike-off excessive sand from the top of mould after ramming.

Swab:

It is a small fibre brush. It is used for moisturizing the edges of the pattern before removing it from the mould. Otherwise, the pattern would carry some of the moulding sand with it, thus requiring major repair of the cavity.

Gate cutter:

It is used to cut a feeding gate to connect the runner hold with the mould cavity.

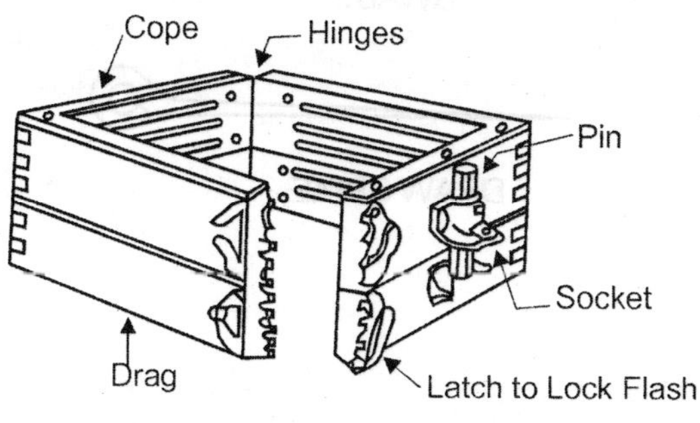

Figure 1.17 Snap Flask

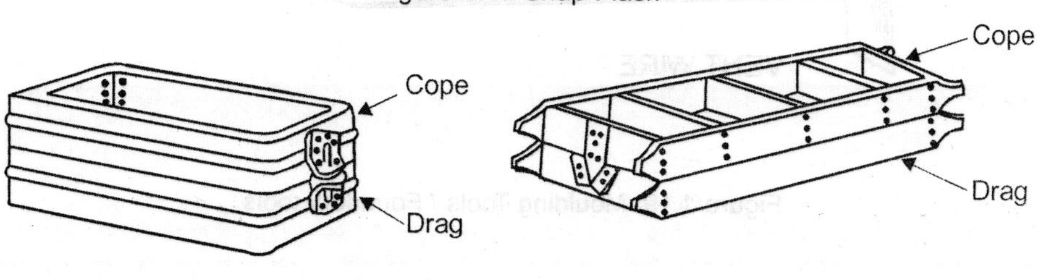

Figure 1.18 Two Part Box Flask

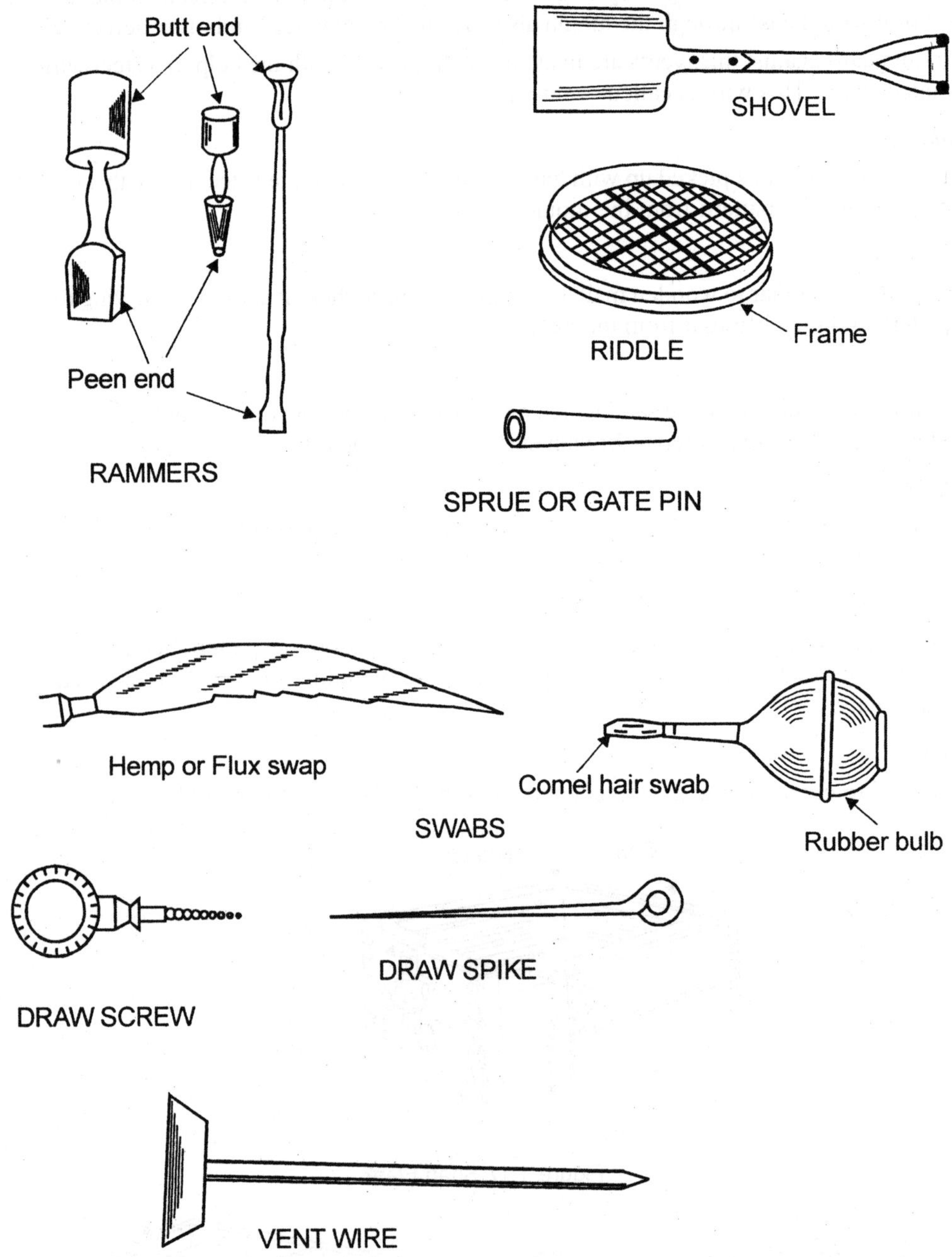

Figure 1.19 Moulding Tools / Foundary tools

1.6 PATTERN MAKING ALLOWANCES

Patterns are not made to the correct sizes of the castings. This extra sizes given on the pattern is called pattern allowance. Pattern allowances are given for the purpose of compensating the metal shrinkage, to provide extra metal for machining, to avoid metal distortion, for easy withdrawal of pattern from mould and for rapping.

If allowances are not given on the pattern, the castings will become smaller (ie) undersize.

The various types of pattern allowances are:

1. Shrinkage allowance
2. Machining or finish allowance
3. Draft allowance
4. Distortion or camber allowance
5. Rapping allowance

1.6.1 Shrinkage Allowance:

When molten metal is poured into the mould it will solidify and cool. During cooling, the metal will shrink and reduce in size. To compensate this, the pattern is made larger than the required casting. This extra size given on the pattern for metal shrinkage is called shrinkage allowance. If shrinkage allowance is not given, the casting will become smaller. Shrinkage allowance for various metals are:

Cast iron, Melleable Iron	.....	10 mm per metre
Steel	.....	20 mm per metre
Brass, Cu Aluminium	.....	15 mm per metre
Zinc, Lead	.....	25 mm per metre

1.6.2 Machining or Finish Allowance:

All the castings are to be machined to finish the rough surfaces and to remove excess metal. For this purpose the pattern is made larger than the required casting. This extra size added to the pattern for machining purpose is called machining or finish allowance. In general, a machining allowance of 3 mm is given for ferrous metals, iron and steel. For non-ferrous metals like brass and aluminium a machining allowance of 1.5 mm is given.

1.6.3 Draft Allowance:

Figure 1.20 Draft Allowance

When a pattern is removed from the mould, the edges of the mould may be damaged. This can be avoided by giving a taper on the vertical surfaces of the pattern. This slight taper given on the vertical sides of the pattern is called draft allowance. The draft usually given in each case depends upon,

(a) Length of the vertical side

(b) Intricacy of the pattern

(c) Method of moulding.

1.6.4 Distortion or Camber Allowance:

Some castings, because of their shape and size may warp or distort during cooling. This is due to uneven metal thickness. To avoid this, the shape of the pattern is slightly bent in the opposite direction. For example, a casting of U shape may distort and the legs become divergent, instead of being parallel. To avoid this, the legs are slightly bent convergent. So after cooling, the legs will become parallel and correct.

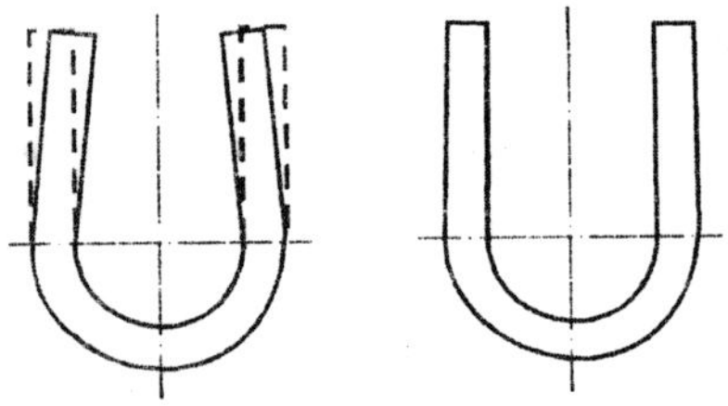

Figure 1.21 Camber Allowance

1.6.5 Rapping Allowance:

Before removing the pattern from the mould, the pattern is shaked from side to side. This is called rapping. So the mould cavity may become large. To avoid this, the pattern is made slightly smaller. This allowance given on the pattern is called rapping allowance. This is a negative allowance.

1.7 MOULDING SAND

1.7.1 Types of Moulding Sand:

Sand is the principal moulding material in the foundry shop. It is used for all types of casting, non-ferrous metals, iron or steel. This is because sand possesses vital properties required for moulding. It can withstand the high temperature of the molten metal. It has high permeability i.e, it allows gases and air to escape from the mould when metal is poured. These properties can be adjusted by varying the composition of sand.

Moulding sand has three constituents. They are:

1) Sand, 2) Binder and 3) Additive

a) Sand:

Sand contains a silica, clay and moisture. Silica is the main constitutent of sand. Silica has 80 to 90 percent silicon dioxide. Silica gives refractoriness to the sand.

Clay is another constituent of sand. Clay gives bonding strength to the sand. Most sand has 5 to 20 percent clay. Moisture is the water added to the sand. It gives the binding action. In general 2 to 8 percent water is added to the sand.

b) Classification of Moulding Sand:

Moulding sand is classified as follows:

i) Natural sands, ii) Synthetic sands, iii) Special sands.

i) Natural Sands:

These are also called green sands. These sands are available at river beds. These contain 80 to 90% of silica, 5 to 10% of alumina or clay and small percentages of lime and magnesia. The clay present in the sand acts as a bond between sand grains. Natural sand is also prepared by crushing and milling soft yellow sand stone.

Natural sand is less costly. It is easily available. It has a wide range of grain sizes and shapes. The only limitation of this sand is that it may get fused with metal. It has high coefficient of expansion. Natural sand is generally used for making light castings in ferrous and non-ferrous metals.

ii) Synthetic Sands:

Synthetic sands are prepared from various ingredients. These are prepared with natural sand with some clay as base. Washed sand without any clay is also used as base. The required properties of moulding sand can be obtained by adding required quantities of bentonite, water and other materials like oxide of iron, calcium and magnesium.

Synthetic sand is used in mechanized production, machine moulding and high pressure moulding. This sand has the following advantages over natural sands.

1) More uniform grain size.

2) Easy mouldability.

3) Higher refractoriness.

4) Easy control of properties.

iii) Special Sands:

Special sands are moulding sands having specific qualities such as refractoriness, high heat conductivity and low expansion. Using special sands, good quality casting with fine surface finish can be produced. The following are the most often used special sands.

1) Zircon sand ($Zr\ Si\ O_4$): It is a green colored sand mainly used for making cores for brass and bronze castings. It is also used as facing sand. It is a fine sand having high heat conductivity, good refractoriness and high density.

2) Olivine sand ($Mg\ Fe\ Si\ O_4$): It is a green colored sand. It has moderate refractoriness. It is mainly used for non ferrous castings of intricate shapes.

3) Chamotte sand: This is made from clay by crushing. It has low heat transfer and low expansion. It is mainly used for heavy iron and steel castings.

4) Chromite sands: These sands have high heat conductivity and good refractoriness. These sands are used for making castings where chilling is required. They are also used for facing sands in steel casting.

1.8 PROPERTIES OF MOULDING SAND

Proper moulding sand must possess six properties. It must have porosity, flowability, collapsibility, adhesiveness, cohesiveness or strength, and refractoriness. The properties are determined, not only by the chemical composition, but by the amount of clay matter in the sand, by its moisture content, and lastly by the shape and size of the silica sand grains.

1.8.1 Porosity (or) Permeability:

Molten metal always contains a certain amount of dissolved gases, which are evolved when the metal freezes, Also, the molten metal, coming in contact with the moist sand, generates steam or water vapour. If these gases and water vapour evolved by the moulding sand do not find opportunity to escape completely through the mould they will form gas holes and pores in the casting.

The sand must, therefore, be sufficiently porous to allow the gases or moisture present or generated within the moulds to be removed freely when the molten metal is poured. This property of sand is called porosity or permeability.

Factors that Influence Permeability of Moulding Sand:

★ Quality and Quantity of Clay and quartz in the mixture.

★ Moisture content.

★ Degree of Compactness.

1.8.2 Flowability:

Flowability of moulding sand refers to its ability to behave like a fluid so that, when rammed, it will flow to all portions of a mould and pack all-round the pattern and take up the required shape. The sand should respond to different moulding processes. High flowability is required of a moulding sand to get compacted to a uniform density and to obtain good impression of the pattern in the mould. Good flowability is very essential where energy for compaction during ramming is transmitted through the sand mass as in machine moulding. Flowability increases as clay and water content increase.

1.8.3 Collapsibility:

After the molten metal in the mould gets solidified, the sand mould must be collapsible so that free contraction of the metal occurs, and this would naturally avoid the tearing or cracking of the contracting metal.

1.8.4 Adhesiveness:

The sand particles must be capable of adhering to another body, i.e., they should cling to the sides of the moulding boxes. It is due to this property that the sand mass can be successfully held in a moulding box and it does not fall out of the box when it is removed.

1.8.5 Cohesiveness or Strength:

This is the ability of sand particles to stick together. Insufficient strength may lead to a collapse in the mould or its partial destruction during conveying, turning over or closing. The mould may also be damaged during pouring by washing of the walls and core by the molten metal. The strength of moulding sand must, therefore, be sufficient to permit the mould to be formed to the desired shape and to retain the shape even after the hot metal is poured in the mould.

This property of sand in its green or moist state is known as green strength. A mould having adequate green strength will retain its shape and will not distort or collapse even after the pattern is removed from the moulding box.

The strength of sand that has been dried or baked is called dry strength. It must have the strength to withstand erosive forces due to molten metal, and retain its shape.

The strength of Moulding Sand depends on:

- ★ Grain size.
- ★ Shape of sand particles.
- ★ Moisture content.
- ★ Density which varies with contents of clay binders in sand mixture.

1.8.6 Refractoriness:

The sand must be capable of withstanding the high temperature of the molten metal without fusing. Moulding sands with a poor refractoriness may burn on to the casting. Refractoriness is measured by the sinter point of the sand rather than its melting point.

1.9 CORES AND CORE MAKING

Cores are separate shapes of sand that are generally required to form the hollow interior of the casting or a hole through the casting. Sometimes cores are also used to shape those parts of the casting that are not otherwise practical or physically obtainable by the mould produced directly from the pattern. The core is left in the mould in casting and is removed after the casting.

1.9.1 Core Requirements:

Cores are subject to severe conditions since after pouring the mould they are surrounded on all sides by molten metal. Consequently, they must possess some special characteristics which are as follows:

1. Cores must be strong enough to retain its shape without deforming, to withstand handling and to resist erosion and deformation by metal during filling of the mould.
2. Cores must be permeable to allow the core gases to escape easily.
3. Cores should be highly refractory in nature to withstand high temperature of the molten metal.
4. Cores must be sufficiently low in residual gas-forming materials to prevent excess gas from entering the metal.
5. Cores must be stable with a minimum of contraction and expansion to make a true form of the casting.
6. Cores should be sufficiently collapsible, i.e., they should disintegrate and collapse after the metal solidifies, to minimise strains on the casting and to facilitate removal of the core from the casting during shakeout.

1.9.2 Core Sands:

The ingredients of core sands are sand and binder. Core sands are usually silica, but zircon, olivine, carbon and chamotte sands are used. Sand that contains more than 5 percent clay cannot be used for cores. Excessive clay reduces not only permeability but also collapsibility.

The type of sand to be used depends on the size of the core and also on the pouring temperature of the metal. The important factors to be considered are the size, shape, and distribution of sand grains, the clay content, and the mineralogical composition of clay particles. The smaller the grain size of the sand used in cores, the smoother will be the surface of the casting made in contact with these cores. Furthermore, sand with rounded grains is more satisfactory for cores than with angular grains for higher values of permeability.

Core binders:

Core sand has no natural bond, as almost pure sand is used for preparing cores. Hence some other materials are added to the sand to act as binders which cement the sand particles together before and after the cores are baked. They impart strength, and sufficient degree of collapsibility in addition to giving good permeability, ample refractoriness and other qualities that cores should possess.

Core binders most commonly used are of three general types: (a) binders that harden at room temperature, (b) binders that require baking to harden, and (c) clays. Various commercial binders are available in the market which consist mainly of oils, cereals, dextrine, resins, sulphite-liquor, molasses and protein.

The action of thermoplasting binders such as rosin and pitch depends on the amount of heat which liquifies and disperses the binder in the sand. Rosin is a form of resin and is obtained by distillation and extraction from pinewood. Thermosetting resin binders are also becoming common owing to their high strength, low gas formation, collapsability, and resistance to moisture absorption. The resin binders usually favoured are of phenol, urea, and furan.

Core oils, as mentioned earlier, are more popular as they are very economical and produce better cores. The chief ingredients of these core oils are vegetable oil, for instance, linseed oil and corn oil. Sometimes, specially processed mineral oils are also added to achieve special properties. They are mainly used in making what are called oil sands for cores.

Oil sands can be used for almost any sand casting application. The following is a typical formulation that contain cereal and a small amount of clay (bentonite); this mixture, which is mixed in a muller, has proved satisfactory for the casting of gray iron and several other metals.

Sand (by weight)	95.80 percent
Cereal flour	1.01 percent
Core oil	1.17 percent
Water	1.86 percent
Bentonite	0.16 percent

Oil sands are very popular because:

1. They are easy to use for core making.
2. An oil-sand core is more collapsible than a clay bonded core known as loam core.

3. The green and dry strength of the oil sand mixture can be controlled by quite simple variations in the proportions of dextrine and oil respectively.
4. The baked cores are very hard and not easily damaged in handling or during closing of the moulds.

1.9.3 Core Making:

Core making consists of the following operation:

(1) Core sand preparation,

(2) Core moulding,

(3) Core baking, and

(4) Core finishing.

(1) Core sand preparation:

The first consideration in making a core is to mix and prepare the sand properly. The mixture must be homogeneous so that the core will be of uniform strength throughout. The core sands are generally mixed in (1) roller mills, and (2) core mixers. In the case of roller mills, the action of the muller and ploughs give a uniform and homogeneous mixing. Roller mills are suitable for core sands containing cereal binders, whereas the core sand mixer is suitable for all types of core binders.

(2) Core moulding:

Cores are then made manually or with machines. Normally, a core box is required for the preparation of cores. Green sand cores are made by ramming the sand mixtures into boxes, the interiors of which have desired shapes and dimensions (Figure 1.22). The methods used to ram core are usually done by machines. Core-making machines can be broadly classified as (1) core blowing machines, (2) core ramming machines, e.g., jolting, squeezing, slinging. The principles that they apply for making a sand mould is same to making a sand core. Except for low-volume production, or very large cores, the sand mixtures is compacted into the core box by core blowing that fills the core sand into the core box by compressed air. The degree of compactness necessary to produce a satisfactory core depends on the type of binder used and on the size and shape of the core.

Fragile and medium size cores are often reinforced with steel wires and rods so that they will have sufficient strength to resist the forces to which they are subjected within the moulds. In large cores, perforated pipes or arbors are used. In addition to giving the strength, they also serve as a large vent.

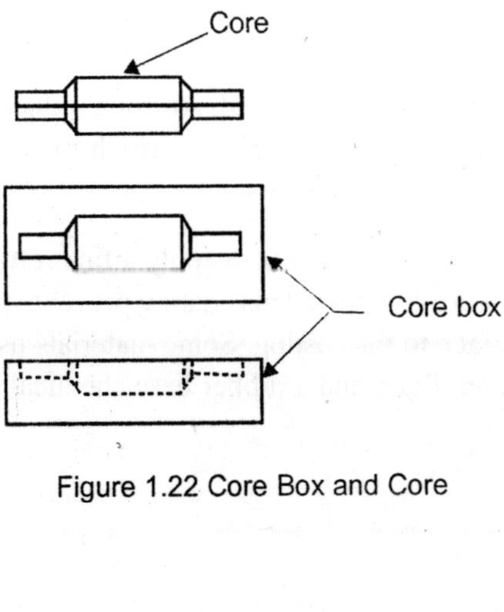

Figure 1.22 Core Box and Core

(3) Core Baking:

After the cores are prepared and placed on metal plate or core carriers, they are baked to remove the moisture and to develop the strength of the binder in core ovens at temperature from 150°C to 400°C, depending on the type of the binder used, the size of the cores, and the length of baking time. The core plates called driers are usually perforated to permit the circulation of gases and to lessen the sticking of the cores to the supports. As a rule, one or more vents are provided in cores to assist in the discharge of gases from the interior of the core prints. According to the kind of production, the core-drying ovens are classified as:

1. Core ovens.

   (i) batch type.

   (ii) continuous type.

2. Dielectric bakers.

(i) Batch type ovens are used for baking variety of cores in batches. The cores are generally placed either in drawers or in portable racks which are finally placed in the ovens.

(ii) Continuous type ovens are preferred for a high rate production of small and medium-size cores. The core racks move slowly through these ovens on a continuous chair or rail. The loading and unloading is continuous, with backing time controlled by the rate of travel of the conveyor. Because the oven is operated steadily at a fixed temperature and because all cores are in the heated zone for the same length of time, only a fairly narrow range of sizes can be baked properly.

Core drying ovens are usually heated by coal, coke, oil, gas or electricity.

2. Dielectric bakers can bake cores in a small fraction of the time required for baking the same cores in conventional ovens. Dielectric heating is employed in modern core ovens for high quality cores made from resin binders. The material to be heated dielectrically is placed between the parallel cement bonded asbestos plates or electrodes and a high frequency current is passed through it. The main advantage of these ovens is that they are faster in operation and a good temperature control is possible with them.

(4) Core Finishing:

After the baking operation, cores are smoothed. All rough places and unwanted fins are removed by filing. Some cores are made in two or more pieces which must be assembled usually by pasting together with dextrin or other water-soluble binders.

The last operation in the making of a core is to apply a fine refractory coating or core wash to the surface. This is sometimes called core-dressing. This coating prevents the metal from penetrating into the core and provides a smoother surface to the casting. Some materials used for core washes include finely ground graphite, silica, mica, zircon, flour, and a rubber-base chemical spray. Coatings may be applied to the core surfaces by brushing, dipping, or spraying.

1.10 TYPES OF CORES

The cores used in foundries are classified according to their shape and their position in the mould. The common types of cores, illustrated in Figure.1.23, are described below.

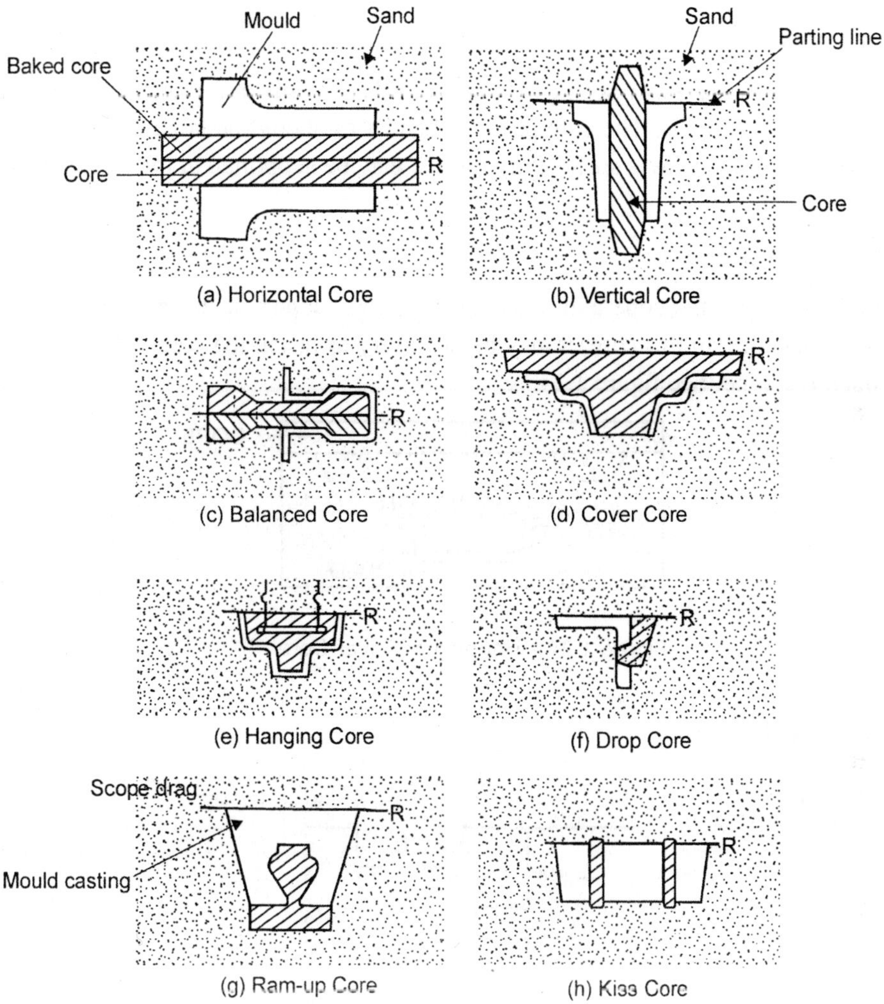

Figure 1.23 Different Types of Cores

1.10.1 Horizontal Cores:

The most common type is the horizontal core. The core is usually cylindrical in form and is laid horizontally at the parting line of the mould. The ends of the core rest in the seats provided by the core prints on the pattern. See figure 1.23(a).

1.10.2 Vertical Cores:

This is placed in a vertical position both in cope and drag halves of the mould. Usually top and bottom of the core are provided with a taper, but the amount of taper on the top is greater than that at the bottom.See figure 1.23(b).

1.10.3 Balanced Cores:

When the casting is to have an opening only on one side and only one core print is available on the pattern, a balanced core is suitable. The core print in such cases should be large enough to give proper bearing to the core. In case the core is sufficiently long, it may be supported at the free end by means of a chaplet.See figure 1.23(c).

1.10.4 Hanging and Cover Core:

If the core hangs from the cope and does not have any support at the bottom of the drag, it is referred to as a hanging core. In this case, it may be necessary to fasten the core with a wire or rod that may extend through the cope. See figure 1.24.

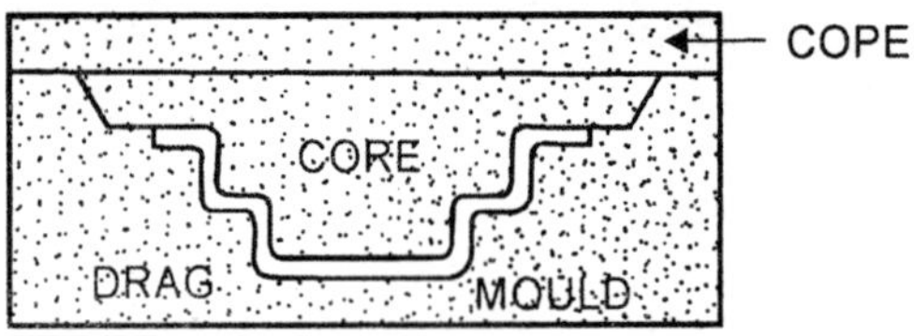

Figure 1.24 A Hanging or Cover Core

1.10.5 Wire Core/Kiss Core:

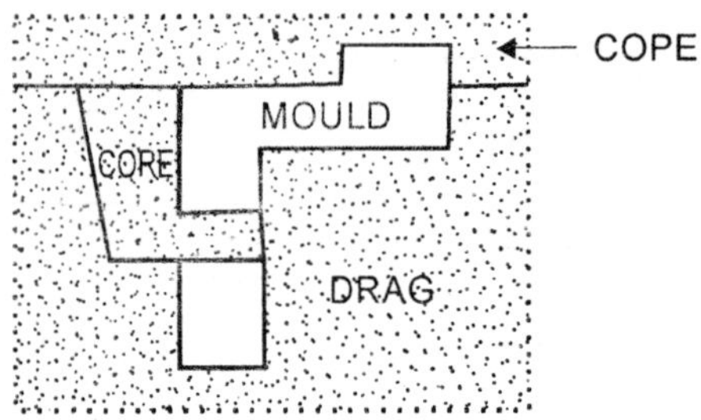

Figure 1.25 A Stop-Off Core

Figure (1.25). It is also known as a stop-off core. It is employed when a hole is desired to be produced in the casting at such a position that its axis falls either above or below the parting line. In such a core, its one side remains flush with the inner surface of the mould and the core, thus, acts as a stop-off. The back surface of this core is provided with enough taper for its easy location. Many other names, like tail core and saddle core, etc., are also given to this type of core according to its shape and use.

Apart from the above there are a few other types also, but they are not so commonly used. One such rarely used type is a ram up core, which is embedded in the mould. Such a need arises when placement of core is not possible after ramming. Similarly, when such a pattern is to be used which carries no core prints, the core is held between cope and drag simply due to the pressure put by the former. It is known as a Kiss core.

1.11 CORE OVEN - TYPES

1.11.1 Introduction:

Classified as "batch type" or "continuous type", core ovens are available in a variety of sizes. Continuous ovens are those through which the cores move slowly on a conveyor. The loading and unloading is continuous, and the baking time is controlled by the rate of travel of the conveyor. Thus, continuous ovens are most suitable for cores of approximately the same size. In batch-type ovens no movement of cores, batches, or parts occurs. Gas, oil, or electricity may be used for heating, and the temperatures should be closely controlled by suitable controlling instruments. Temperatures of around 450°F are used, depending upon the binder. Burners or heating elements should be properly spaced and air-circulated to help to keep all locations within a furnace at about the same temperature. Since moisture from cores will saturate the air inside an oven, some air is continuously exhausted by a blower and replaced by new air from outside the oven.

1.11.2 Dielectric Core Baking:

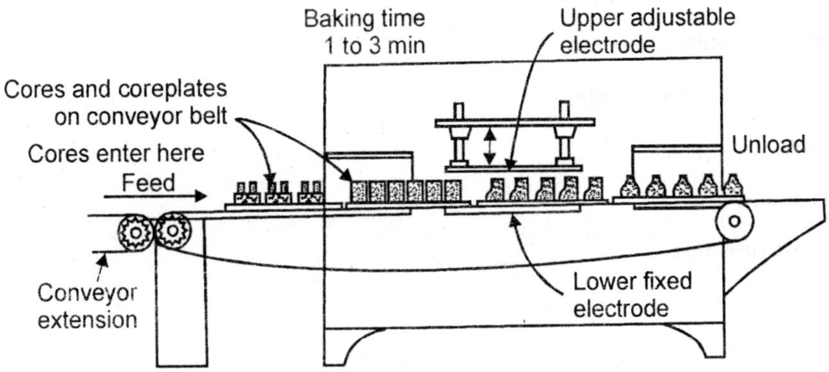

Figure 1.26 Dielectric Core-Baking Oven

Cores may be rapidly baked by dielectric heating. Dielectric heating should not be confused with induction heating. Induction heating is used for heating materials which are conductors of electricity, such as metals, and it is accomplished in a continuously changing magnetic field. Dielectric heating, on the other hand, is used for materials which are nonconductors of electricity. An alternating electric field is established between two parallel plates which function as an electrical condenser. The material to be heated is placed between these parallel electrodes. When the high-frequency alternating electric current, which alternates at around 15 million times per second, is turned on, heat is created within the molecules. Cores are heated uniformly throughout, with their interiors being heated just as rapidly as their exterior surfaces. A core binder, such as linseed oil, which requires oxidation, cannot be used to advantage here. The thermosetting synthetic-resin binders, which cure at around 250°F and which do not require oxidation, are well suited for dielectric heating. Small-sized cores can be baked in 30 seconds, while larger cores require no longer than a few minutes for complete baking. There is also less possibility that overbaking or under-baking will occur.

1.12 METHODS OF SAND TESTING

1.12.1 Introduction:

i) The need for systematic evaluation of the working qualities of moulding sands has led to the development of a wide range of sand control tests.

ii) A moulding sand is supposed to possess many properties (as discussed earlier) for its efficient functioning. These properties depend upon grain shape, size, distribution, and the content and types of binders, additives and moisture.

iii) Sand tests indicate the moulding sand performance and help the foundrymen in controlling the properties of moulding sands.

iv) Sand testing controls the moulding sand properties through the control of its composition.

v) Production of sound castings largely depends upon uniform and good quality of moulding sands.

vi) Sand control tests are performed on the sand which has been prepared and is ready to be transferred to the moulding section,

vii) Various sand control tests are:

1.12.2 The Tests on Moulding Sands:

i) Moisture content test.

ii) Clay content test.

iii) Grain fineness test.

iv) Permeability test.

v) Strength tests,

1. Green and Dry compression.
2. Green tensile.
3. Green and Dry shear.
4. Transverse (bending).

vi) Hot strength test.

vii) Refractoriness test.

viii) Mould hardness test, etc.

1.12.2.1 Testing of Moulding Sand:

i) Moisture content test:

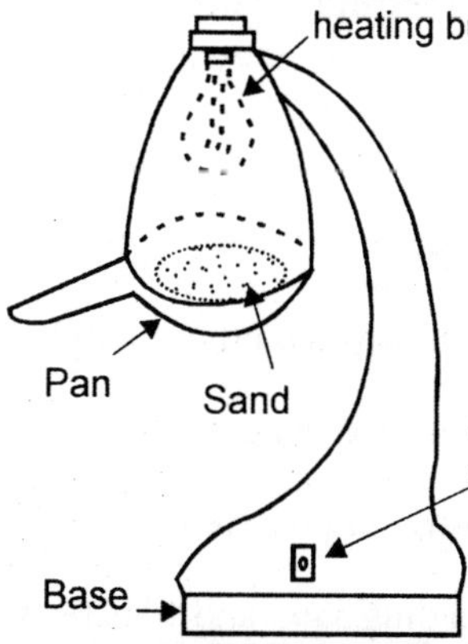

Figure 1.27 Moisture Determining Apparatus.

i) Water is one of the most important factors which controls properties of moulding sands.

ii) Low moisture content in the moulding sands does not develop strength properties.

iii) High moisture content decreases permeability and adds to other problems associated with moulding operation.

iv) Moisture content can be determined by the following methods:

1. Using direct reading moisture teller:

   This makes use of the reaction.

   $CaC_2 + 2H_2O = Ca(OH)_2 + C_2H_2$

   Calcium carbide reacts with the moisture content of the moulding sand and generates acetylene gas. The pressure of C_2H_2 provides a direct reading of the water content on the pressure gauge.

2. Using Electrode probe gauge.

3. Employing measurements of microwave absorption in compacted sand samples.

4. Using infrared heating. (Figure.1.27).

   Steps Involved

   i. 20 to 50 gms of prepared sand is placed in the pan and is heated by an infrared heater bulb for 2 to 3 minutes.

ii. The moisture in the moulding sand is thus evaporated.

iii. Moulding sand is taken out of the pan and reweighed.

iv. The percentage of moisture can be calculated from the difference in the weights of the original moist and the consequently dried sand sample.

ii) Clay content test:

i. Clay is responsible for bonding sand particles together.

ii. Clay influences strength, permeability and other moulding sand properties.

iii. The clay content of the moulding sand is determined as follows:

1. Dry thoroughly a small quantity of prepared moulding sand.
2. Seperate 50 gms of dry moulding sand and transfer the same to a wash bottle.
3. Add in it 475 cc of distilled water and 25 cc of 3% NaOH solution.
4. Using a rapid sand stirrer, agitate the whole mixture for about 10 minutes.
5. Fill the wash bottle with water up to the mark indicated on the same.
6. After the sand etc., has settled for about 10 minutes, siphon out the water from the wash bottle.

   Clay is dissolved in water (due to stirring action) and gets removed along with the same.
7. To the sand thus left in the wash bottle, add more water, stir the constituents again and let the sand settle.
8. Repeat step 7 till the water over the settled sand is clean.

   This assures that the whole of the clay has been removed from the sand.
9. Dry the settled down sand.
10. The clay content can be determined from the difference in weights of the initial (step 2) and final (step 9) sand samples.

iii) Grain Fineness test:

i. Grain fineness test determines the

1. Grain size,
2. distribution, and
3. grain fineness.

ii. The apparatus required for determining grain fineness (Figure.1.28) consists of a number of standard sieves mounted one above the other, on a power-driven shaker. The shaker vibrates the sieves and the sand placed on the top sieve gets screened and collects on different sieves depending upon the various sizes of grains present in the moulding sand.

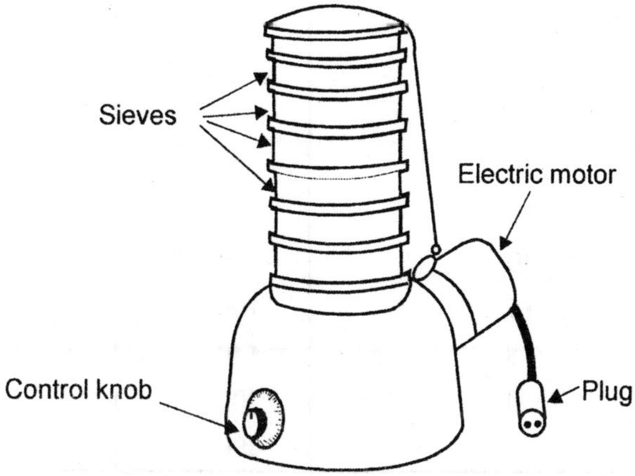

Figure 1.28 Grain Fineness Tester

iii. There are eleven standard sieves mounted one above the other and under the bottom-most sieve is placed a pan.

iv. The top sieve is the coarsest and the bottom-most sieve is the finest of all the sieves.

v. The in-between sieves are placed in order of fineness from top to bottom.

vi. A sample of dry sand out of which clay has been removed is placed in the uppermost sieve; sand is vibrated for a definite period of time, and the amount of same retained on each sieve is weighed and the percentage distribution of grains is computed.

vii. To obtain the AFS grain fineness number, each percentage is multiplied by a multiplier (i.e., factor). The resulting products are added and divided by total percentage of sand grains retained.

viii. ***Table.1*** Illustrates the method of calculating AFS fineness.

Table. 1.

Sieves-Mesh	Percent of sand retained	Factor	Product of col 2 and col 3
1	2	2	4
6	0	3	0
12	0	5	0
20	1	10	10
30	2	20	40
40	3	30	90
50	9	40	360
70	20	50	1000
100	30	70	2100
140	15	100	1500
200	8	140	1120
270	4	200	800
Pan	2	300	600
Total	**93**	—	**7620**

$$\text{AFS grain fineness number} = \frac{\text{Sum of products}}{\text{Total sum of the percentages of sand retained on pan and each sieve}}$$

$$= \frac{7620}{93} = 82$$

iv) Permeability test:

i. Permeability is that property of moulding sand which permits the escape of steam and other gases generated in the mould during hot metal pouring.

ii. Permeability depends upon.
   - — Grain size
   - — Grain shape
   - — Grain distribution
   - — Binder and its content
   - — Degree of ramming
   - — Water content of the moulding sand.

iii. Since permeability is the property of rammed sand, a standard sized sand specimen is first rammed by a specimen Rammer (refer Figure.1.29) and is then used in the permeability tester (Figure.1.30) for finding its (i.e, sand) permeability.

iv. Permeability of the sand specimen prepared as above, is determined by passing a given volume of air through the sand.

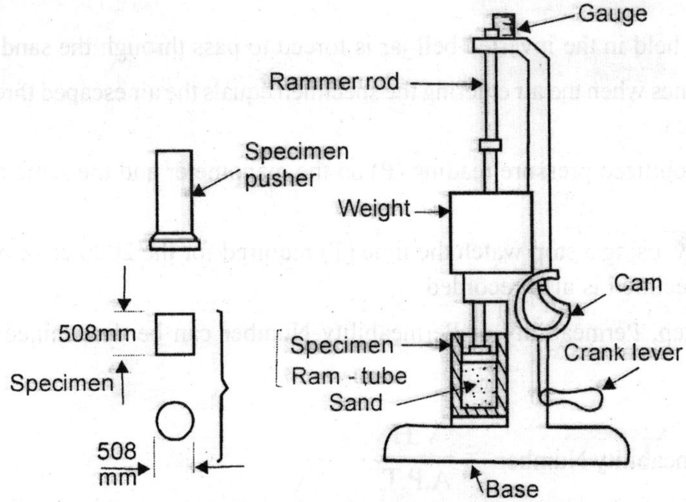

Figure 1.29 Sand Specimen Rammer

v. The permeability tester (Figure.1.30) consists of,

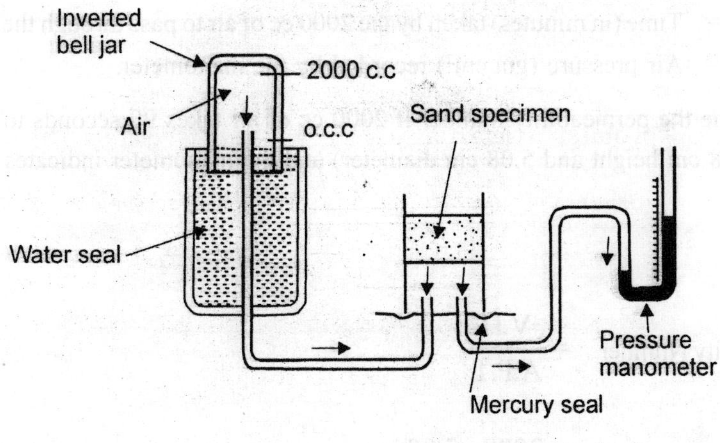

Figure 1.30 Permeability Tester

(1) An inverted bell jar, which floats in a water seal and can permit 2000 cc of air to flow.

(2) Specimen tube, to hold the sand specimen.

(3) A manometer (water gauge) to read the air pressure.

Sand permeability can be determined by TWO methods,

1. Standard method.
2. Rapid (shop) method.

Standard Method:

i) 2000 cc of air held in the inverted bell jar is forced to pass through the sand specimen.

ii) A situation comes when the air entering the specimen equals the air escaped through the specimen (to atmosphere).

iii) This gives a stabilized pressure reading (P) on the manometer and the same can be read on the vertical scale.

iv) Simultaneously, using a stop watch the time (T) required for the 2000 cc of air to pass through the sand of specimen is also recorded.

v) As the next step, Permeability or Permeability Number can be determined by the following relation.

$$\text{Permeability Number} = \frac{V.H}{A.P.T}$$

Where, V = Volume of air passed through the specimen = 2000 cc

H = Height of the specimen = 5.08 cm

A = Area of the specimen = 20.268 cm^2

T = Time (in minutes) taken by the 2000 cc of air to pass through the sand specimen.

P = Air pressure (gm/cm^2) recorded by the manometer.

Example 8.1: Determine the permeability number if 2000 cc of air takes 90 seconds to pass through a standard specimen (5.08 cm height and 5.08 cm diameter) and the manometer indicates an air pressure reading of 5 gm/cm^2.

Solution:

$$\text{Permeability Number} = \frac{V.H}{A.P.T}$$

$$= \frac{2000 \times 5.08}{\frac{\pi}{4}(5.08)^2 \times 5 \times 1.5} = 67\,(\text{Approx.})$$

v) Strength Test:

i. The strength of a foundry moulding sand is determined by:

— Compression

— Tensile

— Shear, and

— Transverse Tests.

ii. All these types of strength tests are used for sand control purpose, of course

iii. The most commonly used test is that of compression strength

iv. Specimen for compression, tension, transverse and shear testings can be made on the sand specimen tester (Figure.1.29) using different attachments.

v. Figure.1.31 shows the sand strength testing apparatus.

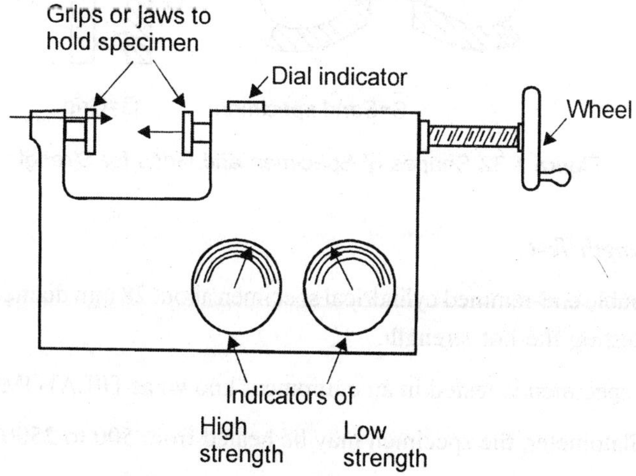

Figure 1.31 Strengths Testing Equipment

1. The specimen is held between the grips.
2. Hand wheel when rotated actuates a mechanism which builds up hydraulic (oil) pressure on the specimen.
3. Dial indicator fitted on the tester measures the deformation occuring in the specimen.
4. There are two (manometer) indicators. One is meant for use when testing low strength sands (say ordinary moulding sands) and the other for relatively high strength core sands.
5. Each indicator has three scales; one for reading compressive strength and the remaining two for recording tensile (or transverse) and shear strengths respectively.
6. The shape of the specimen for measuring compressive, shear, tensile and transverse strengths along with the shape of the specimen grips have been shown in figure 1.32.

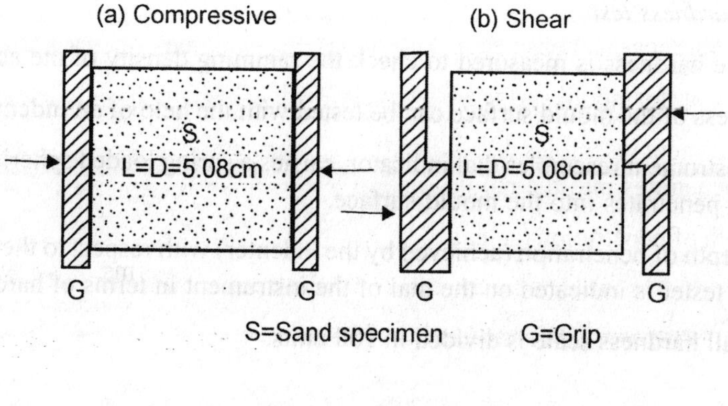

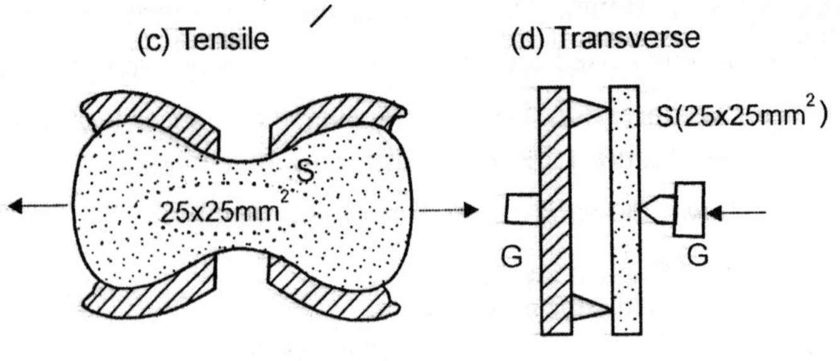

Figure 1.32 Shapes of Specimen and Grips for Strength Testings

vi) Hot Strength Test:

i. A double end-rammed cylindrical specimen about 28 mm diameter and 50 mm long is prepared for testing the hot strength.

ii. The specimen is tested in an equipment known as DILATOMETER.

iii. In dilatometer, the specimen may be heated from 500 to 2500°F and tested in compression.

vii) Refractoriness test:

i. Susceptibility of moulding sand to sintering can be studied with the help of a resistance heated curved platinum strip resting against the periphery of the cylindrical sand specimen.

ii. Refractoriness can also be tested as follows:

1. Prepare a cylindrical specimen of sand.
2. Fire it for two hours at 1550°C.
3. Observe the appearance and the dimensional changes, if any, (occured) in the specimen.
4. A good refractory sand retains specimen shape and shows very little (7.0% or less) expansion.
5. A less refractory-sand specimen will shrink and distort.

viii) Mould hardness test:

i. Surface hardness is measured to check the ramming density of the actual sand mould.

ii. Hardness of the mould surface can be tested with the help of an Indentation hardness tester.

iii. The instrument resembles dial indicator, carries a spring loaded spherical indenter (plunger) which penetrates into the mould surface.

iv. The depth of penetration (achieved by the indenter) with respect to the flat reference surface of the tester is indicated on the dial of the instrument in terms of hardness units.

v. The full hardness scale is divided in 100 units.

1.12.3 Tests on Core Sands:

i) Green strength test.

ii) Permeability test.

iii) Baked strength test.

iv) Hot strength test.

v) Retained strength test.

vi) Core hardness test.

vii) Moisture content test, etc.

1.12.3.1 Testing of Core Sand:

Already mixed and prepared core sand is used for carrying out sand control tests.

i) Green strength tests: (like tension, compression, transverse and shear) and

ii) Permeability tests: on the core sands are carried out in the same manner and using the similar equipments as employed for strength testings of moulding sands.

i. Core sand mixtures possess low green strength as compared to moulding sands and employ a vertically acting compression testing machine.

ii. Green compressive strength of core sand mixtures may normally range between 0.5-1.5 psi, (0.035 to 0.105 kg/cm^2).

iii) Baked strength test:

i. It is conducted on baked cores.

ii. The test determines, whether baking has developed optimum properties of the binder or not.

iii. Transverse (bending) and tensile tests are commonly carried out on baked cores.

iv. Cores are subjected to a controlled baking cycle and tested for tensile and transverse tests in a manner similar to that employed for moulding sands.

v. The baked tensile strength of cores may go up to 20 kg/cm^2.

iv) Hot strength test:

i. Cores are baked before testing.

ii. Hot strength of core sands is tested in the same manner as that of moulding sands.

iii. Dilatometer may be employed for the purpose.

v) Retained strength test:

i. Core sands exhibiting lower retained compression strength are comparatively easier to disintegrate during shake out operation.

ii. A standard core sand baked cylindrical specimen is heated to above 1000°F.

iii. Specimen is held at that temperature for about 10-12 minutes and is permitted to cool down to room temperatures.

iv. The cooled specimen is then tested for its compressive strength.

v. A specimen taking more than 0.70 kg/cm^2 compression will be difficult to shake out.

vi) Core hardness test:

i. It involves testing the hardness of both green and baked cores.

ii. Green core hardness can be tested by using the mould hardness tester having a larger indenter loaded with a light spring; this is because of the low strength of the cores as compared to that of moulds.

iii. The tester employed for finding baked core hardness uses a knife-edge indenter (because of the hardness of the baked cores) which is spring loaded and scratches the core surface.

iv. The tester is slowly drawn over a flat baked core surface and the hardness is directly read from the dial of the tester.

vii) Moisture content test:

i. Since some ingredients of core oils being volatile, evaporate at temperatures at which moisture evaporates, therefore it is difficult to estimate the moisture content by normal heating procedure as employed for finding the moisture content of the moulding sands.

ii. However, the moisture content of core sands can be found out by reacting the same (i.e., water of the core sand) with calcium carbide (refer section 8.2a).

1.13 MOULDING MACHINES

Hand moulding is a slow process. It is suitable only for making few castings. For producing large number of similar castings, moulding is done by moulding machines.

Moulding machines are used for the following reasons:

1. Producing moulds at faster rate.
2. Producing accurate moulds.
3. Reducing the production cost.
4. Getting better quality castings.

Moulding machines will do the following operations:

1. Ramming the moulding sand.
2. Removing pattern form mould.

1.13.1 Types of Moulding Machines:

i) Squeezer machine

a) Top squeezer machine

b) Bottom squeezer machine

ii) Jolt machine

iii) Sand slinger

i) Squeezer Machine:

In a squeezer machine the moulding sand in the flask is squeezed between the machine table and squeezer head.

a) Top Squeezer Machine

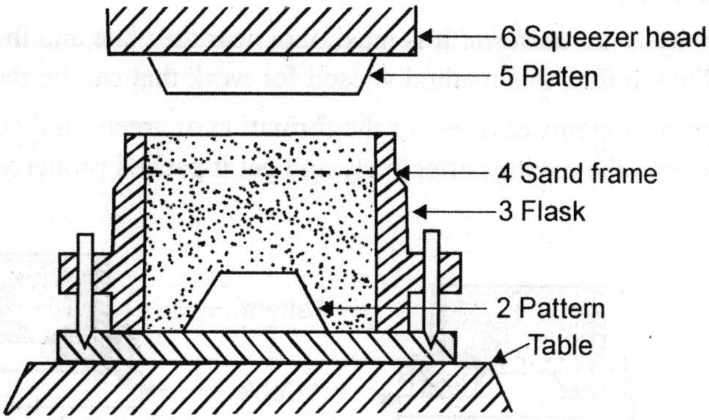

Figure 1.33 Top Squeezer Machine

A top squeezer machine is shown in figure.1.33. A mould board is clamped on the table. The flask is placed on the mould board. The pattern is placed inside the flask. The sand frame is placed over the flask. Moulding sand is filled up and levelled. The table is raised by table lift mechanism against the squeezer head. The platen enters the sand frame and presses the sand lightly. After the squeezing is over, the table comes to the starting position.

This machine is used for small size work (upto 15 cm depth). The sand is rammed more densely on the top of the mould than nearer the pattern. This is the disadvantage of the squeezer machine.

b) Bottom Squeezer Machine

Here the pattern is placed on the mould board. The mould board is clamped on to the table. The flask is placed on a frame and filled with sand. The table with the pattern is raised up against the squeezer head. Thus the flask with the pattern is squeezed between the squeezer head and the table. Then the table returns to its bottom position.

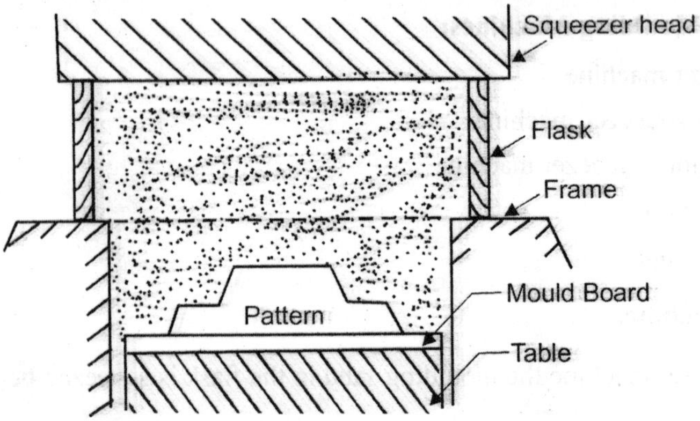

Figure 1.34 Bottom Squeezer Machine

Limitations of Squeezing :

★ Sand density is not uniform. It is maximum near the plate and then fails, gradually towards the pattern. Due to this, this method is used for work that can be moulded in shallow flasks.

★ If the pattern contains cavities for the formation of green sand cores, squeezing does not make sand flow into the cavities effectively and get it packed properly.

ii) Jolt Machine:

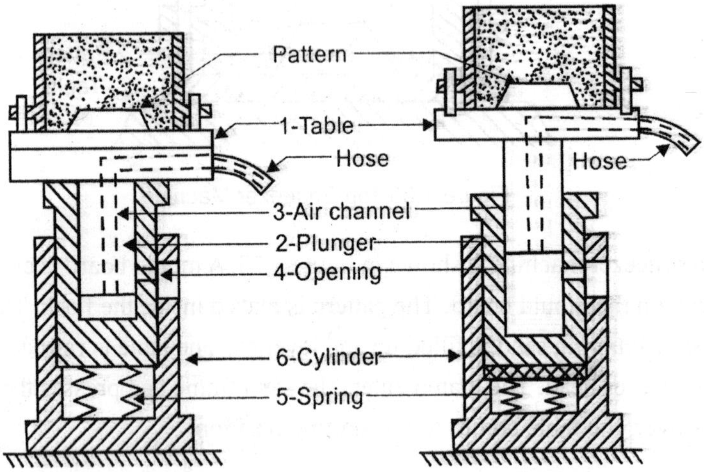

Figure 1.35 Jolt Machine

A jolt machine is shown in figure.1.35. The pattern is placed in the flask on the table. The flask is filled up with moulding sand. The table with the flask is raised to about 80 mm and suddenly dropped. The table will be raised by plunger when compressed air enters under the plunger through channel. The table will be dropped suddenly when the air escapes through the opening. This sudden dropping of table from a height makes the sand pack evenly around the pattern. Springs are used to cushion the table blows and to reduce the noise and vibration. This type of machine is mainly used for horizontal surfaces. Operation is noisy because of jolting. The moulding flask should be strong as it is lifted and dropped.

Drawbacks of Jolt Machine:

★ The density of the sand in the mould is not uniform.

★ High level of noise is produced by Jolt machine in operation.

★ Considerable amount of load acts on the foundation.

iii) Sand Slinger:

A sand slinger is shown in figure.1.36. In this machine, moulding sand is thrown into the box by an impeller. The impeller head has a housing. The blade rotates with a very high speed inside the housing. The sand is delivered to the impeller through the opening by conveyor buckets. The impeller head throws the sand through the outlet with a high velocity. Sand at the rate of 500 kg. per minute is thrown from the impeller over the pattern. So the sand gets rammed around the pattern. The density of the sand can be controlled by the speed of the blade. This machine is used for large moulds. Sand slinger machine gives uniform ramming. Ramming is quick and initial cost of the machine is high.

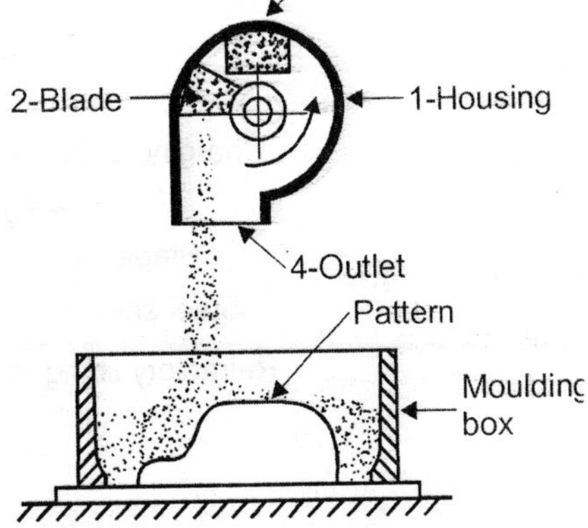

Figure 1.36 Sand Slinger

1.14 MELTING FURNACES

1.14.1 Furnaces - Classification:

Various types of melting furnaces are used in foundry shop. The type of furnace used depends upon the type of metal and the quantity of metal to be melted.

The metal melting furnaces used in foundary are:

1. Cupola furnace . . . (for cast iron)
2. Open hearth furnace . . . (for steel)
3. Crucible furnace . . . (for non-ferrous metal)
   (a) Pit type furnace
   (b) Coke fired stationary furnace
   (c) Oil fired tilting furnace
4. Pot furnace . . . (for non-ferrous metal)
5. Electric furnace . . . (for steel)
   (a) Direct arc furnace
   (b) Indirect arc furnace
   (c) Induction furnace

1.14.2 Cupola:

Cupola furnace is used for producing cast iron. A cupola furnace is shown in (Figure 1.37).

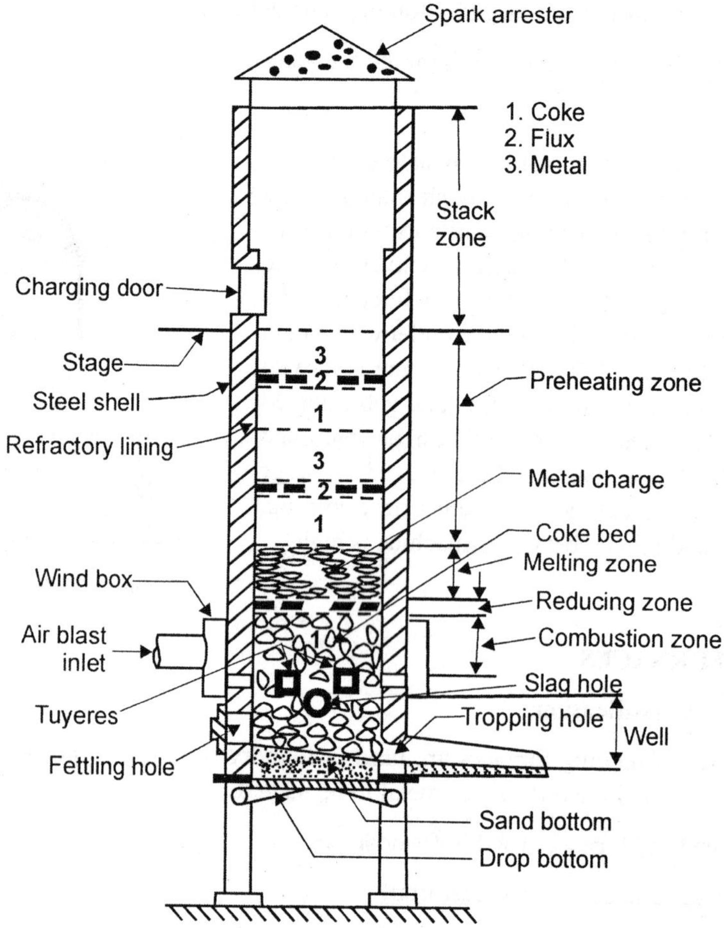

Figure 1.37 Cupola

Construction:

It is a vertical, cylindrical shell made of 10 mm thick steel plate. The furnace is lined with refractory bricks inside. The bottom of the cupola is closed by two bottom doors. A sand bed is prepared over the bottom doors. The sand bed is laid sloping towards the tap hole. Molten metal comes and stays on this sand bed (hearth). The legs of the cupola are set in concrete. There is a tap hole for taking out molten metal. The tap hole is closed by a plug made of clay. There is slag hole in the shell above the tap hole. Slag floating over the molten metal is removed through the slag hole. There are openings called tuyeres, 1 metre above the bottom. Air for combustion of fuel is sent through the tuyeres. There is a wind box and blower for the supply of air into the furnace. There is a charging door and charging platform halfway to the cupola. Metal and fuel are charged into the furnace through the charging door.

Preparation:

The slag and waste from previous melting is cleaned. Broken bricks are repaired. The bottom doors are closed. A sand bottom is prepared sloping towards tap hole. The height of sand bed is to be about 200 mm. A tap hole is formed and lined with clay. A slag hole is prepared. The cupola is dried thoroughly.

Firing:

Oiled waste and wooden pieces are placed at the bottom and fire is started. Now air is supplied into the furnace. When the wood starts burning, coke is charged in several portions. When the coke burns, more coke is added upto the tuyere level. The blast is turned off. Again coke is added upto the level of bed charge. When the coke bed burns for half an hour, the charging is done.

Charging and Melting:

Pig iron and iron scrap are charged into the furnace through the charging door. Then coke is charged. Alternate charges of iron and coke are made. Coke is used as a fuel. Lime stone is added to the charge to remove impurities. It will help in thorough mixing of molten metal. The ratio of pig iron to coke is 10:1. The ratio of pig iron to limestone is 25:1. The cupola is fully charged. The iron is soaked for one hour. Then the blast is turned on. Within few minutes molten metal will begin to collect at the sand bed. When enough molten metal is melted, clay plug is removed and metal is collected in ladles. From the ladles, molten metal can be directly poured into moulds. The slag floats on the molten metal and is tapped out through the slag hole. The furnace should be charged again to the full level. At the end of the melt the cupola is shut off by stopping the air blast. The remaining molten metal is removed. The bottom doors are opened by pushing out the drop. The wastes drop down and they are quenched by water.

Advantages of Cupola :

★ Continuous operation.

★ Low cost of operation.

★ Easy to operate.

★ Composition and temperature of melt can be controlled.

1.14.3 Crucible Furnace

Crucible furnaces are used for melting non-ferrous metals. The capacities range from 30 to 150 kg. The type of crucible furnace are

i) Pit furnace, ii) Coke fired stationary furnace, iii) Oil fired tilting furnace.

i) Pit Furnace:

Pit furnace is shown in figure 1.38. The crucible is placed in a pit below the floor level. It is fired with coke. The charge to be melted is placed in the crucible. Coke is packed around the crucibles. Natural draft is provided by a tall chimmey. Many crucibles can be placed in a single pit. After the metal is melted, the covers are removed. The crucibles are lifted out with the help of tongs and are taken to the pouring place. This furnace is used for melting non-ferrous metals.

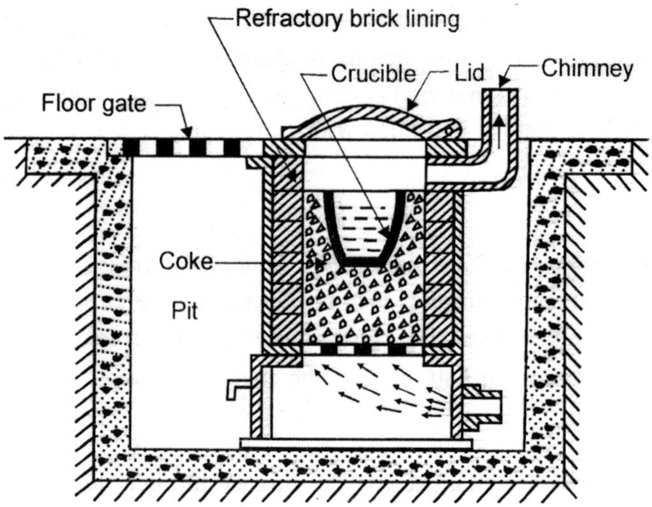

Figure 1.38 Pit Furnace

ii) Coke Fired Stationary Furnace:

This furnace is used for melting small quantities of non-ferrous metals. The furnace is erected above the floor level. The crucible is placed in the heating chamber. The heating chamber is lined with refractory material. Coke is used as the fuel. Forced draft is used. A blower is used for the supply of air. (Refer figure 1.39).

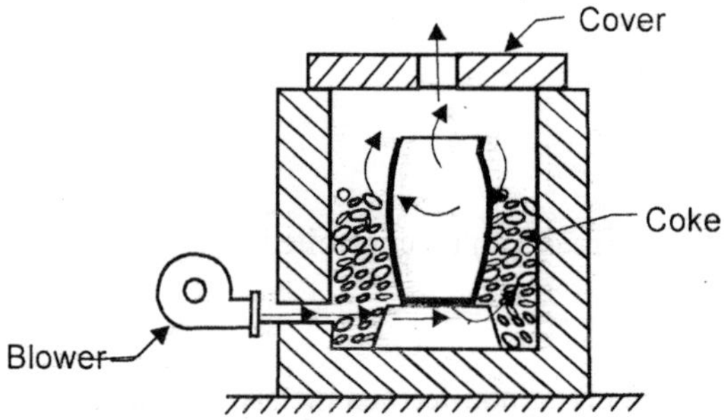

Figure 1.39 Coked Fired Crucible Furnace

iii) Oil Fired Tilting Furnace:

This furnace is shown in figure.1.40. This furnace is used for melting small quantities of non-ferrous metals. The furnace is mounted on two pedestals above the floor level. For pouring molten metal, the furnace is rotated by a geared hand wheel. This furnace is fired with oil. A drum of kerosene or waste oil is placed at a height of 6 m above the floor level. Oil and air are admitted with pressure through a nozzle. The crucible is placed in the heating chamber and is heated by the flame. Oil fired furnace has higher melting rate. The furnace can be stopped whenever needed. The temperature can be controlled easily. Oil or gas fired furnace require lesser floor space. It gives lesser pollution.

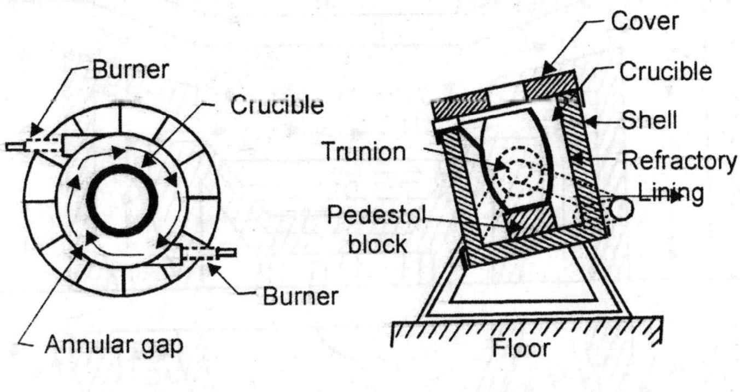

Figure 1.40 Tilting Crucible Furnace

iv) Pot Furnace:

Pot furnace is used for melting non-ferrous metals like aluminium, magnesium, zinc, lead and tin. The pot is made of cast iron or steel. The furnace may be fired by gasoline oil or coke. A coke fired pot furnace is shown in figure.1.41. The molten metal is taken out from the pot using ladles. Normally the capacity of the pot furnace is around 500 kg. Pot furnaces are used in permanent mould casting processes and die casting processes.

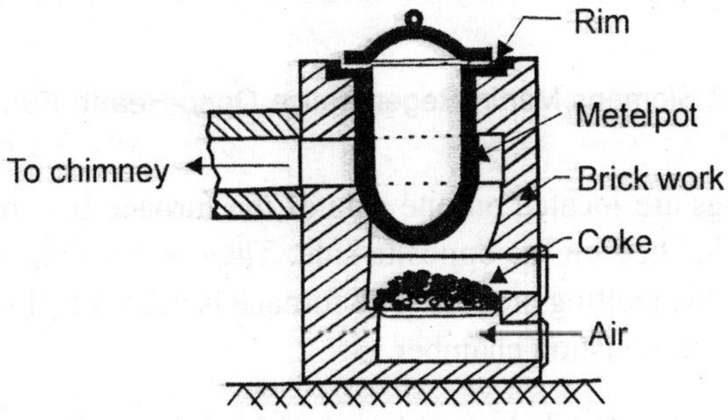

Figure 1.41 Pot Furnace

1.15 OPEN HEARTH PROCESS

It is based on the regenerative principles of heating, i.e., very high temperature required for steel making are obtained by preheating the gaseous fuel and air by the outgoing products of combustion.

Construction:

Open hearth furnace consists of a shallow hearth about 12 metres long, 4.5 metres wide and 50 cm deep lined with a highly refractory material (acid or basic) depending upon the kind of metal charged and the type of steel to be made. A slightly arched roof of refractory brick and brick side walls supported by heavy metal stays are provided.

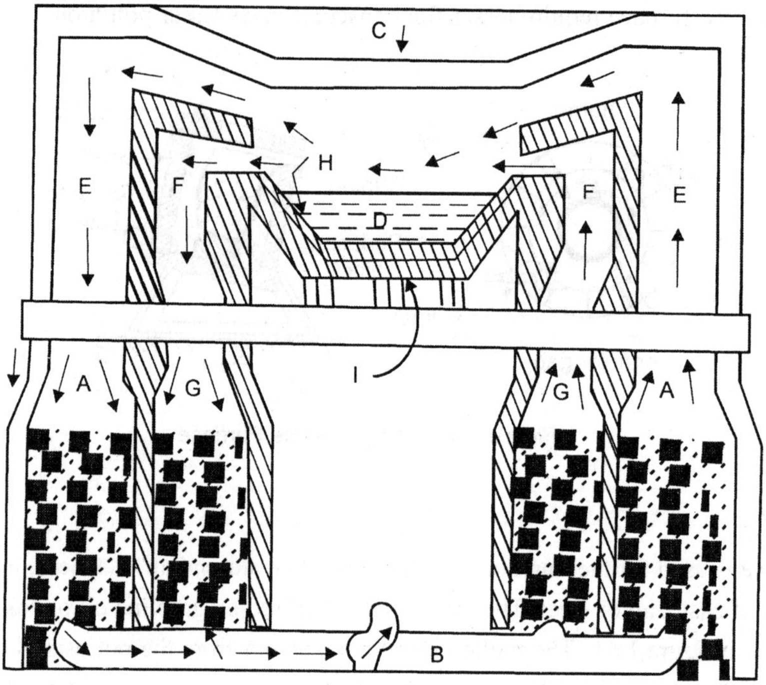

A, A – Hot air regenerative chambers; B – Gas and air pipe; C – Fire brick lining; D – Molten metal bath; E, E – Air ducts; F, F – Gas flues; G, G – Gas regenerative chambers; H – Hearth lining; I – Steel girders.

Figure 1.42 Siemens-Martin Regenerative Open-Hearth Furnace

Several charging openings are located on one side of the furnace for convenience in loading the metal charges. The tap hole is located on the opposite side. The combustion chambers, located at each end, are used alternately during the melting process. The furnace is heated by burning gas and air that has been heated before it enters the combustion chamber.

Regenerative chambers are located below each end of the furnace. These chambers are filled with fire brick laid in a checker-board pattern to absorb the heat from the gases discharged from furnace hearth. This heat will be returned to the incoming gas and air, when the flow is reversed. Waste gases are exhausted into the atmosphere through a stack after they have passed through the regenerative chamber.

The flame is directed over the top of the metal to prevent oxidation. The metal is heated by indirect radiation from the roof and direct radiation from the flame. Tests are made during the melting process to determine the condition of the metal. When the process has been completed, the metal is tapped into ladles to pour.

Working:

Gaseous fuel used may be producer gas, natural gas, coke oven gas or a mixture of coke oven gas and blast furnace gas.

Gas and air flow through one of the pair of regenerators and absorb heat from them. These are then led through the uptake flues into the hearth and get ignited there. Products of combustion produced by ignition sweep over the bath containing molten pig iron and transfer their heat to it.

The combustion products then leave the furnace through the ports on the other side, pass through the other pair of regenerators transfering most of their heat to the checker bricks and finally escape through the chimney to the atomsphere.

When sufficient heat is stored in the checkers, the furnace is reversed i.e, the gas and air now enter the pair of regenerators on the other side, become preheated and combustion products leave the hearth by ports and regenerators through which gas and air was passing previously. The furnace is generally reversed every 20 to 30 minutes.

(a) Acid Open-Hearth Process:

Like acid bessemer process, it requires low phosphorus pig iron as raw material. Lining used is silca or fire clay and so, flux like lime stone etc. cannot be used, for reducing phosphorous and sulphur. These impurities, therefore, cannot be removed by this process.

For steel making, this process is rarely used, as the corresponding acid bessemer process is much more simpler and economical.

(b) Basic Open-Hearth Process:

For producing steel from high phosphorus pig irons this process is employed. Lining is of some base material like dolomite, magnesite etc. and so flux like limestone can be used for reducing phosphorus and sulphur.

The hearth may either be of fixed or tilting type for the metal to flow out easily. Its capacity varies between 100-200 tonnes. It is made of fire bricks at the bottom, covered by magnesite bricks upon which another layer of calcined magnesite mixed with basic open-hearth slag is put.

The furnace charge consists of solid or molten pig iron, scrap iron ore (for affecting oxidation) and limestone as a flux.

The ports should be designed at such an angle that (i) the gases do not directly impinge on the bath and also (ii) the roof may not be unduly heated. To prevent excessive oxidation of the metal and permit good mixing, the air port is usually placed above the gas port.

Advantages of Open-hearth process over Bessemer process:

i) The temperature and the composition of the bath are under much better control as the ore acts as an oxidising agent and heat is applied externally.

ii) This method has greater flexibility as greater variety of raw materials can be employed.

iii) Due to addition of ore and scrap, it gives much increased output from the same amount of pig iron.

iv) Basic open-hearth process permits the use of any pig iron. It can eliminate phosphorus completely whereas in Bessemer process, there are limitations.

(c) Duplex Process or Special Open-hearth Process:

It consists of acid bessemer converter in conjunction with a basic open-hearth furnace.

Advantages:

i) High silicon and high phosphorus pig iron can be converted into good steel by this method.

ii) It increases the life of open-hearth furnace as there is no silica present which may react with the lining.

iii) It also cosiderably reduces the time of open-hearth heat and labour.

1.16 AIR FURNACE (OR REVERBERATORY FURNACE)

i) An air furnace is an acid-lined reverberatory type furnace used for the production of Malleable iron and high-test gray iron castings.

ii) An air furnace (Figure.1.43) consists of a long, rectangular structure having removable arched roof sections (called Bungs) over a shallow hearth made up of refractory sand dampened with clay wash.

iii) It resembles open hearth furnace except for few differences namely,

— It does not have any regenerative chambers for preheating the incoming air.

— Metal is charged through the bungs.

— Temperature is less than that in the open hearth furnace.

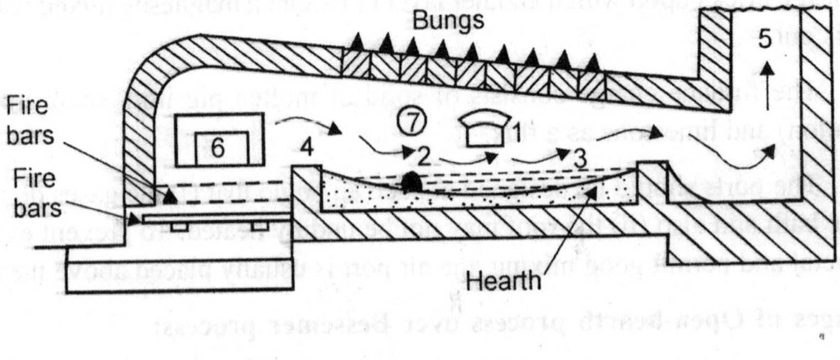

Figure 1.43 Air Furnace

iv) Oil or pulverized bituminous lump coal is used as the fuel for heating and melting the metal.

v) Air and fuel are blown through one end of the furnace so that the flame passes over the metal charge lying in the hearth of the furnace.

The flame and hot gases (as in open hearth furnace) heat up the air furnace roof and walls. The heat reflected and radiated from the roof and walls is utilised for melting and superheating the metal charge.

— Acid slag above the molten metal protects it from direct exposure to the flame and furnace atmosphere.

vi) An air furnace is charged by removing bungs and then dropping the charge (i.e., scrap, pig iron, etc.) through the hole; the bungs are then replaced and sealed with clay.

vii) The entire charge i.e. entire amount of raw materials is put into the furnace at a time, melted in one batch and drawn off when the correct composition (found after analysis) and temperature have been obtained.

Difference between cupola and air furnace:

— In an air furnace, time is available for analysing samples of molten iron and thus proper additions, such as pig iron or ferro alloys, can be made if necessary. In air furnace, considerable amount of control can be exercised over the chemical composition and hence the quality of the melt.

— Whereas, cupola is continuously tapped and thus there is no time for analysing and controlling the chemical composition of the melt before tapping.

viii) Unlike cupola, in air furnace only the flame and not the fuel comes in contact with the molten iron and thus transfer of certain elements such as carbon and sulphur from the coke and into the melt is eliminated.

ix) Since an air furnace permits good control of metal compositions, it is often used in a duplexing operation, with a cupola which (quickly and efficiently) melts and supplies molten cast iron and the same is then transferred to an air furnace for refining and adjusting its composition.

x) A melting (metal to fuel) ratio in air furnace is of 2:1 but that of 1:1 is also not unusal.

— Since this ratio is generally 10:1 or 8:1 in cupola, a cupola involves lower working costs as compared to an air furnace.

xi) Since cupola provides a small but continuous supply of cast iron, which simplifies its disposal throughout the foundry, a large number of small castings can be easily produced.

— On the other hand an air furnace supplies molten metal in large quantities in batches (at intervals of time), therefore it facilitates the production of large castings such as rolls etc.

xii) Besides the working cost, the initial cost of air furnace is also higher than that of a cupola and for this reason air furnace does not find wide applications except for producing malleable iron castings.

xiii) The capacity of air furnaces varies from 5 to 50 tons per heat; a heat implies the amount of metal melted at one time.

1.17 ROTARY MELTING FURNACE

i) Rotary furnaces got originated in Germany but they are now being used also in other countries of Europe, U.S.A., etc.

ii) Since air furnace (discussed earlier) for making malleable cast iron is difficult to operate and wasteful in fuel, a rotary furnace is an improvement over the same.

iii) A rotary furnace consists of a horizontal cylindrical steel shell lined with refractory material and mounted on rollers (Figure. 1.44) for rotating or rocking purposes. The cylindrical shell revolves completely at about 1 r.p.m.

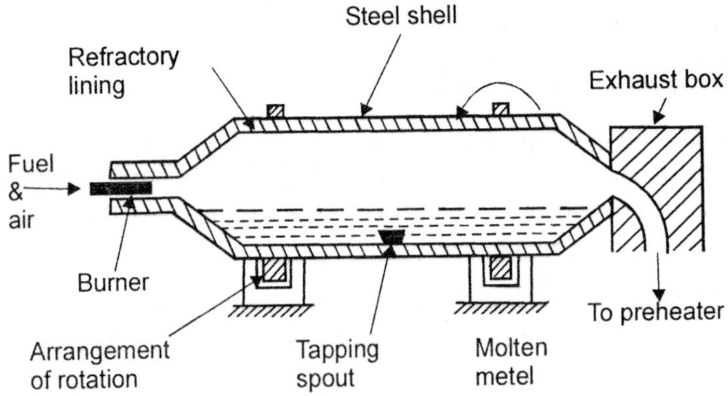

Figure 1.44 Rotary Furnace

iv) Due to rotation of the furnace, the metal gets heat from the walls and is melted more efficiently and faster. The melting time is reduced to about half that of a stationary furnace.

— The first and the subsequent heats for melts of gray cast iron need about 2 to 3 hours and $1^1/_2$ to 2 hours respectively.

v) The melting (metal-to-fuel) ratio is 5:1 for pulverized coal and 6:1 for oil fuel.

vi) The refractory lining may be renewed after 200-300 heats.

vii) Rotary furnace is charged from one of its conical ends after removing burner or exhaust box temporarily.

viii) The burner at one end burns the fuel and the resulting high temperature flame melts and superheats the metal charge lying in the shell.

ix) Since molten metal does not come in contact with the fuel (such as coal) there is no danger of carbon or sulphur pick up.

x) The furnace atmosphere can be controlled within limits.

xi) The slag floating on the molten iron, protects the metal and its constituent elements (like Si, Mn etc.) from oxidation.

Thus the melting losses in a rotary furnace are lower than in cupola. The melting losses in a rotary furnace are of the order of about 2%.

xii) Unlike cupola, in a rotary furnace, a sample of molten metal can be drawn before pouring and analyzed as regards its chemical composition which subsequenty can be accurately controlled by additions if required.

xiii) Charge in a rotary furnace can be super-heated to high temperatures.

xiv) Cheap and light scrap like steel clippings or cast iron borings which cannot be used in cupola, may be added in the charge without affecting the melt quality. This offsets the higher melting cost of rotary furnace when compared to a cupola.

xv) After the melting is complete, the initially plugged tap hole is opened and the furnace is rotated slowly until the tap hole reaches the level of molten metal in the furnace, when the same (i.e, metal) can be poured into a ladle.

xvi) In case the tap hole is not provided in the cylindrical body, the liquid metal can be tapped from the other (i.e., Figure 1.33, right hand) end of the furnace. Besides tapping, this end is used for charging, slagging and evacuating the products of combustion.

xvii) The products of combustion may be sent to a preheater or recuperator unit where they may heat the cold air to be used subsequently in the rotary furnace.

— The temperature of this heated air reaching at the burner is adjusted to suit the moisture content, volatile content, and fineness of the pulverized coal.

xviii) A rotary furnace may have a capacity ranging from 1 to 50 tons depending upon the requirements of a consumer.

xix) Rotary furnaces find applications in producing high-duty cast irons having alloying elements such as Mo, Ni, Cr, etc. which if added in cupola will suffer a loss.

xx) Metals previously melted in cupola can be held and super-heated in rotary furnaces.

xxi) Fuel costs of cupola and air furnace do not differ much in case pulverized coal is used for the later.

1.18 BESSEMER CONVERTER

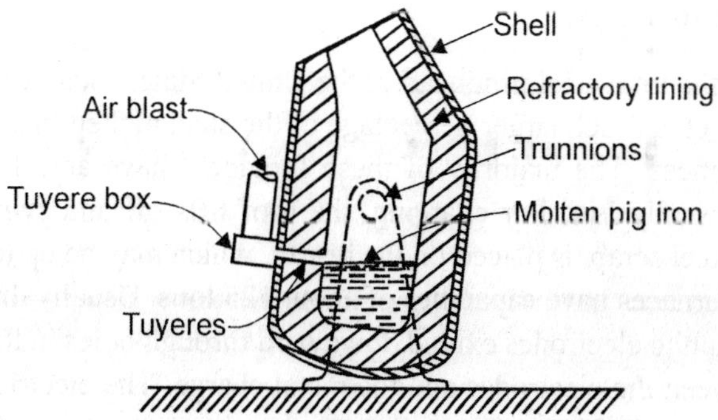

Figure.1.45 Bessemer converter

Similar in construction to the side-blow converter shown in figure 1.45, the bessemer converter differs in that the tuyeres are located in the bottom instead of at the side. The bessemer process for making steel depends upon the ability of oxygen in the air to oxidize silicon, manganese, carbon, and some phosphorus more readily than it oxidizes iron at high temperatures. Sulphur cannot be oxidized and remains in the steel.

The chemical reactions are the same as those used in the open-hearth process. The converter is filled with molten pig iron by tilting it to about the horizontal position. Then a blast of air is turned on through the tuyeres in the bottom, and the converter is rotated to its upright position. An air-blast pressure of 20 to 25 psi is sufficient to keep the metal from entering the tuyeres. The chemical reactions of the elements with oxygen from the air generate sufficient heat to raise the temperature of the metal several hundred degrees. The carbon oxidizes to carbon monoxide, which burns at the mouth of the converter with a long flame. When this flame shortens enough to indicate that the carbon content has been sufficiently lowered, the operator rotates the converter to its emptying position and turns off the air blast. Today, instead of relying upon the operator's judgment, a photoelectric cell is used to determine precisely the end of the blow.

Phosphorus will not be removed in a converter unless a basic slag is used. In the United States, only acid-lined bessemer converters are employed. Bessemer converters have capacities ranging from 20 to 40 tons and are generally considered capable of producing about 1 ton of steel per minute of operation. Since the initial cost of the equipment is relatively low, this is an economical method for making some steels.

1.18.1 Side-blow converter:

In foundries which produce steel castings, the side-blow (Tropenas) converter is more suitable than the bessemer converter. The air blast from tuyeres, located at the side, impinges on the surface of the molten pig iron, oxidizing the undesirable elements. An air-blast pressure of about 5 psi is used. Here the carbon is burned largely to carbon dioxide, creating more heat than with the bessemer process. This produces higher-temperature molten steels (3200 to 3300°F), which are needed to pour steel castings in thin sections. Side-blow converters are normally smaller than bessemer converters, holding from 1 to 4 tons of steel.

1.19. ARC FURNACE

1.19.1 Electric direct-arc furnaces:

About 7 percent of the total steel produced in the United States today can be attributed to electric direct-arc furnaces. However, a much larger percentage of the steel melted in steel foundries is produced in electric direct-arc furnances. The majority of these furnaces have acid linings, since sulphur and phosphorus contents are usually low. For charging, the roof lifts off and swings clear of the furnace, while the charge, usually steel scrap, is placed on the hearth, which may be up to 22 feet in diameter. The largest electric direct-arc furnaces have capacities of about 125 tons. Usually three-phase electric current is used, and three large graphite electrodes extend downward through holes in the roof. Heat is created by electric arcs obtained between the electrodes and the metal charge. The electric current from any one of the electrodes arcs to the metal charge, flows through the charge, and arcs back to one of the other electrodes. The temperatures at the three arcs thus produced are about 11000°F. This readily melts the metal scrap underneath the electrodes. A pool of molten metal forms underneath the electrodes, which in turn helps to melt the remaining metal from the bottom upward.

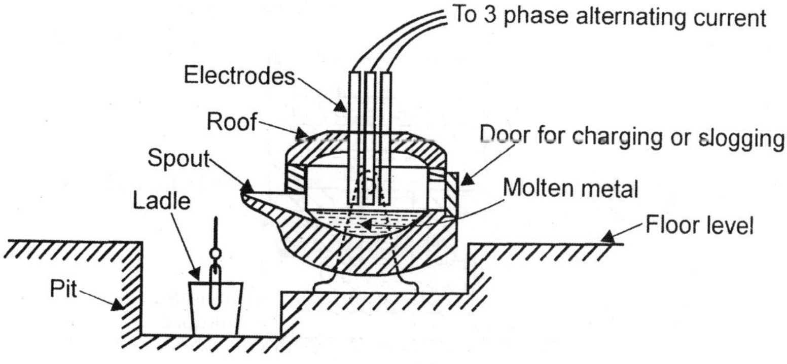

Figure.1.46 Electric direct-arc furnaces

The proper are distances from the ends of the electrodes to the metal below are adjusted automatically by maintaining a predetermined voltage across the arc gaps. For a 50-ton furnace, the electric current in each of these arcs is about 25,000 amperes with a voltage of about 250 volts. Spare roofs are repaired and kept on hand, since the life of a roof may be about 100 heats.

Electric direct-arc furnaces have several important advantages. The electric arc is a source of pure heat without any contamination from elements in fuels, and the rate of heat application can be closely controlled. Thermal efficiency is high, about 70 percent, but the higher cost of electric power may tend to counteract this advantage. The furnace atmosphere above the molten metal can be controlled. Most alloying elements, such as chromium, nickel and tungsten, can be recovered from the scrap with negligible losses. Steel may be made from pig iron and steel scrap using the same methods as those described for the open-hearth furnace. In other cases, when the charge consists almost entirely of steel scrap, steel is made by only melting and refining. An oxidation stage followed by a deoxidation stage is not necessary. In general, higher-quality carbon steels and alloy steels required in larger quantities are made in electric direct-arc furnaces.

1.19.2 Electric indirect-arc furnaces:

The indirect-arc furnace differs from the direct-arc type in that the arc is formed between two graphite electrodes, and no electric current passes through the metal charge. A common type of electric indirect-arc furnace is shaped like a barrel and mounted on rollers. The rollers allow the furnace to rock during the melting, so the metal will obtain additional heat from the side walls. This rocking provides faster melting, increases the life of the refractory lining, and stirs the metal. The two electrodes are introduced through the two opposite ends of the furnace and project to near the centre, where the arc is formed. With the larger furnaces both the arc distance between the electrodes and the rocking action are automatically controlled once they are set. For tapping, the furnace is tilted by operator control farther than for rocking to allow the metal to flow out of the taphole. Capacities of rocking electric indirect-arc furnaces range from a few pounds up to 2 tons. The temperatures obtained may not be as high as those obtained with direct-arc furnaces, but cast iron and non-ferrous metals can be readiiy melted.

1.19.3 Core-type induction furnace:

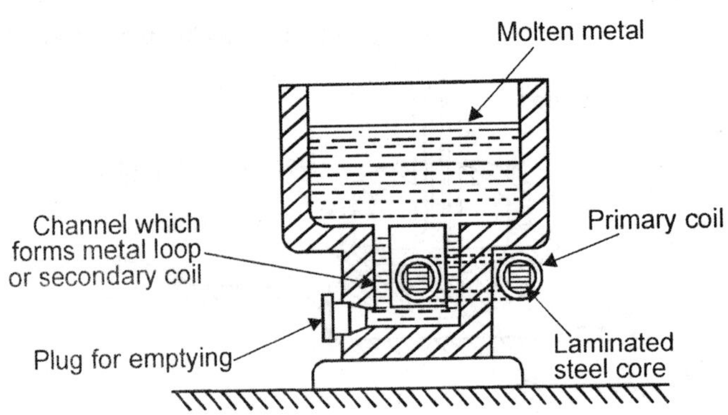

Figure 1.47 Core-type induction furnace

This type in the past also been known as the "low-frequency induction furnace", since it has used commonly supplied a-c frequency, or 60 cycles per second. Induction furnaces create heat by electrical resistance as electrical currents are induced within the metal charge itself. The principle of operation of a core-type induction furnace is the same as that of an ordinary transformer. The primary coil consists of many turns, and the secondary coil has one turn, which is a loop of molten metal. The voltage induced in this loop, or channel, is low, and the current is high, being thousands of amperes.

Since both primary and secondary coils encircle a suitable laminated iron core, this type of furnace is known as the "core-type induction furnace". The channel of molten metal connects to the reservoir above, which holds the metal charge. During operation, molten metal circulates in the channel stirring the metal in the reservoir. Operation of this type of furnace must be started by first filling the channel with molten metal, which must be obtained from some other furnaces. This will form a closed secondary circuit.

During periods when the furnace is temporarily not in operation, such as during nights and weekends, molten metal may remain in the channel. The electrical current is adjusted and controlled automatically to keep this metal in the molten condition. This metal should not solidfy, since it cannot be remelted by the heat created in the secondary loop.

A furnace of this type is somewhat restricted to one alloy, since changing alloys requires emptying the furnace completely, cleaning it, and starting it up again with the new molten alloy. Melting us rapid and clean, and the metal can be accurately controlled in temperature and composition. Stirring ensures uniformity of the metal. The operation is economical with high thermal efficiencies up to 80 percent. This furnace is used for the melting of cast irons and most nonferrous alloys. It serves well as a holding furnace for die casting and permanent-mould casting, because the temperature of the metal can be so easily controlled. In one installation producing malleable cast iron, the low-frequency induction furnaces used hold 2½ tons of metal, and their melting rate is 1,200 pounds per hour. Capacities of core-type induction furnaces range upto about 5 tons.

1.19.4 Coreless-type induction furnace:

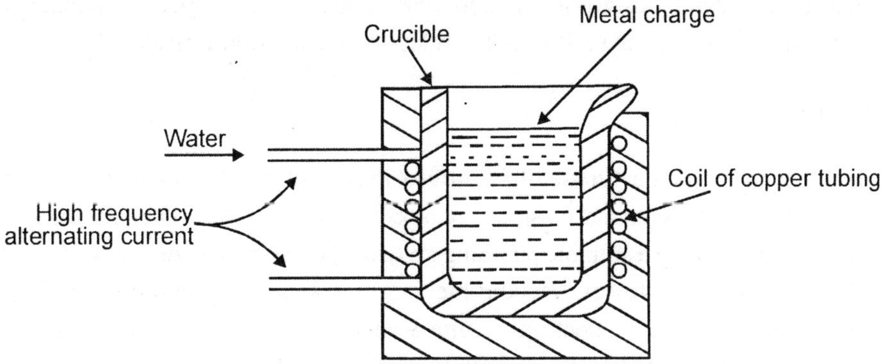

Figure. 1.48 Coreless-type induction furnace

In the past this type has been also known as the "high-frequency induction furnace". In order to obtain a simpler type of induction furnace, eliminating the need for the channel of molten metal and the laminated iron core, the coreless-type induction furnace was developed. In it, a suitable crucible for holding the charge is closely surrounded by a heavy coil. When an alternating electric current rent is passed through this coil, all of the space inside the crucible is subjected to the resulting, continually changing magnetic field. Electric currents induced within the metal of the charge create heat by electrical resistance. In order to increase the rate of energy transfer from coil to metal charge, the use of relatively high frequencies has been necessary, until quite recently. The ordinary line frequency of 60 cycles per second could not be used satisfactorily. With recent developments it is now possible to melt metal in the larger coreless-type induction furnaces using ordinary 60 cycle electric current. With 60 cycle electric current, one of the disadvantages still to be overcome is the violent stirring of the molten metal which occurs.

Two methods are used for obtaining higher-frequency alternating electric current. Spark-gap converters provide frequencies of 20,000 to 80,000 cycles per second, and these are used for the smaller furnaces with ratings from 3 to 40 kilowatts. Larger furnaces use motor-generator sets, providing frequencies within the range of 500 to 10,000 cycles per second. Lower frequencies are used for the larger-capacity furnaces. Other methods for obtaining high frequencies are also possible. Thermal efficiencies of coreless-type induction furnaces are lower than those of the core-type induction furnaces because of energy losses within the spark-gap converter or motor-generator set. Thermal efficiencies up to 60 percent are obtained. Melting is fast, and the rate of energy input is readily controlled. The melting is clean without contamination, and the atmosphere above the melt can be controlled easily. A stirring action within the molten metal ensures complete homogeneity. No warm-up period is needed with these furnaces, and the change from one alloy to another is much more easily accomplised than with the core-type induction furnace. Because of the higher initial costs of these furnaces and their auxiliary equipment, the coreless-type induction furnace is usually restricted for the melting of high-quality metals and of smaller amounts of metal. It is especially useful in the laboratory. Capacities range from 3 pounds upto 12 tons, and almost any metal or alloy can be melted in a coreless-type induction furnace.

1.19.5 Vacuum melting:

The coreless-type induction furnace has proved itself suitable for vacuum melting, because it operates independently of pressure, and its size may be conveniently small. The furnace is enclosed in a vessel which can be evacuated by a suitable pumping system. Provision is made for pouring the metal into desired moulds while inside the vessel. Vacuum melting may also be accomplised by bombarding the metal to be melted with a stream of electrons within a vacuum chamber. This method can be used for purifying metals, since impurities with lower boiling points can be eliminated by vaporizing them and drawing them off with the vacuum system.

Vacuum melting is done largely to obtain higher-purity metals. From steels, the removal of oxygen, nitrogen, and hydrogen is possible. Gaseous impurities can also be removed from other metals and alloys. Titanium is arc-furnace-melted in either a vacuum or an inert atmosphere of pure argon or helium to prevent the absorption of oxygen, nitrogen and hydrogen. Small percentage of these impurities produce hardness and reduce ductility.

1.20 SPECIAL CASTING PROCESS

So far, the details of science of sand casting processes have been discussed. But sand casting is not suitable nor economical in many applications where the special casting processes would be more appropriate. In the following topics the details of some of the commonly used special casting methods would be described.

1.21 GRAVITY DIE CASTING METHODS

This is also called as Permanent Mould Casting. Moulds made of sand are destroyed after casting. So the moulds can not be reused. So for producing a large number of castings, a permanent mould is necessary.

A permanent mould is made of heat resisting cast iron, alloy, steel, graphite or other suitable material. So the same moulds can be used again and again after each casting.

Permanent moulds are made of two halves for easy removal of castings. Pouring cup, runner and riser are provided in the mould halves itself. The two halves are hinged on one side. They are closed and clamped tightly before pouring the metal. Almost all metals can be cast in this mould. Zinc, Copper, Aluminium, Lead, Magnesium and Tin alloys are most often cast in this method.

This method is suitable only for components of simple shapes and design and uniform wall thickness. If necessary, mould halves can be cooled by circulating water or providing fins at the periphery.

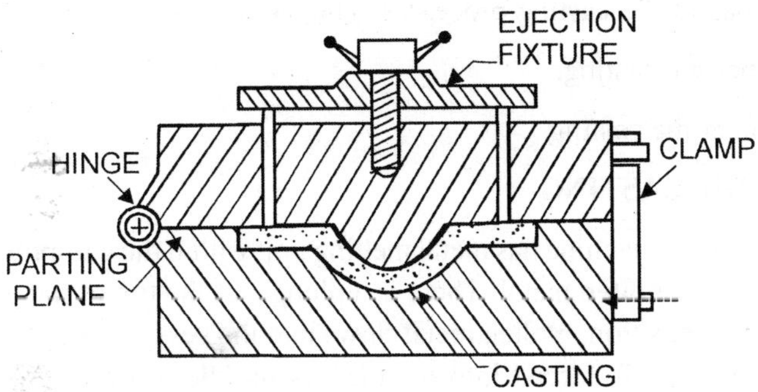

Figure. 1.49 Gravity Die Casting

Advantages:

1. Accurate castings can be obtained.
2. Good surface finish is obtained.
3. Less wastage and rejection.
4. Less production cost.
5. Castings are free from defects.
6. Moulding is made of metal. So the heat from the casting is conducted quickly. Hence, fine grained structure is obtained in the casting.

Limitations:

1. Large size casting can not be produced as the metal mould becomes costlier.
2. Removal of casting from the mould is difficult.
3. Complicated shaped castings can not be produced easily.

1.22 PRESSURE DIE CASTING

Pressure die casting is a process in which molten metal is forced under pressure into a permanent metal mould. The metal mould is called the die. This is made of two parts. One parts is stationary and the other is movable.

Die casting is suitable for casting lead, magnesium, tin, zinc, aluminium alloys and brass.

The die casting is done as follows:

1. The molten metal is forced under pressure into the assembled die.
2. The die is water cooled. So the molten metal cools down and becomes solid immediately.
3. The die is opened. Then the finished casting is ejected by pins.

There are two types of die casting processes. They are:

1. Hot chamber die casting.
2. Cold chamber die casting.

1.23 HOT CHAMBER DIE CASTING

Hot chamber die casting machine has metal melting unit within the equipment. There is a gooseneck vessel which is submerged in molten metal. There is a plunger at the top of the goose neck vessel. When the plunger is in the upward position, molten metal flows into the vessel through a port. When the plunger comes down, the molten metal is forced into the die. As the die is water cooled, there is immediate solidification. Then the dies are seperated and the finished casting is removed by ejectors. Afterwards the dies are assembled. The plunger goes up and the cycle is repeated. The plunger is actuated by hydraulic means. The operating pressure is 15 (MN/m^2.)

Hot chamber die casting is suitable for casting of metals such as zinc, tin and lead.

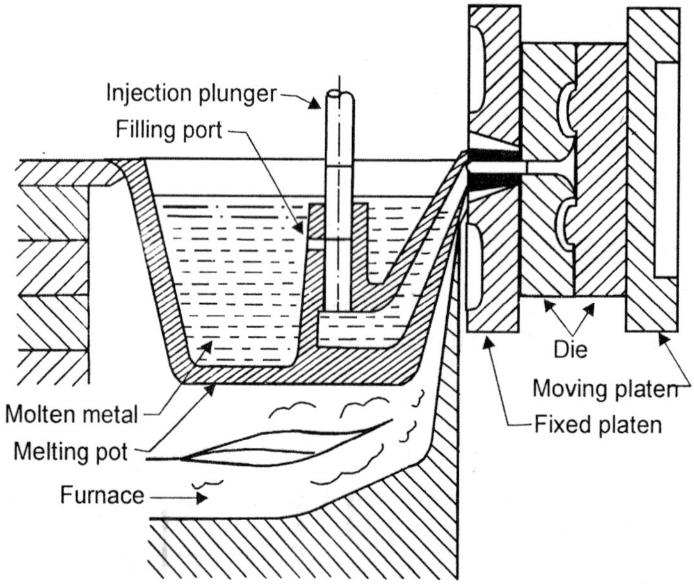

Figure. 1.50 Hot Chamber Die Casting

1.24 COLD CHAMBER DIE CASTING

In cold chamber process. The metal melting unit is not integral with the machine. The metal is melted in a separate furnace and brought to the machine for pouring. The process is shown in figure 1.51.

There is a plunger operated hydraulically. The molten metal is poured into the injection cylinder. Then the plunger moves to the left and forces the molten metal into the die cavity. As the die is water cooled, there is immediate solidification. Then the dies are separated. The finished casting is removed by ejectors. Then the die is assembled. The plunger goes to the right and the cycle is repeated.

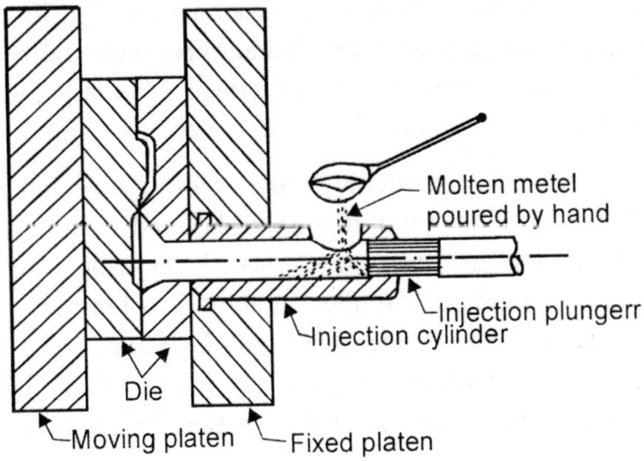

Figure. 1.51 Cold Chamber Casting

Application of die casting:

Automobile parts like fuel pump, carburetor body, horn, heater, wiper, and crankcase, washing machine parts, ash tray, lamp base, toys, camera parts, locks, clocks, steering wheel hub, textile machine parts are made by die casting method.

Advantages:

1. Rate of production is very high (700 casting per hour)
2. Castings have very good surface finish.
3. Very accurate castings are obtained.
4. The die has long life (75000 castings)
5. Less floor space is sufficient.
6. Labour cost is low.
7. Very thin castings can be made (0.5mm)
8. Waste of metal is very low.
9. Cost per casting is less.
10. Casting defects are less.

Disadvantages:

1. Only non-ferrous metals can be cast.
2. Heavy castings can not be cast.
3. Die casting is suitable only for mass production of small castings.

1.25 CARBON-DIOXIDE MOULDING

The idea of using carbon-dioxide as a hardening medium for sand moulds was conceived as long back as 1698, but its advantageous commercial application could be established only in 1948 by Alfred Augstein in Czechoslovakia. Since then it has gained wide popularity owing to its various advantages. Later researches in its application have further enhanced its utility.

This process is basically a hardening process for moulds and cores. The sand mixture used in this process consists of pure dry silica sand, free from clay, and sodium silicate liquid base binder. In order to have a high permeability the sand used should not be very fine. The sand is mixed with 3.5 to 5 percent sodium silicate liquid base binder in a muller for about 3 to 4 minutes. The moisture content in the mixture should not exceed 3 percent. Some other additives like wood, flour, coal dust, pitch or graphite etc., may also be added in suitable proportion (1% max,) to improve the collapsibility of the mould or core. The prepared mixture can then be packed around the pattern in the mould or in the core box for core making either manually or by machines.

After packing the sand, CO_2 gas is forced into the mould or core at a pressure of about 1.41 kg/cm^2. This process of forcing the gas into the sand mass is called gassing. This CO_2 reacts with the sodium silicate to form silica gel, which binds the sand grains together to provide the strength and hardness to the mould core.The chemical reaction which takes place inside can be represented by the following simple equation:

$$\underset{\text{(sodium silicate)}}{Na_2SiO_3 \times H_2O + CO_2} \quad = \quad \underset{\text{(sodium carbonate)}}{Na_2CO_3} \qquad \underset{\text{(silica gel)}}{SiO_2 \times H_2O}$$

Immediately after gassing, the moulds are ready for pouring. So are the cores for being placed in the moulds. Overgassing should, however, be avoided as the same is likely to deterioate the sand. The method of gassing should be simple, quick and capable of distributing the gas uniformly throughout the sand mass. The gassing process usually takes 15 to 30 seconds only. A few typical methods of gassing a mould are illustrated in Figure.1.52.

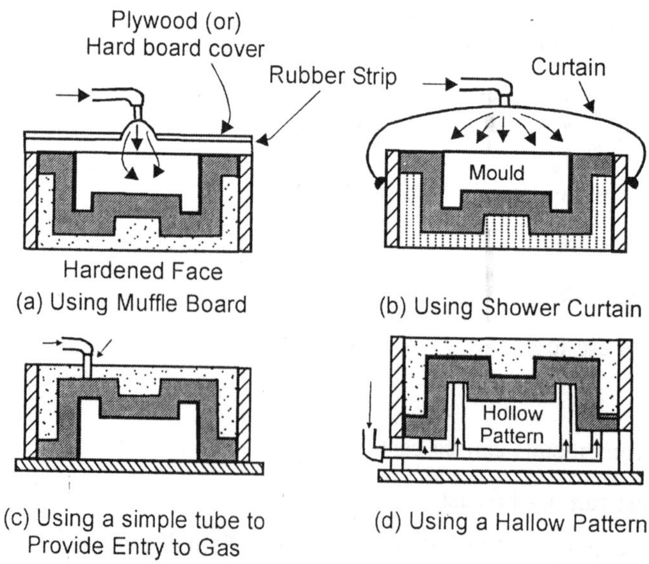

Figure. 1.52 Gassing Methods

Evacuation Technique:

For most effective and efficient gassing of cores and moulds it is necessary to evacuate completely the air present in the pores of the rammed sand to allow CO_2 to replace this air in the pores. For this, a suitable arrangement is always required to be made. In this, one side of the mould or core box carries a gas tight seat to which is connected the CO_2 hose. The other side of the mould or the core box is connected to vacuum pump by means of a pipe, through a reservoir. When the pump operates, it sucks the air from the sand and CO_2 replaces the same by entering from the other side.

The hardness of moulds and cores can be further increased by exposing them to free atmosphere for a short while after gassing. Sometimes a little heating at about 200°C also helps in increasing the hardness. Both metallic and wooden patterns can be used in the process. The varnish coating on the wooden patterns causes sticking of sand on the pattern surface. It can be prevented by withdrawing the pattern before gassing. But this will effect the dimensional accuracy. Therefore, a better remedy will be to apply a coat of cellulose lacquer over the pattern surface to prevent the said sticking of sand. A more effective alternative is to provide a zinc coat of about 0.05 to 0.13 mm thickness over the pattern surface, followed by spraying a layer of aluminium or brass of about 0.25 mm thickness and finally rubbing it with a smooth emery paper. The last remedy of course, is costlier.

Main Advantages:

1. It is a very speedy operation and, therefore, heavy orders at short notice can be easily handled.
2. A very high skill is not required to the operators.
3. No core and mould baking equipment is generally required.
4. The use of core plates is totally avoided.
5. Cores and moulds can be stored for longer periods, if required.
6. Same sand may be used for both moulds as well as core-making.
7. Less floor space is required because more frequent pouring can be done.
8. On account of higher dimensional accuracy, the rejection is less and requirement of further machining is minimised.
9. Use of heavy cranes is normally not required since the moulds are not to be shifted to ovens for baking.
10. Hollow core can be easily produced, effecting economy in the cost of core mix.
11. Higher permeability and better flowability of sand is achieved.
12. It can be used for both ferrous and non-ferrous castings.

1.26 CENTRIFUGAL CASTINGS

The process of centrifugal casting is also known as liquid forging. It consists of rotating the mould at a high speed as the molten metal is poured into it. Due to the centrifugal force the molten metal is directed outwards from the centre, towards the inside surface of the mould, with considerable pressure. As a result of this a uniform thickness of metal is deposited all along the inside surface of the mould, where it solidifies, and the impurities being lighter remain nearer to the axis of rotation. This process enables the production of castings with greater accuracy and better physical properties as compared to sand castings. It also enables the production of distinct surface details and dense metal structure. Although many different shapes can be cast through this process, but those with symmetrical shapes are best suited for it. The better physical properties of the castings are the result of proper directional solidification of the metal inside the mould. It is achieved because the denser (or cold) metal is automatically forced towards the outer side of the casting by the centrifugal force, whereas the hotter metal remains on the inner side of the casting to provide the required feeding of metal during solidification. The centrifugal casting methods can be classified as follows:

i) True Centrifugal Casting.

ii) Semi-Centrifugal Casting.

iii) Centrifuging.

i) True Centrifugal Casting:

The main features of a true centrifugal casting are that the axis of rotation of the mould and that of the casting are the same. Also the central hole through the casting is produced by the centrifugal force without the use of the central core. The axis of rotation of the mould may be horizontal, vertical or inclined at any suitable angle between 70° and 90°. End cores are usually employed at the two ends of the mould to prevent the molten metal from being thrown out at the ends. A few examples involving the application of this method are the hollow cast iron pipes, gun barrels, bushings, etc.

A typical horizontal true centrifugal casting machine is illustrated in Figure.1.53. It is shown having a large cylindrical mould of casting cast iron pipes. Similar equipment can be used for casting other cylindrical items. The mould consists of an outer metallic flask provided with a rammed sand lining inside. The mould is rotated between two sets of rollers as shown. The bottom rollers are mounted on a shaft driven by a variable speed motor mounted at one end. Pouring in the mould is done through a pouring basin formed on the body of a trolley. Initially, during pouring, the mould is rotated at a slow speed. After the pouring is over, the mould is rotated at a very fast speed to effect even distribution of the metal all along the inside surface of the mould and proper directional solidification. After solidification, the flask is replaced by a new one and the process repeated. Wall thickness of the casting is controlled by the volume of molten metal poured into the mould. Pouring temperatures range between 1482° to 1649°C.

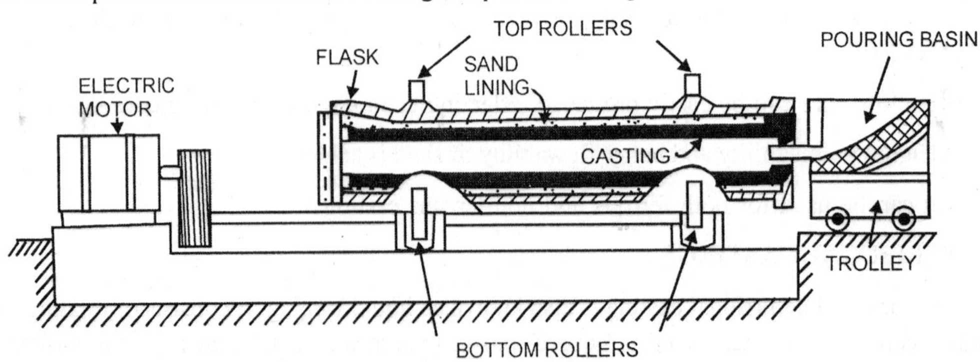

Figure. 1.53 Horizontal True Centrifugal Casting Machine

For sucessful casting, the application of correct spinning speeds is necessary. Slow speeds will not allow the molten metal to adhere to the inside surface of the mould and too high speeds will develop high stresses in the casting. The main factors effecting the selection of a proper speed are the size and metal of castings. Depending upon these factors the speed requirements may vary from 50 to 3,000 revolutions per minute.

There are two common horizontal axis true centrifugal casting processes:

1. De Lavaud Process.
2. Moore Sand Spun Process.

1. De Lavaud Process:

It employs a steel mould, provided with a refractory spray inside, which is completely surrounded by cooling water. The whole unit is mounted on wheels so that it can travel along an inclined track. The mould is rotated by an electric motor, capable of providing variable speeds. A hydraulic cylinder and plunger are used for moving the unit along the track. Initially the mould is brought to the raised end of the track and revolved. Pouring is started and as the same proceeds the mould starts travelling down the track at a constant speed. Thus a uniform layer of metal is deposited all along the inside surface of mould through a helical movement of the molten metal. After solidification, the casting is withdrawn and the process repeated for the next casting.

2. Moore Sand Spun Process:

This process employs a sand lined mould, driven as usual by variable speed electric motor. The mould does not travel but continues to rotate in its position. One end of the mould carries a tilting mechanism through which it can be raised or lowered. The metal is poured at this end in its raised position. Initially the mould is rotated at a slow speed and is gradually lowered to horizontal position as the pouring continues. After the horizontal position is achieved and the pouring stopped, the speed of rotation is increased and maintained till solidification. After that the rotation is stopped, casting removed and the mould made ready for next casting.

Vertical and inclined axes:

Vertical or inclined axes of rotation for moulds are adopted generally for short length castings. A general defect noticed in casting produced in these positions is that the produced central hole is not truly cylindrical. Instead of this it is a parabolical hole. However, this defect can be minimised considerably by adopting high spinning speeds. It is important that during pouring the molten metal should be directed towards the centre of the mould bottom. Permanent metal moulds, sand-lined moulds or graphite moulds can be used in the process. The main advantage of adopting these axes of rotation is the convenience in metal pouring and ejection of casting.

ii) Semi-Centrifugal Casting:

This process, which is also known as profiled centrifugal casting, is widely used for relatively large castings which are symmetrical in shape, such as discs, pulleys, wheels, gears, etc. In this method the mould is rotated about a vertical axis and the metal poured through a central sprue. It is not necessary to cast only one mould at a time. Several moulds can be stacked together, one over the other, and fed simultaneously through a common central sprue, as shown in Figure 1.54. This provision increases the rate of production considerably. The centrifugal force is used to feed the metal outwards to fill the mould cavities completely. The centre of the castings is usually solid, but, if required, a dry sand central core may be used to produce the central hole. The speed of rotation of these moulds is much lower than that in true centrifugal casting. With the result, the pressure developed is too low and the impurities are not directed towards the centre as effectively as in true centrifugal casting. The speed of rotation of these moulds is such that a linear speed of about 180 meters per minute is obtained on the outer edge of the casting. The moulds used may be of green sand, dry sand, metal or any other suitable material.

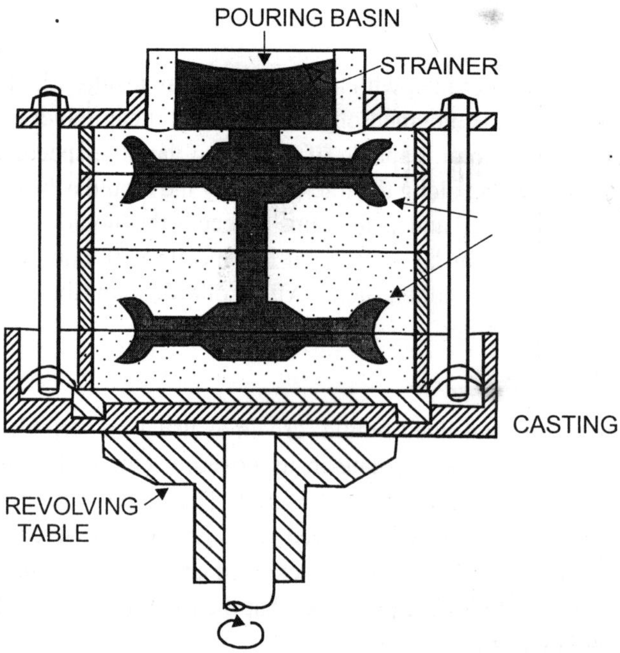

Figure 1.54 Section Through A Semicentrifugal Casting Machine

Advantage of Semi-Centrifugal Casting:

★ Casting shapes can be more complicated than true centrifugal casting.

Product Application:

Disk shaped parts like

★ Wheels,

★ Sheaves,

★ Pulleys,

★ Fly wheels.

★ Turbocharger diaphragm etc. can be made.

iii) Centrifuging

This is also sometimes known as pressure casting. It mainly differes from true centrifugal casting methods in that, unlike the later two, the axis of rotation of the moulds do not coincide with each other, as the moulds are situated at a certain distance from the central vertial axis of rotation all around the same. Shapes of castings do not carry any limitations in this method and a variety of shapes can be cast. A number of small mould cavities are made around a common central sprue and connected to the same through radial gates. For a higher rate of production the stacked moulds can be used with advantage. As in semi-centrifugal method, in this method also the mould assembly is rotated about a vertical axis and centrifugal force used to force the molten metal from the central sprue into the mould cavities through the radial gates. Sectional view through a typical mould for centrifuging is shown in Figure 1.55.

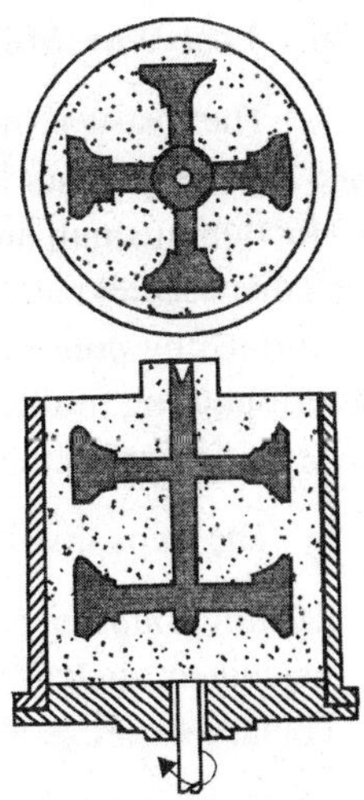

Figure. 1.55 Section through a Mould Showing the Centrifuging Process

Advantages and Disadvantages of true centrifugal casting:

Advantages:

1. The castings produced are very sound, having a clean metal as most of the impurities are collected on the inner surface and can be removed later on through machining.
2. The percentage of rejects is very low.
3. The castings have dense metal with very good mechanical properties.
4. An effective directional solidification from the outer surface towards the inner is invariably achieved unless the wall thickness may, of course, require the use of some exothermic material to prevent shrinkage defects.
5. The requirement of a core to produce a central hole is avoided in most of the cases.
6. The need of separate gates and risers is totally eliminated.
7. Even minute surface details on castings may be easily obtained.
8. The production rate is sufficiently high.
9. Thin sections and intricate shapes can be easily cast.
10. Inspection of castings is considerably simplified on account of the fact that the defects, if any, are normally found to exist on the surface and not within the casting.

Disadvantages:

1. All shapes cannot be cast through this process.
2. The complete equipment requires a heavy initial investment.
3. Its maintenance also is quite expensive and its operation needs the employment of skilled labour.

1.26.1 Lost-Wax Method:

The lost-wax method, sometimes, also called simply as 'Precision-investment casting' has been used for many years by jewelers and dentists. Since world-war-II, the method has been adapted to the production of small and precise industrial castings. Basically, the method involves the use of expendable (heat disposable) pattern surrounded with a shell of refractory material to form the casting mould. Castings are formed by pouring molten metal in the mould cavities created by melting out the pattern. Since the pattern pouring molten metal in the mould cavities created by melting out the pattern. Since the pattern made of wax is melted out and gets destroyed, that is, why the name 'lost-wax method'.

The steps involved in this method are:

i) *Making a master pattern* of the part to be cast. The pattern is usually made of a metal that can be easily machined, such as brass, aluminium alloy or steel, or a fusible alloy (alloy of tin, lead and bismuth). It is made oversize to compensate for wax and metal shrinkage.

ii) *Making a composite die* to the master pattern (by casting) for casting the wax/plastic patterns. The die-material is a low melting-point alloy like bismuth alloys or even aluminium and cast iron. The master dies can be directly made from die-blocks of steel by machining the cavities.

iii) *Making of wax/plastic patterns.* For this, the two halves of the master die are clamped together and molten wax/plastic is injected into the die cavity under pressure. The molten wax is slightly above its melting point and the injection pressure is about 4 bars with the dies preheated to about 65°C. For plastic materials (Polystyrene, Polythenes etc.,) the injection pressure is of the order of 35 bars with higher temperature.

iv) *Assembling the wax patterns to a wax gating system.* Several small wax patterns are assembled together to a wax gating system connected to a central sprue by wax welding with the aid of heated tools. In this manner, a cluster or "tree" of wax patterns is formed (Figure 1.56).

v) *Investing the wax patterns:* The wax patterns are invested in two stages. Firstly, a thin coating (about 1mm) of primary investment slurry is made around the wax patterns by dipping these in the slurry. This slurry is made by mixing extremely fine silica sand with a water / ethyl silicate or gypsum solution. This slurry coating being indirect contact with the surface quality of the castings. After this primary coating has dried sufficiently, the final investment layer consists of coarser and less expensive slurry. The final investment moulds may be either solid type or shell type. A solid type mould is formed by placing a metal container type flask over the cluster of patterns and then pouring a hard-setting moulding material into the flask (Figure 1.56(c)). Shell type investment moulds are made by dipping the cluster of patterns in a ceramic slurry and the procedure is repeated until the required thickness of the mould or shell (about 6.5 mm) is obtained. The common refractory used for investment material is silica. The binder is gypsum or water-based sodium silicate.

vi) *Melting out the wax patterns and baking the mould:* The finished mould is dried in air for 2 to 3 hours and then baked in an oven for 2 hours to melt out the wax. At a temperature of 100 to 120°C, the wax melts and the moulds are inverted so that all or most of the wax will run out of the sprue.

vii) *Melting the metal and pouring the mould:* The mould of flask is transferred to a drying furnace where it is first held at 150°C and then gradually heated to 800 to 900°C. This wil! vaporize any remaining wax in the mould. The preheated mould also ensures that the molten metal will completely

fill the mould

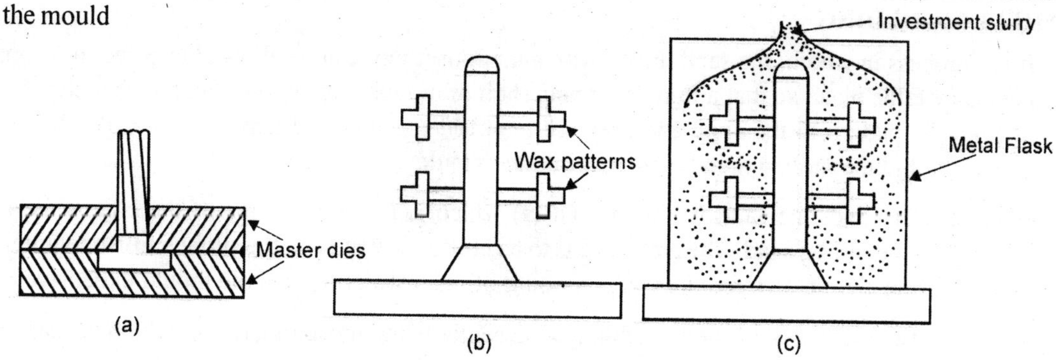

Figure. 1.56 Lost-wax method

cavity. The preheated mould is poured with molten metal which fills the cavity and is allowed to cool and solidify. Another way to fill the cavity is to place the mould on a vacuum table and drawing a vacuum through the slurry which will help suck the metal right into the mould or by placing the mould in a centrifugal casting machine and pouring the metal in this manner.

viii) *Shaking out the castings,* removing gates and feeders. After cooling, the fragile mould material is broken away freeing the castings. The gates and runners are removed in the normal way.

Advantages of lost-wax method:

1. Intricate details can be cast.
2. Undercuts and other shapes, which would not allow the withdrawal of a normal pattern, are easily obtained.
3. More than one casting can be made at a time.
4. Unmachinable alloys (heat resisting steels or nimonic alloys) can be cast.
5. The surface is very smooth (1.5 to 2.25μm) and there is no parting line
6. High accuracy can be obtained and so that much of complicated and costly machining can be eliminated.

Limitations:

1. The process has the limitations in use of and location of holes.
2. The process is involved and thus expensives.
3. The parts are limited in size to a few kg.

Product applications:

1. Parts for computers and data processing equipment.
2. Parts for food and beverage machinery.
3. Parts for machine tools and accessories, scientific instruments, sewing machines.
4. Parts for aerospace industry, in aircraft engines, frames, fuel systems and instruments.
5. Parts for movie cameras and projection.
6. Rock drill thread chaser holder blocks.
7. Nozzles, buckets, vanes and blades for gas turbines.

1.27 SHELL MOULDING

It is a process in which, the sand mixed with a thermosetting resin is allowed to come into contact with a heated metallic plate, so that a thin and strong shell of mould is form the pattern. Then the shell is removed from the pattern to the cope and drag are removed together and kept in a flask the necessary back up material and the molten metal is poured into the mould.

Generally, dry and fine sand (90 to 140 GRN) which is completely free of the clay is used for preparing the shell moulding sand. The grain size is to be chosen depends on the surface finish required on the casting. Too fine a grain size requires large amount of resin which makes the mould expensive.

The synthetic resins used in shell moulding are essentially thermosetting resins, which get hardened irreversibly by heat. The resins most widely used, are the phenol formaldehyde resins. Combined with sand, they have very high strength and resistance of heat. The phenolic resins used in shell moulding usually are of the two-stage type, that is, the resin has excess phenol and acts like a thermoplastic material. During coating with the sand the resin is combined with a catalyst such as hexa-methylene-tetramine (hexa) in a proportion of about 14 to 16% so as to develop the thermosetting characteristics. The curing temperature for these would be around 150°C and the time required would be 50 to 60 s.

Typical sand compositions for shell moulding are given in the table 2.

Additives may sometimes, be added into the sand mixture for good surface finish and avoid thermal cracking during pouring. Some of the additives used are coal dust, pulverised slag, manganese dioxide, calcium carbonate, ammonium boroflouride and magnesium silicofluoride. Some lubricants such as calcium stearate and zinc stearate may also be added to the resin sand mixture to improve the flowability of the sand and permit easy release of the shell from the pattern.

The first step in preparing the shell mould is the preparation of the sand mixture in such a way that each of the sand grain is thoroughly coated with resin. To achieve this, first the sand, hexa and additives which are all dry, are mixed inside a Mueller for a period of 1 min. Then the liquid resin is added and mixing is continued for another 3 min. To this cold or warm air is introduced into the muller and the mixing is continued till all the liquid is removed from the mixture and coating of the grains is achieved to the desired degree.

Table 2 Shell sand compositions in percentages

Casting material	Silica Sand	Zircon Sand	Resin	Additives
1 Low carbon and alloy steels	63	30	5	2
2 Medium and high carbon steel	--	96	3	1
3 Grey cast iron	60	35	4	1
4 Brass and bronze	90	--	6	4
5 Aluminium alloys	95	--	4	1
6 Brass and bronze	--	95	4	1

Since the sand resin mixture is to be cured at about 150°C temperature, only metal patterns with the associated gating are used. The metal used for preparing patterns is grey cast iron, mainly because of its easy availability and excellent stability at the temperatures involved in the process. But sometimes additional risering provision is required as the cooling in shell moulds is slow.

The metallic pattern plate is heated to a temperature of 200°C to 350°C depending on the type of the pattern. It is very essential that the pattern plat is uniformly heated so that the temperature variation across the whole pattern is within 25 to 40°C depending on the size of the pattern. A silicone release agent is sprayed on the pattern and the metal plate. The heated pattern is securely fixed to a dump box, as shown in figure 1.57(a), wherein the coated sand in an amount larger than required to form the shell of necessary thickness is already filled in.

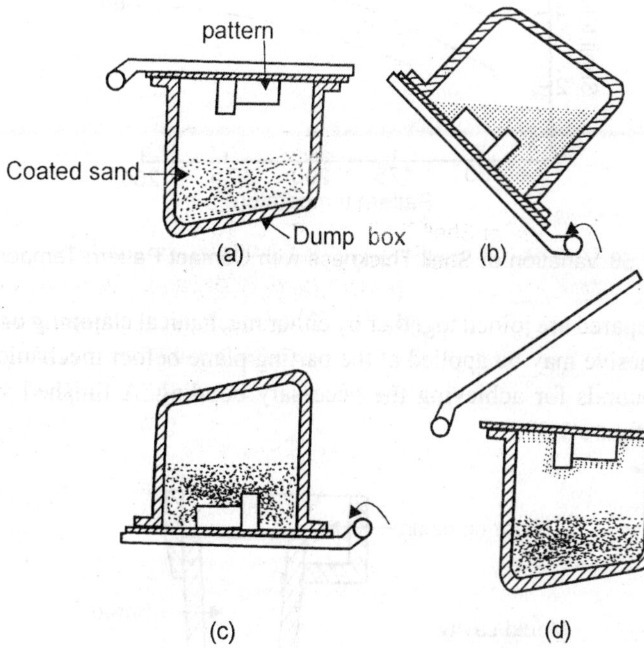

Figure. 1.57 Shell Moulding Procedure

Then the dump box is rotated as shown in figure 1.57(b) so that the coated sand falls on the heated pattern. The heat from the pattern melts the resin adjacent to it thus causing the sand mixture to adhere to the pattern.

When a desired thickness of shell is achieved, the dump box is rotated backwards by 180° so that the excess sand falls back into the box, leaving the formed shell intact with the pattern as in figure 1.57(d). The average shell thickness achieved depends on the temperature of the pattern and the time the coated sand remains in contact with the heated pattern. Figure 1.58 shows typical shell thicknesses that can be obtained with various pattern temperatures and contact times. The actual shell thicknesses required depends on the pouring metal temperature and the casting complexity. This may normally be achieved by trail and error method.

The shell along with the pattern plate is kept in an electric or gas fired oven for curing the shell. The curing of the shell should be done as per requirements only because over-curing may cause the mould to break down as the resin would burn out. The under-curing may result in blow holes in the casting or the shell may break during handling because of the lack of strength.

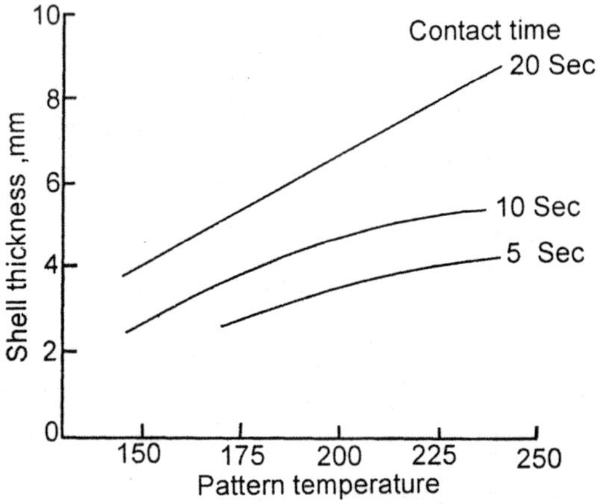

Figure. 1.58 Variation of Shell Thickness with Contact Pattern Temperature

The shells thus prepared are joined together by either mechanical clamping or by adhesive bonding. The resin used as an adhesive may be applied at the parting plane befoer mechanical clamping and then allowed for 20 to 40 seconds for achieving the necessary bonding. A finished shell mould ready for pouring, is presented in figure 1.59.

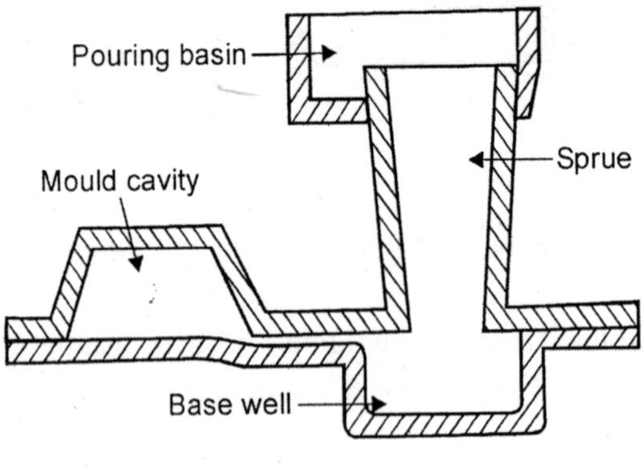

Figure. 1.59 Shell mould ready for pouring

Since the shells are thin, they may require some outside support so that they can withstand the pressure of the molten metal. A metallic enclosure to closely fit the exterior of the shell is ideal, but is too expensive and therefore impractical. Alternatively, a cast iron shot is generally preferred as it occupies any contour without unduly applying any pressure on the shell. With such a backup material, it is possible to reduce the shell thickness to an economical level.

Advantages:

1. Shell mould castings are generally more dimensionally accurate than sand castings. It is possible to obtain a tolerance of ± 0.25 mm for steel castings and ± 0.35 mm for gray cast iron castings under normal working conditions. In the case of close toleranced shell moulds one may obtain it in the range of ± 0.03 to ± 0.13 mm for specific applications.

2. A smoother surface can be obtained in shell castings. This is primarily achieved by the finer size grain used. The typical range of roughness is of the order of 3 to 6 microns.

3. Draft angles which are lower than the sand castings are required in shell moulds. The reduction in draft angles may be from 50 to 75% which considerably saves the material costs and the subsequent machining costs.

4. Sometimes, special cores may be eliminated in shell moulding. Since the sand has high strength the mould could be designed in such a manner that internal cavities can be formed directly with shell mould itself without the need of shell cores.

5. Also, very thin sections (upto 0.25 mm) of the type of air cooled cylinder heads can be readily made by the shell moulding because of the higher strength of the sand used for moulding.

6. Permeability of the shell is high and therefore no gas including occur.

7. Very small amount of sand needs to be used.

8. Mechanisation is readily possible because of the simple processing involved in shell moulding.

Limitations:

1. The patterns are very expensive and therefore are economical only if used in large scale production. In a typical application, shell moulding becomes economical over sand moulding above 15000 pieces because of the higher pattern cost.

2. The size of the casting obtained by shell moulding is limited. Generally castings weighing upto 200 kg can be made, though in smaller quantity castings upto a weight of 450 kg were made.

3. Highly complicated shapes cannot be obtained.

4. More sophisticated equipment is needed for handling the shell mouldings such as those required for heated metal pattern.

Applications:

Cylinders and cylinder heads for air cooled IC engines, automobile transmission parts, cast tooth bevel gears, brake beam, radome hubs, track rollers for crawler tractors, transmission planet carrier, steel eyes, gear blanks, chain seat bracket, refrigerator valve plate, small crank shafts are some of the common applications of shell mould castings.

1.27.1 Precision Investment Casting

This is the process where the mould is prepared around an expendable pattern. The first step in this process is the preparation of the pattern for every casting made. To do this, molten wax which is used as the pattern material is injected under pressure of about 2.5 MPa into a metallic die which has the cavity of the casting to be made. The wax when allowed to solidify would produce the pattern. To this wax pattern, gates, runners and any other details required are appended by applying heat.

To make the mould, the prepared pattern is dipped into a slurry made by suspending fine ceramic materials in a liquid is allowed to drain off from the pattern. Dry refractory grains such as fused silica or zircon are "stuccoed" on this liquid ceramic coating. Thus a small shell is formed around the wax pattern. The shell is cured and then the process of dipping and stuccoing is continued with ceramic slurries of gradually increasing grain sizes. Finally when a shell thickness of 6 to 15 mm is reached, the mould is ready for further processing. The shell thickness required depends on the casting shape and mass, type of ceramic and the binder used.

The next step in the process is to remove the pattern from the mould, which is done by heating the mould to melt the pattern. The melted wax is completely drained through the sprue by inverting the mould. Any wax remnants in the mould are dissolved with the help of the hot vapour of a solvent, such as trichloroethylene.

The moulds are then pre heated to a temperature of 100 to 1000°C, depending on this size, complexity and the metal of the casting. This is done to reduce any last traces of wax left off and permit proper filling of all mould sections which are too thin to be filled in a cold mould.

The molten metal is poured into the mould under gravity, under slight pressure, or by evacuating the mould first. The method chosen depends on the type of casting.

Other pattern materials used are plastics and mercury in place of wax. In the process called 'Mercast', the mercury is kept under -57°C where the mercury is frozen. The complete mould preparation is to be undertaken at a temperature below -38°C. The main advantage of mercury as a pattern material is that it does not expand when changed from solid to liquid state as wax. But the main disadvantage is keeping the pattern at such low temperature the shell of necessary thickness is already filled in.

1.28 CONTINUOUS CASTING

The process consists of continuously pouring molten metal into a mould, which has the facilities for rapidly chilling the metal to the point of solidification, and then withdrawing it from the mould.

Experimental work and research has proved that there are many opportunities for saving in the continuous casting process of metals. Following processes are typical of those in use or in the process of development.

(i) Reciprocating Mould Process of Continuous Slab Casting

In this process, a reciprocating water-cooled copper mould is used, the down stroke being synchronized with the discharge rate of the slab. Molten metal is poured into the holding furnace and is discharged to the mould through a 12 mm tube at the rate of 10,000 kg per hour. Molten metal is distributed across the mould from a submerged, horizontal crosspiece, the level of the metal being held constant at all times.

Pouring rate of molten metal is controlled by a needle valve through the top of the holding furnace. As the metal becomes chilled in the lower part of the mould, it is discharged at a constant rate and enters the withdrawing rolls. These are synchronized with the downward movement of the mould and are mounted just above a circular saw which cuts the slab to required lengths.

Brass slabs produced by this process are further processed, by cold rolling, into sheets and strips.

(ii) Asarco Process for Continuous Cast Shapes

This process differs from other continuous processes in that the forming die or mould is integral with the furnace and there is no problem of controlling the flow of metal. Metal is fed by gravity into the mould from the furnace as it is continuously solidified and withdrawn by the rolls below.

An important feature of this process is the water-cooled, graphite-forming die, which is self-lubricating, has excellent thermal shock resistance and is not attacked by copper-base alloys. The upper end, being in the molten metal, acts as a riser and compensates for any shrinkage which might take place during solidification, while simultaneously acting as an effective path for dissipating evolved gases. These dies are easily machined to shape. Products may be produced from 10 mm to 130 mm in diameter. Multiple production from a single die permits casting the small section rods.

In starting the process, a rod of the same shape as that to be cast is placed between the drawing rolls and inserted into the die. This rod is tipped with a short length of the alloy to be cast. As the molten metal enters the die, it melts the end surface of the rod, forming a perfect joint. Casting cycle is then started by the drawing rolls and the molten metal is continuously solidified as it is chilled and withdrawn from the die.

When the casting leaves the furnace, it ultimately reaches the sawing floor where it is cut to desired length while still in motion. A tilting receiver takes the work and drops it to a horizontal positions, and from there it goes for inspection and straightening operations.

For phosphorized copper and many of the standard bronzes, the process has proved successful. Alloy compositions may be produced with satisfactory commercial finish as rounds, tubes, squares, or special shapes. Physical properties are superior to permanent mould and sand castings.

(iii) William Continuous Casting Process for Steel

This process, developed for continuous casting of carbon and alloy steels, utilizes thin walled brass moulds having cross-sectional areas upto 280 square cm. These moulds are perferably oval in cross-section.

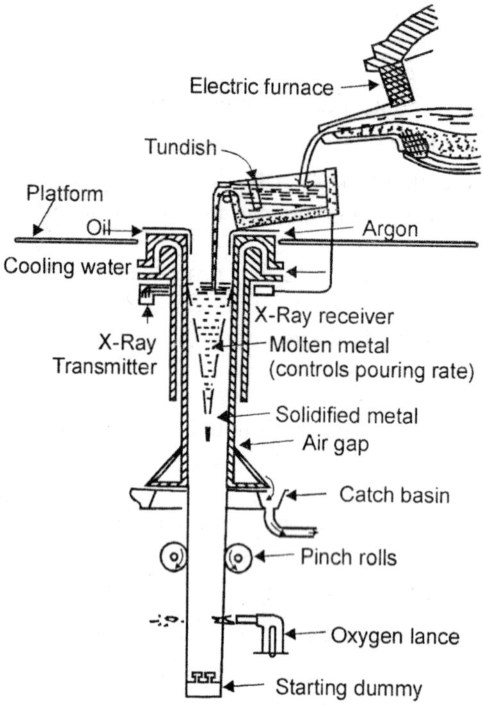

Figure. 1.60 Continuous Method for Casting Steel

A small stream of metal is poured into the mould from an electric holding furnace a rate controlled by the metal level in the mould. The mould must necessarily be constructed of a material having a high heat conductivity and one which is not easily wetted by the liquid metal.

Rapid mould cooling is essential for the success for this process, and results in:

i) Improved mould life.

ii) Less segregation.

iii) Smaller grain structure, and

iv) A better surface.

Actually the metal next to the mould wall solidifies only a few centimetres below the top surface and shrinks slightly from the mould sides. As the cast section leaves the cooled moulds, it passes through a section that controls the rate of cooling and then to the drawing and straightening rolls.

Below this point, it is cut to length by an oxy-acetylene torch and finally lowered to a horizontal position. Steel blooms and billets produced by his process, have good crystalline structure, little segregation, uniform section, and a size close to that required for many rolling mills.

(iv) Alcoa Direct-Chill Process for Continuous Casting of Aluminium Ingots

This process consists of pouring molten aluminium from a holding furnace through a refractory trough into shallow, stationary moulds, the bottoms of which rest on a hydraulic elevator. When the metal at the bottom of the mould becomes chilled, the elevator drops at a rate of 4 to 125 cm per minute. As the ingots descend, they are sprayed with water to complete their solidification.

The shallow moulds used are made in sizes of 30 by 90 and 30 by 120 cm and are rectangular in shape. Other sizes and circular shapes can be cast if desired. Length of the ingots is regulated to give a convenient size for rolling and is usually 3-5 metres.

This process leaves a rough surface on the ingots which must be removed by milling before they can be further processed in the rolling mill. Most of the aluminium used in the United States is cast by this process.

(v) Direct Casting of Sheet (Hazelett Process)

Much work has been done in the attempt to cast various metals into sheet form between water-cooled rolls, as is done in the manufacture of glass by the continuous method.

So far this process has not proved commercially successful because the metal solidifies faster at the edges of the strip than at the centre with the result that it builds up a greater thickness at the edges than is desired. This forces the cooling rolls apart and changes the sheet thickness.

Also, in casting solid-solution alloys, the process is affected by segregation problems, resulting ultimately in wide variations of metal analysis. Some progress has been made in the elimination of these difficulties by increasing rolling speeds and reducing the contact area that the metal has with rolls.

Applications of Continuous Casting Processes

The continuous casting process has been used for several years for casting non-ferrous metals. It has been recently used for steel, and for preparing rounds, squares, pipes, tubes, sheets, etc.

When fully developed, this process is claimed to have many advantages over the old methods of rolling since heavy equipment like ingot moulds, blooming mills, etc., are completely dispensed with. Due to lower production costs, steels may also become cheaper.

1.29 INVESTMENT CASTING

Introduction

i) Investment mould casting process is also known as the LOST WAX PROCESS, or PRECISION CASTING.

ii) The term *Investment* refers to a clock or special covering apparel.

iii) In investment casting, the clock is a refractory mould which surrounds the precoated wax pattern.

iv) The term *investment casting* is used to describe a group a processes in which moulds are produced from liquid refractory slurries. These, containing finely divided materials, give the mould a fine surface texture which is subsequently transmitted to the castings.

v) Investment casting processes are of two types, one using expendable and the other based on permanent pattern.

vi) Investment casting forms an expendable pattern of wax in a die which in turn is employed to form a mould in an investment material. After the mould of investment material is set, the wax (pattern) is melted or burned and the molten metal is poured in the cavity thus formed.

Procedure steps in the Investment Casting Process

1. Producing a die for making wax patterns

- Dies may be made either by machining cavities in two or more matching blocks of steel or by casting a low melting point alloy around a (metal) master pattern.
- For long production runs, steel dies are most satisfactory. They are machined from the soild blocks by die sinking and are assembled in the tool room. The dies thus formed, achieve the highest standard of accuracy and have considerable longer life.
- Dies of low melting point alloys are made by casting and require a master pattern or metal replica of the final casting.
- The master pattern is given an allowance for subsequent contractions of pattern and metal, up to 2%.
- The master pattern is used to produce two halves of the die or mould by embedding it in plaster or clay and casting one die half at a time by pouring a low melting point alloy such as that of bismuth or lead.
- Die halves are sent for necessary machining and drilling the gate through which wax is to be injected for preparing expendable patterns.
- Cast dies are more economical than sunk dies for short production runs.

2. Making of expendable patterns and gating system

- An expendable pattern may be made out of plastic, tin, fozen mercury (Mercast process) or wax; the wax being more commonly used.
- Waxes employed are blends of beeswax, carnauba, ceresin, acrawax, paraffin and other resins usually obtained as proprietary mixtures.
- Wax patterns are produced using wax-injection machines. Wax at 150–170°F is injected into the die (halves clamped in position) at a pressure ranging from 7 to 70 kg/cm^2 [Figure 1.61(a)].
- Small shallow vents cut in the parting surface of the die provide adequate venting.

- Gates and sprues are formed in the same manner as the wax patterns and are attached to the patterns and are attached to the pattern assembly.
- A pattern is produced for each casting to be made. Individual wax patterns thus produced may be wax-welded together to build a larger assembly to
  i) enable many small castings to be poured in one group for economy.
  ii) compensate for teh solidification shrinkage of the metal so as to produce sound castings.
  iii) form a system of runners and feeders that allows metal to flow into each casting cavity.

3. *Precoating the pattern assembly*
   - The wax pattern assembly is dipped into a slurry of a refractory coating material.
   - A typical slurry consists of 325-mesh silica flour suspended in ethyl silicate solution.
   - Wax pattern assembly is, next, sprinkled with 40 to 50 AFS silica sand and is permitted to dry.
   - The slurry used for precoating imparts a fine textured surface at the mould-metal interface and sprinkled sand serves to key the precoat to the regular investment.

4. *Investing the wax pattern assembly for the production of moulds* [Figure 1.61(d)]
   - The precoated wax pattern assembly is then invested in the mould.
   - Investment moulds may be formed by either Solid moulding or Shell moulding.
   - **Solid Moulding**
     i) The wax pattern assembly is first placed in a metal flask.
     ii) A specially formulated ceramic slurry (the investment) is then poured into the flask and is allowed to harden around the wax pattern assembly.
     iii) The flask may be vibrated so that the investment material settles and air bubbles rise away from the pattern.
     iv) The investment hardens after about 8 hours of air drying.
     v) A typical investment moulding mixture consists of 91.2% sand, 33.8% water, 6.5% primary calcium phosphate and 2.3% MgO, 300-mesh.
   - **Shell Moulding**
     i) It consists of coating the wax pattern assembly by dipping the same in a (ceramic) slurry made of 325-mesh refractory flour (fused silica, alumina etc.) and liquid binders (colloidal silica, sodium silicate etc.) and immediately covering or dusting it with a 20 to 70-mesh refractory stucco.

ii) After the (first) dip has dried, the process is repeated so that each dip and dusting form another layer of ceramic on the pattern assembly.

iii) A number of dips are given in order to build a shell thickness of the order of 6 to 12 mm.

iv) The first coating imparted to the wax pattern assembly is composed of very fine particles to produce a good surface finish whereas the following coats are coarser to build up the required shell thickness.

5. *Removing wax pattern from the investment mould* [Figure 1.61(d)].

a) *Solid moulds are placed upside down in progressive furnaces.*

First of all, the wax pattern is melted and the wax is drained from the mould.

As the mould progresses through the furnace, it experiences high temperatures and gets cured in about 16 hours.

The oven which melts wax is kept at a temperature of 200 to 400°F whereas the curing or burn out oven works at a temperature of the order of 1300° to 1900°F.

b) *Wax patterns from shell moulds may be removed in the following ways:*

i) The mould may be plunged upside down directly into a furnace kept at about 1400 – 2000°F.

ii) The mould may be kept inverted in a an autoclave and heated quickly to about 300°F under pressure.

iii) Since wax has a coefficient of expansion about 10 times that of usual investment mould material, the forces created by heating may damage some mould. For this reason, instead by heating, wax pattern may be removed by dissolving wax in a solvent such as trichloroethylene. It is more effective in removing wax from the investment shell moulds.

After removing the wax, the shell can be quickly fired at the desired temperature.

6. *Pouring metals into the moulds* [Figure 1.61 (f)]

- The metal to be poured may be melted in a coreless induction or some other furnace and brought in small ladles to the preheated moulds for pouring.
- Moulds are preheated (before getting poured) to about 1900°F depending upon the metal to be poured example for light alloys and for steels, the preheated mould temperatures should be of the order of 400 to 800°F and 1200 to 1800°F respectively.
- Moulds are preheated because,

i) Preheating vaporizes any remaining wax in the moulds,

ii) Metal may flow more easily and fill every detail of the moulds,

iii) Expansion of the mould cavity helps compensating for the solid shrinkages of the castings and the wax patterns.

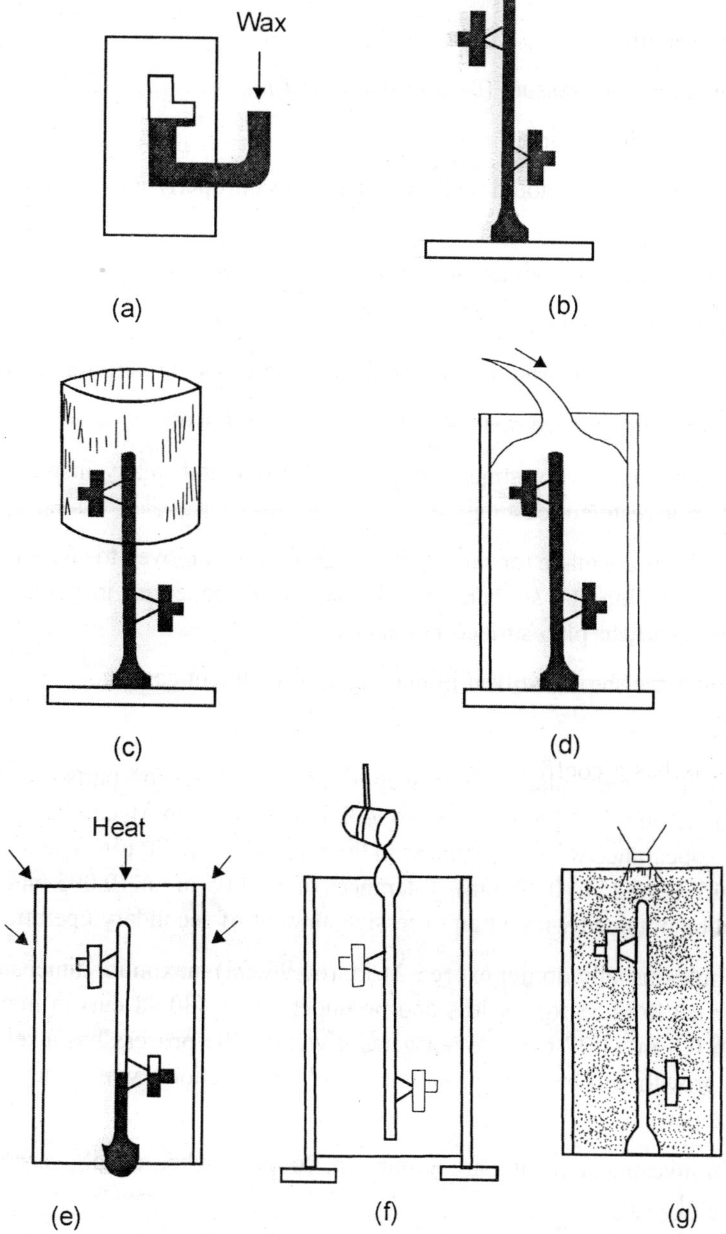

(a) Wax injected into die to make pattern, (b) Patterns have been gated to centra sprue, (c) Placing metal flask around the pattern assembly, (d) Investing the wax pattern assembly, (e) Removing wax pattern from investment mould, (f) Pouring molten metal into the mould, (g) Removing casting from the mould by breaking mould material.

Figure 1.61 Steps involved in making investment casting

- Preheated moulds may be filled with molten metal,
  - i) under gravity,
  - ii) under direct air pressure (0.35 to 0.7 kg/cm^2),
  - iii) under centrifugal force.

Molten metal forced into the mould under pressure flows properly into sharp details, this sections and intricate shapes.

7. *After they are solidified, the castings are removed from the mould.*
8. *Cleaning, Finishing and Inspection*
   - Each casting is separated from the assembly and the gates etc., are removed.
   - The cast part is sand blasted and (normally) it is ready for use.
   - Since the standards for investment castings often demand, many inspection operations are carried out on investment castings.
   - Visual inspections are made for surface faults, gages are employed to check casting dimensions, X-rays determine internal soundness and Fluorescent penetrant inspection is carried out to reveal cracks, surface pits, surface porosity etc.

Figure 1.61 shows the steps involved in making an investment casting.

Advantages and limitations

Investment casting is particularly advantageous for small precision parts of intricate design that can be made in multiple moulds. Thin walls down to 0.030 in. (0.76 mm), can be cast readily. Surfaces have a smooth matte appearance with a roughness in the range of 60 to 90 μin. The machining allowance is about 0.010 to 0.015 in. (0.25 to 0.38 mm). Tolerances of ± 0.005 in./in. (0.005 mm / mm) are normal; closer tolerances can be obtained without an excessive amount of secondary operations.

Investment moulds usually do not exceed 24 in. (60.96 cm) maximum dimension. Preferably, the casting should weigh 10 lb (4.53 kg) or less and be under 12 in. (30.48 cm) in maximum dimension. Sections thicker than 0.5 in. (1.27 cm) are not generally cast. The process has a relatively low rate of solidification. Grain growth will be more pronounced in larger sections, which may limit the toughness and fatigue life of the part.

Metals used in investment casting are aluminum, copper, nickel, cobalt, carbon and alloy steels, stainless steels, and tool steels.

1.30 CERAMIC PROCESS

The Ceramic or Shaw process, developed in England by Clifford and Noel Shaw, is somewhat similar to the investment casting process in that a ceramic slurry is poured over a pattern. In this case, however, the pattern, made out of plastic, plaster, wood, metal, or rubber, is reusable. The slurry hardens on the pattern almost immediately and becomes a strong green ceramic of the consistency of vulcanized rubber. It is lifted off the pattern while it is still in the rubber like phase. The mould is ignited with a torch to burn off the volatile portion of the mix. It is then put in a furnace and baked at 1800°F (982°C), resulting in a rigid refractory mould. The mould can be poured while still hot. The Shaw process finds its biggest advantage in being able to mould steels to very precise dimensions. Steel castings weighing over 100 lb (45.35 kg) have been made by this process. On small castings, tolerances may be held to within 0.001 in./in., but on larger parts the tolerance is generally about 0.010 in./in. (0.25 mm/mm). The surface finish varies from 80 to 120 μin. Pattern to finish casting can be done in 2 hr. The major tonnage of Shaw cast tooling goes into drop-forging dies, but the process is also used in making plastic moulds, die-casting moulds, glass moulds, stamping dies, and extrusion dies.

1.31 FULL MOULD CASTING

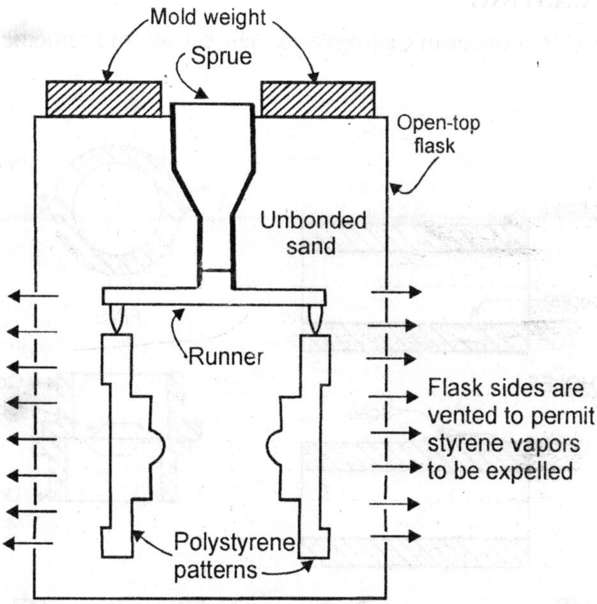

Figure 1.62 A pouring cup is placed around the exposed, uncoated EPS sprue top. weights are placed on top of the mould to counteract the metalostatic casting forces.

Full mould casting is also known as lost foam casting and EPS casting, for evaporative polystyrene patterns. The process begins with making a pattern out of expandable polystyrene. The patterns are produced by blowing the preexpanded beads into a mould cavity and heating them with steam to fuse the individual beads into a solid mass.

After the patterns are made, they are aged (warm forced air is often used in production to obtain the required pattern shrinkage). The patterns are then assembled, if necessary, by means of air-set or hot-melt adhesives. It is typical to assemble may foam patterns to a single gating system (cluster), similar to that of investment casting.

The cluster of patterns is coated with a special permeable washed, dried, and invested in a flask for pouring. The permeable wash prevents the metal from contacting the sand in the flask, while allowing the gas generated by vaporization of the pattern to escape through the loose unbonded sand that surrounds the pattern. Since the pattern is not removed from the mould but is replaced by the molten metal, the term full-mould is used.

During investment in the flask, the loose sand must be compacted. This is accomplished with a mechanical compaction table. Once the foam patterns have been invested and the flask properly compacted, a pouring cup on basin is set around the exposed, uncoated EPS sprue top (Figure. 1.62). Molten metal is then poured directly onto the foam, which vaporizes and is replaced by a casting that will accurately duplicate the finest details of the foam pattern.

After the casting has cooled, the flask is dumped and the loose sand flows away from the casting. Because of the gentle nature of removal, most of the wash coating remains on the casting and is easily removed by direct sand blasting.

1.32 DEFECTS IN CASTING

Various defects that occur in castings and their causes and remedies are tabulated in the following pages.

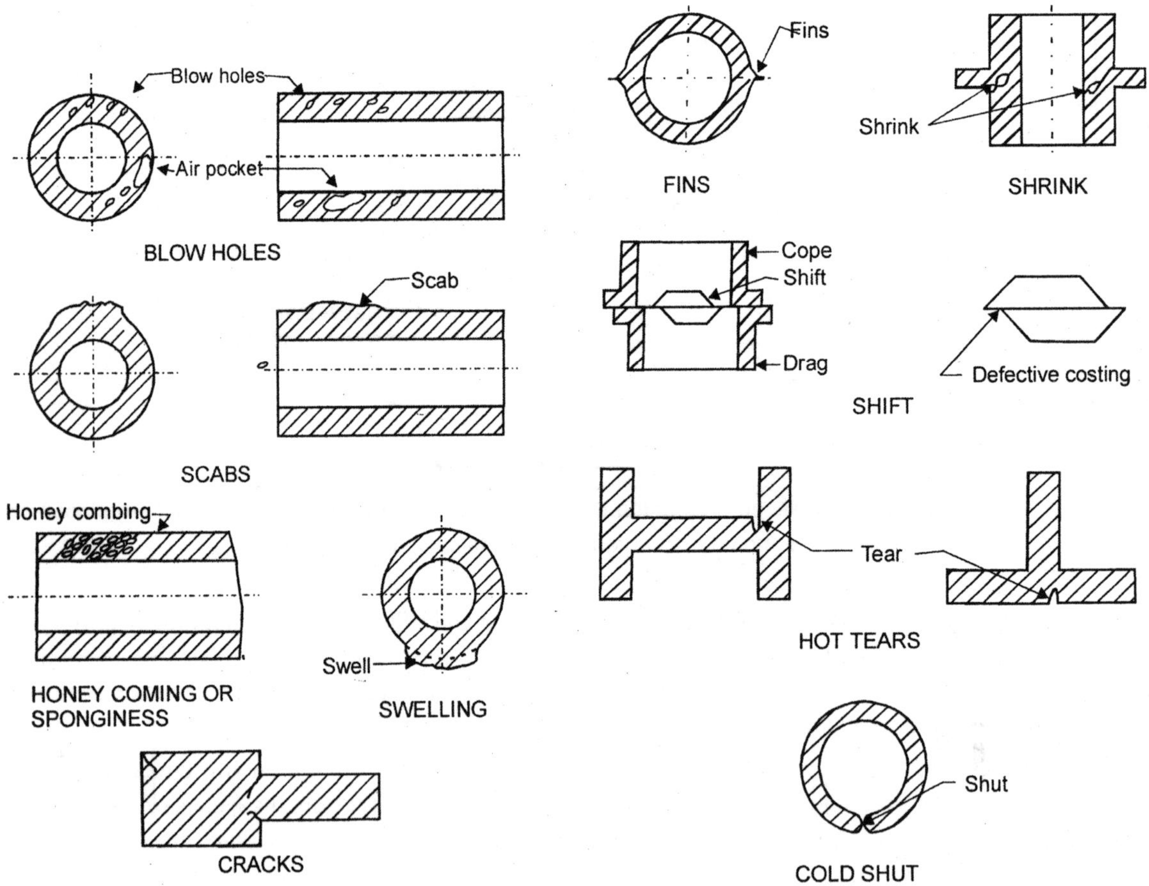

Figure 1.63 Defects in Casting

S.No	Defects	Causes	Remedies
1	BLOW HOLES: Gases and steam are formed when molten metal is poured.If these gases could not out, below holes come will result in the casting.	1.high moisture in sand. 2.very hard ramming 3.improper vent holes 4.low porosity of moulding sand	1.control moisture in moulding sand. 2.ram properly. 3.provide enough vent holes. 4.increase the porosity of sand
2	SCAB: Erosion or breaking down a portion of mould and the recess filled with metal.	1.low thermal shock resistance of sand. 2.uneven ramming of sand. 3.high velocity pouring.	1.sand should have more thermal resistance (add suitable additives). 2.ram properly. 3.pour smoothly.
3	SHIFT: Mismatching of casting sections.	1.misalignment of pattern pieces. 2.misalignment of cope & drag. 3.misalignment of core in mould.	1.align pattern pieces properly using dowels. 2.align properly cope and drag using clamps. 3.seat the core properly in coreprint and mould. 4. place weight over the box.
4	SWELL: Enlargement of casting.	1.very soft ramming. 2.no pouring weight during pouring. 3.very quick pouring.	1.increase the strength of ramming. 2.place pouring weight. 3.improve gating. 4.pour steadily.
5	SHRINKS: Shrinkage cavity on casting surface.	1.improper risers. 2.improper use of chills. 3.improper solidification. 4.incorrect pouring temperature.	1.proper risering system depending upon thickness of casting. 2.correct pouring temperature. 3.correct gating design. 4.proper use of chills.
6	HOT TEAR: Internal or external discontinuity in the casting.	1.sudden change in section. 2.no fillets in corner. 3.improper pouring temperature.	1.improve pattern design. 2.provide fillets whenever there is a change of section.

7	COLD SHUT AND MISRUN: When two streams of molten metal meet and do not fusetogether it is cold shut.	1. low pouring temperature. 2. slow pouring. 3. small gates.	1. improve gating system. 2. increase pouring temperature. 3. pour steadily. 4. low fluidity of metal.
8	METAL PENETRATION: Molten metal penetrating the sand. Casting will have very rough external surface.	1. soft ramming. 2. large grains of sand. 3. improper use of mould and core washes.	1. proper ramming. 2. use fine grained sand. 3. give proper facing sand in mould.
9	HARD SPOTS: Some spots on the surface become hard.	1. very quick cooling of surface. 2. metal composition incorrect.	1. cool normally. 2. use carbide splitting alloys like nickel with pouring metal.
10	HONEY COMBING: Number of small cavities present in the surface of casting.	1. improper skimming. 2. moulding sand not strong. 3. faulty pouring. 4. faulty gating.	1. proper skimming to remove slags. 2. hard ramming. 3. steady pouring. 4. correct gating.
11	RUNOUT: Metal leaks out of mould during pouring.	1. faulty moulding. 2. faulty flask. 3. pattern too large for the flask	1. proper size of moulding flask. 2. proper parting line and gating. 3. use larger flask.
12	CRACKS OR WRAPAGE: Small cracks appear on the sharp corners of the casting.	1. very thin sections. 2. sharp corners.	1. improve casting design. 2. avoid sharp corners.

1.33 INSPECTION METHODS

1.33.1 Purpose of Inspection

Purpose of inspection of a casting is to determine the presence of internal as well as external defects and provide complete knowledge of the extent of such defects. Sometimes, a casting possessing a very good surface finish and excellent appearance from outside is found to have internal flaws of machining of pressure testing.

It is, therefore, necessary to provide such method of inspection which may show the condition inside the casting and reveal the extent of the faults and their probable effect upon its final serviceability.

1.33.2 Methods of Inspection

They are broadly classified as:

i) Destructive methods of inspection.

ii) Non-destructive methods of inspection.

In Destructive method, a specimen sample out of a given lot of castings is sectioned into two or more parts on a metal band saw. It is then examined for internal flaws and defects.

Limitations of Destructive method:

i) Accuracy of locating a defect obviously depends on the judgement of the inspector who is to anticipate the real location of the defect.

ii) Selected specimen casting may not be really a representative of all the lot and many defective castings may be accepted.

iii) There is practical limitation to the number of pieces that can be taken or sectioning.

Method of destructive inspection is, therefore, quite unless as a means of checking accidental defects which often occur in large lots despite exercising all care during the process.

In Non-destructive methods, the casting relation its final shape, dimensions and usable state. These methods can disclose the presence and true location of the defects in the casting without destroying it.

Non-Destruvtive methods in regular use are:

i) Visual Examination

ii) Sound or Percussion Test (Ring Tests or Vibration Tests)

iii) Impact test

iv) Pressure test

v) Magnetic and Fluorescent Powder inspection

vi) Radiographic Examination

vii) Electrical conductivity test

viii) Penetrants or Chalk testing

ix) Supersonic test

x) Weight or Displacement testing

xi) Chemical analysis

xii) Metallogaphic test

xiii) Mechanical testing

xiv) Hardness testing

1.33.3 Visual Examination

This is the oldest and most commonly used method. The success of other lately developed methods also depends on visual inspection. The method usually consists of examining the surface of the casting with the unaided eye, a magnifying lens or a low-power microscope. It also comprises checking of thickness by using calipers,gauges, templates and so on. Special calipers may have to be designed to measure internal thickness.

Although the method is very simple and convenient, but it requried great skill on the part of the inspector to locate the defects. Defects detectable by visual inspection are blowholes, rates tails, cold shuts, swells, burnt-in sand, surface porosity, or surface cracks.

Hot tears is an internal defect. But it often expends to the surface. And so it may be defected if observed carefully provided it is not too narrow to be visible to the eye.

Limitations of Visual testing:

Main drawback with this method is that the internal defects cannot be detected. Surface defects can only be observed.

1.33.4 Sound or Percussion Tests (Ring tests or Vibration tests)

This also is a very old method although a number of variations and refinements have been made. Basically, it consists of suitably supporting the part of suspending it by chains etc., permitting it to swing freely and then tapping it with hammer. Weight of hammer blow is so adjusted that vibrations are set up in the casting producing a certain characteristics tone. This tone may or may not be changed in pitch or quality by the presence of discontinuities which tend to change the sound wavelength produced by the blows.

This method detects relatively large discontinuities in an otherwise homogeneous metal. It may be successful in simple shapes and uniform cross sections. This test reflects this soundness of a piece of metal by the quality of the 'ring' it produces. It is an easily conducted but rather crude test when the 'ring quality' is determined by the human ear.

Recently inspection equipment has been developed to 'listen' to the sounds electronically and compare the tone with that of a casting known to be sound.

Limitations of Sound tests:

It is difficult to judge the extent and harmful effect of the defect and locate the fault position.

1.33.5 Impact Tests:

These may be destructive or non-destructive in nature depending on the quality of the casting.

A proper sized hammer is used to strike or fall on certain members of the casting, where defect is suspected. The casting containing harmful defects will break and so will be automatically rejected. Those having no defects will stand the test.

In another form, the casting is dropped from a specified height onto a steel base.

Limitations of Impact Test:

i) This method cannot be used in all cases owing to its very nature which may cause damage to the casting.

ii) It cannot be very reliable and sometimes even the defect-free castings may break.

1.33.6 Pressure Tests:

These are employed to locate the leaks and test the overall strength of the parts viz., valves, fittings, boilers etc. required to carry fluids in service under pressure.

Fluid used in testing may be water, air of steam. Water is generally preferred because it is incomepressible and less danger is involved even if the casting shatters due to pressure. Pressure for testing varies from one and a half times to two times the working pressure.

Pressure is generally applied by means of a small hand pump for safety reasons. A leak if not located immediately, may be detected on the pressure gauge.

Steam test has an advantage because steam can press through smaller holes or openings through which water may not pass readily. Heat of steam may cause minute cracks due to expension. For safety, castings are first tested hydraulically and then under steam, the pressure used being higher in case of water.

For testing pneumaticaly, the casting is immersed in a tank carrying water and air pressure is applied. Air bubbles are formed if there is a leak.

1.33.7 Magnetic and Fluorescent Powder Inspection:

1) Magnetic practicle inspection is used for iron or steel and their alloys possessing magnetic properties. In this method, the casting is first magnetised. Iron fillings are then sprinkled over the path of magnetic field. Filling align themselves in the direction of the lines of force. This distribution of proportional to the strength of the magnetic field.

If a casting is free from defects, particles will be distributed uniformly all over the surface. If a defect exists, the iron fillings jumble round the defect. The reason is that a discontinuity in the casting causes the lines of forces to by-pass the discontinuity and concentrate around the extremities of the defect. The depth at which the defect occurs can also be judged by studying the concentration of the particles. Considerable experience is, however, necessary to judge the defects correctly.

The casting is magnetised by passing an electric current (A.C. or D.C.) through it. A.C. is used when high surface sensitivity is desired and D.C. is preferred where defects are to be located beneath the surface.

Other methods used for magnetising the castings are —

i) By placing the casting between two magnetic poles, and

ii) By placing the casting in a coil carrying D.C. by wrapping a flexible cable round the casting.

Iron powder may be applied either —

i) dry by a hand shaker or bulb blower, or

ii) wet by spraying or pouring over the surface.

When wet, the powder is carried in suspension form in a liquid like kerosene or carbon tetrachloride.

2) Flourescent powder inspection can be used on all materials, but detects only surface defects.

Castings on non-magnetic materials may be immersed in a warm suspension of fluorescent powder in penetrating oil. The suspension penetrates even minute cracks and pores excess suspension is washed away in warm water, and the casting is dusted with a drying powder. This draws some of the suspension from the cracks to the surface, where it fluoresces and is readily visible under ultraviolet light.

3) Dye-penetrant inspection is a variation of fluorescent powder inspection where in a coloured dye is suspended in penetrating oil; rather than fluorescent powder. The use of ultraviolet light is thus obviated, but defective areas are somewhat less readily seen.

Advantages of Magnetic and Fluorescent Powder Inspection:

i) These methods of inspection detect fine surface cracks and flows more readily than radiographic examination.

ii) They are also considerably cheaper than radiography.

1.33.8 Radiographic or X-ray Examination:

Radiography can be utilized for inspecting casting of any metal. Materiallym the techniques are not different from medical investigations.

Radiant energy from a X-ray tube, radium sulphate capsule (radium or gamma-ray radiography), or radioactive cobalt is passed through the casting, or section of the casting, and recorded on a film held against the opposite surface. Defects in the form of voids or cracks are recorded as blackened areas on the film. Since the radiant energy moves more easily through the less dens regions.

Interpreting radiographic films of castings correctly requires considerable skill and experience. Following defects are readily disclosed by radiographic inspection: Sand spots, cracks, internal and external hot tears, unfused chills and chaplets, shrikage, sub-surface gas or pin-hole porosity and discontinuity or collection of discontinuities exceeding about 2 to 3% of the casting thickness.

X-ray Radiography, as compared with radium radiography enjoys:

i) shorter exposure times; usually minutes rather than hours as required with gamma-rays.

ii) better definition.

iii) clearer images.

It is relatively very high in first cost and maintenance having limited manoeuvrability. It requires the casting to be brought to the machine.

Radium Radiography, is particularly applicable when castings are extremely large and inspection has to be done at the site. In addition, many castings can be radiographed simultaneously by placing them in a circle around the capsule and exposing overnight.

Radioactive cobalt has characteristics similar to radium. Radium pills can be rented for unlimited times at low cost.

Electromagnetic waves having low wavelengths varying between 10–6 and 10–10 cm called X-ray are used as a means of inspection. X-ray have properties similar to those of light waves, but have much shorter wavelengths lying outside the range of humans sensitivity. X-rays can defected by a sensitive photographic film. Due to their shorter wavelengths, these waves can penetrate materials that are normally opaque to light. Shorter wavelength is required to penetrate the denser materials.

X-rays are produced by an X-ray tube. Tube carried two copper elements, the cathode and anode which are sealed. Cathode carries an electrically heated filament which generates electrons. Electrons are driven across towards the positively charged anode, when they strike the tungsten target fixed to the anode. Striking electrons cause their kinetic energy to be partly converted into heat which is conducted away through the cooling fins provided on the anode and remaining energy converted into electromagnetic waves called 'X-rays'.

X-rays pass out of the tube through a window in the form of a beam. Their intensity is controlled by controlling the current which passes through the filament. Wavelength of the ray is inversely proportional to the voltage applied between the two poles. Thus, for shorter wavelength, higher voltage has to be applied.

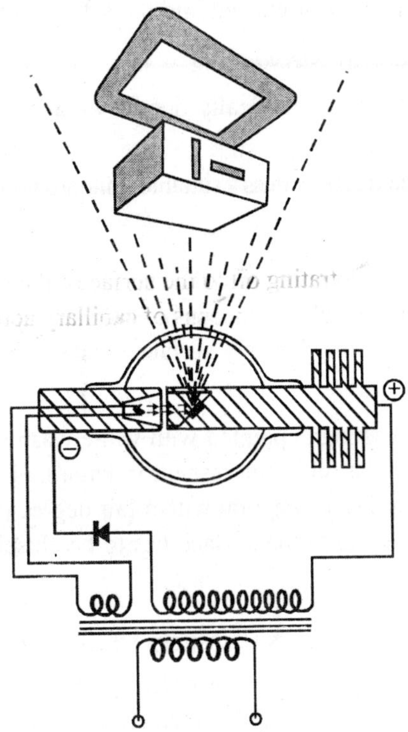

Figure 1.64 X-ray Inspection.

Top - High voltage transformer and X-ray tube, Centre - Casting with internal voids (black), Bottom - Appearance of X-ray film. Blackening of film is proportional to thickness of the internal defect.

The casting under inspection is kept against the X-ray. If there is cavity or a hole, the rays meet less obstruction and so penerate more freely than at the place where metal is more solid and dense. The rays penetrating and emerging out of the casting are absorbed by a photographic plate. Thus, the part of the photographic plate opposite the defect receives more rays and is exposed more than the rest of the plate. This produces a contrasting image on the negative.

For getting more accurate results, special films with an emulsion coating are used. Sometimes, in place of a photographic plate, fluoroscopic screens (made of materials which fluorsence when exposed to X-rays in a dark room) are used. For protecting the viewer from continuous exposure to the rays, the image of the screen is observed in a mirror. The mirror is so placed that the observer is out of the path of X-rays. Voltages for X-ray machines depend on the metal density and thickness. 150,000 volts machine may be used for steel sections upto 25 mm thick.

Gamma rays (like X-rays) emitted during decomposition of radium are used for inspecting casting. Their wave lengths range between $10^{-7.5}$ and $10^{-10.5}$ cm. They can penetrate metals more easily since they are shorter than X-rays. The method needs expensive protective equipment and is not so common due to high cost of radium. See figure 1.64.

1.33.9 Electrical Conductivity Method:

Current is passed through the casting and the same is read on an ammeter. The casting having imperfections offers resistance to the flow of current and the same is shown by a drop in the reading.

Limitations of the Electrical Conductivity Method:

i) The method is difficult to apply practically due to variation in sectional thickness, size and metallurgical structure.

ii) The method cannot be used directly unless a suitable standard is developed for given lot of castings.

1.33.10 Penetrants Method:

In consists of applying a thin penetrating oil to the surface of the casting. It is allowed to stand for sometime so that the oil passes into the cracks by means of capillary action. Oil is then thoroughly wiped and cleaned from the surface. If the casting has any surface cracks, the oil remains in these cracks and tends to seep out.

For detecting the defects, the casting is painted with white wash coat or powdered with talc. It is then viewed under ultra-violet light. The oil is fluorescent in nature, is easily detected under this light. Thus, the defects are clearly revealed. For estimating with a fair degree of accuracy the form and size of the crack the amount of penetrant coming to the surface has to be closely observed.

Applications of Penetrants Method:

i) It is used to detect small surface cracks in castings which cannot be observed with a naked eye.

ii) This method can also reveal those surface defects which are not revealed by radiographic inspection. It is, therefore, often used to supplement the radiographic test.

Limitations of Penetrants Method:

Although this method shows up the finest surface defects in a magnified form, interior defects where the penetrant cannot reach, cannot be detected.

1.33.11 Supersonic or Ultrasonic Testing of Castings:

It is comparatively vary recent development. This method holds very encouraging prospects in near future owing to the high degree of accuracy with which the results are revealed.

Working Principle:

It is based on the principle of reflection of high frequency sound waves. If the surface under test contains some defects, the high frequency sound wave when emitted through the section of the casting, is reflected from the surface of the defect and returns in a shorter period of time. If the section is homogeneous and free from defect, the wave is reflected back after travelling through the whole of the section. In this case, it takes longer time or return to its source.

To detect the lengths of time, an oscillograph is used. The path of the travel of the sound waves is plotted on the oscillograph screen where it can also be measured.

Advantages of Supersonic Testing:

Over other methods is that

i) the defect, even if in the interior, is detected and located accurately,

ii) the dimensions of the defects can also be quickly measured without damaging or destroying the casting.

Applications of Supersonic Testing:

Ultrasonic reflectoscope is used for finding sub-surface defects, example: cracks and shrinkage cavities in large castings.

A flat surface is prepared on the casting against which a quartz crystal is pressed. A thin film of oil insures good contrast. Ultrasonic vibrations are sent through the crystal, and their reflections from any free surface are picked up and recorded on an oscilloscope. Position of the defect within the casting can be found accurately by a calibrated scaling system on oscilloscope.

1.33.12 Dimensional Control:

Initial castings from a new pattern are usually carefully measured in a layout room to ensure whether the casting conforms to blue print specifications. Gages are specially constructed to permit rapid check of particular casting dimensions during a production run.

1.33.13 Smooth Surface Finish:

It is always a desirable characteristic. It is described by a number which is an average of the heights of the peaks and depths of the voltage on a casting surface. The number may be 'simple arithmetical average' or a 'root mean square average'. The two types of average are practically very nearly the same within 10 or 15%.

Casting surface finishes very from about 2250 micromillimetres for smooth moulded plaster castings to over 7500 micro mm for heavy casting made in coarse sand. The degree of surface roughness on as cast surface is determined by visual comparison or by use of a direct-reading surface roughness indicator known as 'Profilometer'.

In profilemeter, a stylus is drawn carefully over the surface. An electronic pick-up translate the simple from the stylus into an average surface roughness reading.

1.33.14 Other Tests:

These fall more in the category of acceptability standards, but which are considered inspection tests, are listed below:

i) Chemical analysis for determining whether composition is within allowable limits.

ii) Metallographic test as a check on grain size, non-metallic impurities, sub-microscopic pinholes type and distribution of phases present in the casting structure, and response to heat treatment.

iii) Mechanical test includes the following:

a) Tensile test for determining ultimate tensile strength, yield strength, elongation and reduction of area.

b) Bend, Notch bend, Impact and Transverse tests for determining ductility and shock resistance.

c) Fracture of the rough casting to check apparent soundness.

d) Pressure testing of as-cast or mechanical unit to pressures in excess of service requirements.

iv) Hardness test. Hardness is determined by measuring resistance of metal of deformation by penetration with a steel ball, or indentor. Hardness reflects ductility and tensile strength of round metal. Rockwell, Brinall and Seleroscope Hardness testers are most useful for checking castings.

SUMMARY

- ***Metal forming processes classified:*** Metal forming processes are classified as: Forging, Casting (Foundry).
- ***Pattern:*** A pattern is a model of the object to the cast. It is used to make a mould cavity to desired shape and size. The pattern is usually made of wood or metal or plastics.
- ***Types of patterns:*** Patterns are classified according to their constructional features. They are Single piece pattern or solid pattern, Split pattern, Match plate pattern, Loose piece pattern, Gated pattern, Sweep pattern, Skeleton pattern, Segmental pattern, Shell pattern, Lagged up pattern etc.
- ***Function of a pattern:*** A pattern is used to make a cavity of required size and shape in the mould.
- ***Basic steps in the Foundry Process:*** 1) Pattern making, 2) Moulding, 3) Metal melting and pouring, 4) Cleaning.
- ***Pattern materials used in foundry:*** Wood, Metal, Plastics or Plaster, Waxes, etc.
- ***Pattern making allowance:*** Patterns are not made to the correct sizes of the castings. Patterns are made slightly larger than the required castings. This extra size given on the pattern is called pattern allowance.
- ***Reasons for providing pattern allowance:*** Pattern allowance is provided for the following reasons: Metals shrink in size when they solidify, Castings require machining and/or other surface finishing operations, Castings tend to warp or bend during solidifaction.
- ***Important pattern allowances:*** i) Contraction allowance (or) shrinkage allowance, ii) Draft allowance, iii) Machine finish allowance or finish allowance, iv) Rapping allowance or shake allowance, v) Distortion or camber allowance.

- ***Shrinkage allowance:*** When the molten metal is poured into the mould, it will solidify and cool. During cooling, the metal will shrink and reduce in size. To compensate this, the pattern is made larger than the required casting. This extra size given on the pattern for metal shrinkage is called shrinkage allowance.

- ***Master Pattern:*** For making a metal pattern, a wooden pattern is used. The wooden pattern is used to make a metal pattern is called Master Pattern.

- ***Factors for selecting pattern material:*** Selection of material for making pattern depends upon the following: i) Quality of the Casting,ii) Intricacy of the component, iii) Surface finish desired, iv)Number of castings to be produced, v) Possibility of design changes.

- ***Mould:*** A mould can be defned as a cavity created in a compact sand mass by using a pattern.

- ***Casting:*** Casting is the one of the processes used for making components of complicated shapes in large quantity. It is the process of producing metal parts by pouring molten metal into the mould cavity of the required shape and allowing the metal solidify. The solidified pieces of metal are known as castings.

- ***Moulding:*** Moulding refers to the process of making a mould by using a Moulding box, Moulding sand, Pattern and Moulding tools.

- ***Important moulding tools:*** Moulding board, moulding box, shovel, trowels, riddle, rammers, sprue pin, draw spikes (or) draw screw, bellows, vent rod, gate cutter, swab, mallet, lifter, strike of bar.

- ***Core:*** Core is a sand mass used to make a cavity or a hole in a casting. The shape of the core is similar to the required cavity in the casting made. Core is made using core sand in core boxes.

- ***Core print:*** A core print is the projection on a pattern. Core prints forms a seat in the mould. The core is supported in the seat formed by the core print.

- ***Core Sand:*** It is a mixture of silica, clay and binders like resin, linseed oil etc.

- ***Materials used for making a core:*** Core is made of core sand, binder and additives, core sand has 90% silica sand. Binders used are linseed oil, resin, bentanite, phenol, urea and water.

- ***Additives used in core:*** Pitch corn-flour, straw, graphite, cow dung and seal coal.

- ***Core making:*** Cores are made using core boxes. Cores are made by hand moulding or using machine like (i) Jolting machine, (ii) Squeezing machine and (iii) Sand slinger.

- ***Important types of cores:*** Horizontal core, Vertical core, Hanging core (or) cover core, drop core, kiss core, balanced core and ram up core.

- ***Types of cores boxes:*** Half core box, slab or rectangular or dump core box, split core box, strickle core box, gang core box.

- ***Function of the core in sand moulding:*** Cores are the materials used for making cavities and hollow projections which cannot be normally produced by the pattern alone. Any complicated contour or cavity can be made by means of cores so that intricate shapes can be easily obtained. These are generally made of sand and are even used in permanent mould. In generals, cores are surrounded on all sides by the molten metal and are therefore subjected to much severe thermal and mechanical conditions and therefore, the core sand should be of higher strength than the moulding sand.

- ***Characteristics of a good core:*** Some special characteristics of good core; green strength, dry strength, refractoriness, permeability, collapsibility, smoothness and low gas emission.
- ***Moulding Sand:*** It is a special type of sand used for making mould. Moulding sand has three constituents. They are 1. Sand, 2. Binder, 3. Additive - Moulding sand is used for making the mould cavity; the mould cavity is made using pattern.
- ***Types of Moulds:*** 1. Green sand moulds, 2. Dry sand moulds, 3. Skin dry moulds, 4. Loam moulds, 5. Metal moulds and 6. Special moulds.
- ***Moulding Sand Additives:*** Additives are added to moulding sand to improve some special and specific properties. 1.Cereal flour, 2.Ground pitch and asphalt, 3. Coal powder and Graphite powder, 4. Wood powder, 5. Molasses and Dextrin, 6. Ferrous oxide and perlite, 7. Fuel oil.
- ***Purpose of using additives in Moulding Sand:*** Additives are added in order to improve toughness and dry strength and to enhace temperature resistance, plasticity and collapsibility of moulding sand.
- ***Important properties of moulding sand:*** porosity or permeability, flowability, adhesiveness, cohesiveness or strength (here strength refers to green strength, dry strength and hot strength), plasticity, refractoriness, yieldability, survivability or serviceability and collapsibility.
- ***Classification of moulding sand:*** Moulding sands are classified as: 1. Natural sand, 2. Synthetic sand and 3. Special sand.
- ***Constituents of Moulding Sand:*** Silica, Kaolinite and bentonite clay and water in proper proportions are the constituents of moulding sand, clay content improves mouldability.
- ***Purpose of providing vent holes in moulding sand:*** Vent holes are pricked in moulding sand to allow the stream and hot gases produced during pouring of molten metal to come out of the mould.
- ***Types of moulding sand:*** Moulding sands are the prime materials there are a number of other sand materials which are also used for some specific purposes. Some of them are, Green sand, Dry sand, Loam sand, Core sand, Facing sand, Parting sand, Baking sand, System sand.
- ***The melting furnances used in foundry are:***

  i) Cupola furnace (for cast iron)

  ii) Open hearth furnace

  iii) Crucible furnace

    a) pit type furnace, b) coke fied stationary, c) oil fired tilting furnace, d) pot furnace

  iv) Electric furnace

    a) Direct arc furnace, b) Indirect arc furnace, c) Induction arc furnace

- ***The normal size:*** of cupola is 1 m to 3 m in diameter and 5 m to 10 m in height.
- ***Zones of cupola furnace:*** are, Crucible zone, Combustion zone, Reducing zone, Melting zone, Preheating zone, Stack zone.
- ***There are three main classes of refractory materials:*** 1) Acid refractories, 2) Basic refractories, 3) Neutral refractories.

- ❖ *Use of flux:* A suitable amount of flux is always added to the metal charge in order to remove the impurities like ash, dirt and sand etc., from the metal.
- ❖ *Cupola specifications:* Its melting capacity, diameter of the shell without lining and with lining, plate thickness, total height, number of tuyers and horse power of the blower. These specifications vary slightly with different manufacturers, but variation is very small.
- ❖ *There are three main types of die-casting machines:* i) Hot chamber die casting machine, ii) Cold chamber die casting machine, iii) Vacuum die-casting machine, iv) Air blown or goose neck type machine.
- ❖ *Operations are not required in I, gravity die casting (or) sand casting:* 1) Opening and closing of the halves, ii) Holding the die at predetermined position during pouring and solidification of the metal, iii) Forcing the molten metal into the die, iv) Insertions are removal of core from the die casting, v) Ejecting the castings fromt the mould.
- ❖ *Shell moulding (casting):* Shell moulding process is otherwise known as croning or c-process or shell process. The process consists of making a mould that has two or more think shell like parts. The shells are moderately hard and smooth.
- ❖ *Merits:* Good surface finish on castings is obtained, High dimensional accuracy, Less labour is required. Dermerits: It needs an expensive, machined metal pattern and hence not economical for small run production suitable for small components only.
- ❖ *Applications of shell moulding process:* Cylinders and cylinder heads for air cooled IC engines, automobile transmission parts, cast tooth bevel gears, brake beam, track rollers for crawler tractors, transmission, planet carrier, steel eyes, gear blanks, chain seat bracket, refrigerator valve plate, small crank shafts are some of the common applications.
- ❖ *Investment casting:* It is also as lost-wax or precision casting unlike shell casting the mould is made of a single piece in this process. Here a pattern of required shape is formed using wax or plastics by pouring or injecting molten wax into a metallic die and allowed to harden. The pattern may be gated together so that several parts can be made or once.
- ❖ *Merits of investment casting:* i) Suitable for small precision parts of intricate design, ii) Parts can be made in multiple moulds, iii) Thin wall castings can be readily cast, iv) Very good surface finish, v) Close tolerance can be maintained, vi) Aluminium, Copper, Nickel, Cobalt, Carbon and alloy steels can be cast.
- ❖ *Die casting:* is a process of casting metals in permanent moulds made of metals called dies. (It should be clearly understood that of all permanent mould castings or gravity die-castings only is called as permanent-mould casting, whereas the other types are called pressure die-casting or simply die-casting)
- ❖ *Pressure die-casting:* is defined as a process in which the molten metal is forced into a permanent mould called die. The molten metal is introduced under pressure.
- ❖ *Types of alloyes cast in cold chamber die casting machine:* The cold chamber machine is ideal for metals such as aluminium alloys which cannot be cast in hot chamber machines due to the ready reactivity of molten aluminium with steel. High melting temperature alloys of the non-ferrous type are also best die cast in cold chamber machines.

- ***Gating system:*** The gating system is defined as a passage through which the molten metal enters the mould cavity. The quality of the castings can be improved by providing proper gating system. Gating system consists of pouring cup, sprue, gate, runner and riser.
- ***Use of moulding process:*** 1) Casting range from 200 gm to 200 kg, 2) For both ferrous and non-ferrous, 3) Good dimensional accuracy is required, 4) Good surface finish is required.
- ***Riser:*** is a reservoir of molten metal present in the casting, through which the hot metal can flow back into the mould cavity when there is a reduction in volume of metal due to solidification.
- ***Feeding range of risers in casting:*** The feeding range of riser is the distance a riser can feed like metal in a casting. Usually to maintain the feeding range of about 4.5 times the thickness for casting.
- ***Important risers:*** 1) Top rises, 2) Bottom risers (or) Blind risers
- ***Important types of gating system:*** The gating system is broadly classified as, i) Top gating, ii) Bottom gating, iii) Side gating, iv) Parting line gating.
- ***Types of top gating:*** i)Wedge types, ii) Pencil gate, iii) Top gate with Strainer core.
- ***Casting defects:*** Blow holes, Cales, Cracks, Swell, Shifts, Shrinks, Hot tear, Cold shut and misrun, Metal penetration, Hard spots, Honey combing.
- ***Types of testing in moulding sand are:*** Moisture content test, Clay content test, Grain fineness and distribution test, Permeability test, Refractoriness test, Strength test, Mould hardness test, Core hardness test.
- ***Main types of Inspection of castings:*** The Inspection methods can be categorized of i) Destructive testing method, ii) Non-destructive testing method.
- ***Non-destructive testing methods are:*** Visual inspection, Inspection for dimensional accracy, Pressure test, Sound test, Impact test, Radiography or X-ray technique, Magnetic particle test, Penetrant test, Ultrasonic test.
- ***Types of core sand test:***

  i) Core hardness test

  ii) Permeability test

  iii) Strength test

  a) Green strength test

  b) Baked strength test

  c) hot strength test

  d) retained strength test.

  iv) Moisture content test

REVIEWS

1. What is meant by manufacturing?
2. How is manufacturing activity related to standard of living?
3. What are the major classifications of manufacturing process?
4. What is casting?
5. What are the metals used for castings?
6. What are the advantages and disadvantages of castings?
7. What are the fundamental steps in making in sand castings?
8. What is the colour code followed in pattern making?
9. What are the materials of patterns and how are they finished?
10. Discuss the various allowances used in pattern making.
11. Discuss briefly different types of pattern.
12. What is the composition of moulding sand?
13. What are the classifications of moulding sands based on clay content?
14. Briefly explain the properties required of moulding sand.
15. Define various moulding sands.
16. Write a short note in mixing and preparation of sand for moulding.
17. Distinguish between green and dry sand mould.
18. Explain how a sand mould for a simple object is prepared.
19. Write a full note on making a green sand mould for a wheel.
20. What is split pattern?
21. What is a core? What is a core box?
22. What is a sprue?
23. What is a gate? What is a gating system?
24. What are the functions of a gating system?
25. Distinguish between hand moulding and machine moulding.
26. Explain pouring and cleaning of castings.
27. Write a short note on inspection procedure for castings.
28. Explain in detail the various defects of castings.
29. Describe a crucible furnace with a neat sketch.
30. What is a tilting furnace?
31. Write a short note on pit furnace.
32. Explain in detail the working of a cupola furnace with a neat sketch.
33. Explain with a neat sketch of Blast furnace.
34. What are the differences between cupola and air furnace?
35. Explain with a neat sketch open hearth furnace.

36. Explain with a neat sketch of pot and electric arc furnace.
37. Explain briefly indirect arc furnace with neat sketch.
38. Write down the principle of operation of high frequency induction furnace.
39. What are the advantages, limitations and applications of high frequency induction furnace.
40. Explain with a neat sketch of core type induction furnace.
41. What is the use ladle?
42. What are the types of ladle used in foundry?
43. Describe briefly hot chamber die-casting with a neat sketch.
44. Explain briefly cold-chamber die-casting process with a neat sketch.
45. Write down the various die material.
46. What are the types of die-casting dies?
47. Write down the product applications of die-castings.
48. Explain with a neat sketch of centrifugal casting.
49. What are the advantages and disadvantages of centrifugal casting?
50. Draw a neat diagram of semi-centrifugal casting and explain briefly.
51. Explain briefly with a neat sketch of continuous casting.
52. Explain Hazelett process.
53. Explain in detail about precision casting.
54. Explain in detail about investment casting.
55. Describe briefly with a neat sketch of shell moulding.
56. What are the factors which influence the choice of casting method?
57. What are the types of gating systems?
58. Explain any three types of gating systems.
59. What is the use of open and blind riser?
60. Write short notes on the following.
    a) Visual examination
    b) Sound or percussion test
    c) Supersonic test
    d) Hardness testing
    e) Magnetic and fluorescent powder inspection
61. Write down the advantages and application of supersonic testing.
62. Explain in detail about foundry mechanization.
63. Explain in detail computer application of metal casting.
64. Explain briefly about design of sand casting.
65. What are the problems encountered with pressure die-casting?

CHAPTER - 2

FABRICATION PROCESSES

2.1 INTRODUCTION

Welding is a process of joining similar metals by the application of heat. Welding can be done with or without the application of pressure. While welding, the edges of metal pieces are either melted or brought to plastic condition. Welding can be done with the addition of filler material or without it. Welding is used for making permanent joints. It is used in the manufacture of automobile bodies, aircraft frames, railway wagons, machine frames, structural works, tanks, furniture, boilers, general repair work and ship building. Almost in all metal working industries welding is used.

2.2 CLASSIFICATION OF WELDING PROCESSES

i) Gas Welding

1. Oxy – acetylene
2. Air – acetylene
3. Oxy – hydrogen

ii) Arc Welding

1. Carbon arc welding
2. Metal arc welding
3. Metal inert gas welding
4. Tungsten inert gas welding
5. Atomic – hydrogen arc welding
6. Plasma arc welding
7. Submerged arc welding
8. Electro – slag welding

iii) Resistance Welding

1. Butt welding
2. Spot welding
3. Seam welding
4. Projection welding
5. Percussion welding

iv) Thermit Welding

v) Solid state Welding

1. Friction
2. Ultrasonic
3. Diffusion
4. Explosive

vi) Newer Welding processes

1. Electron-beam welding
2. Laser welding

vii) Related Processes

1. Oxy-acetylene cutting
2. Arc cutting
3. Hard facing
4. Brazing
5. Soldering

2.3 TYPES OF WELDING

There are two main types of welding process. They are

(a) Plastic welding.

(b) Fusion welding.

In plastic welding, the metal pieces are heated to a plastic state. They are pressed together to make the joint. The plastic welding is also known as pressure welding. Electric resistance welding is an example of plastic welding. In fusion welding the metal at the joint is heated to molten state. Then it is allowed to solidify. Pressure is not applied in this welding process. So it is known as non pressure welding. Gas welding and electric are welding are examples of fusion welding.

2.4 ARC WELDING METHODS/EQUIPMENTS

2.4.1. Metal arc welding and Equipment:

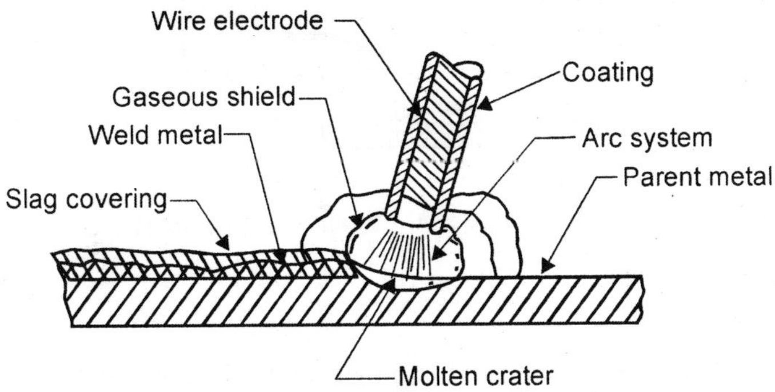

Figure.2.1 Arc Welding (with cooled electrode)

This is simply called are welding. Arc welding is the process of joining two metal pieces by melting their edges by an electric arc. The electric arc is produced between two conductors. The electrode is one conductor and the work piece is another conductor. The electrode and work piece are brought nearer with a small air gap (3 mm approximate).

When current is passed, an electric arc is produced between the electrode and the work piece as shown in figure 2.1. The work piece is melted by the arc. The electrode also melted. Both molten pieces of metal become one. The electrode also supplies additional filler metal into the joint. Temperature of arc is about 4000°C. A transformer or generator is used for supplying the current.

Electrodes used in arc welding are generally coated with a flux. This flux produces a gaseous shield around the molten metal. This prevents the reaction of the molten metal with oxygen and nitrogen in the atmosphere. The flux removes the impurities from the molten metal and form a slag. This slag gets deposited over the weld metal. This protects the weld seam from rapid cooling.

Equipment needed for arc welding are given below:

1. A welding generator (D.C.) or Transformer (A.C.)
2. Two cables-one for work and one for electrode.
3. Electrode holder.
4. Electrode.
5. Protective shield.
6. Gloves.
7. Apron.
8. Wire brush.
9. Chipping hammer.
10. Goggles.

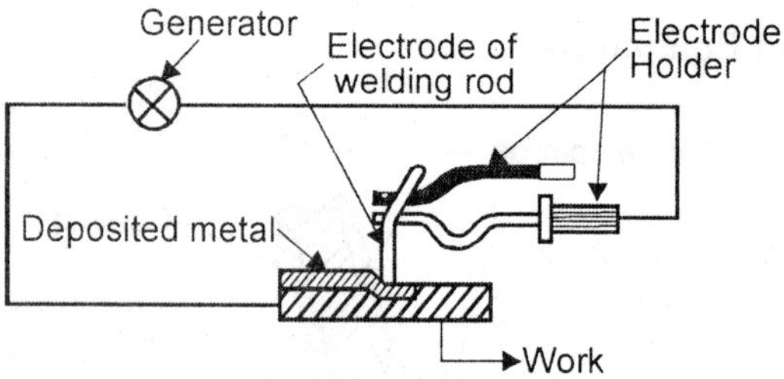

Figure.2.2 Electrical Arc Welding Equipments

Specification:

The specification of D.C. generator and transformer for welding are given below (Refer figure 2.2):

Specification of a typical Generator:

Motor, generator (separately or self excited 3-Phase 50 cycles per second).

Current range: 125 A to 600 A – D.C.

Open circuit voltage – 30 V to 80 V

Arc voltage – 20 V to 80 V

Efficiency – 0.6

Electric consumption – 6 to 20 Kwh\Kg of metal deposit

Power factor – 0.7

Specification of a typical (A.C.) Transformer:

Oil cooled double wound step-down transformer.

Phase – 3-50 cycles per second

Current – 50 A to 400 A

Open circuit voltage – 80 V

Efficiency – 0.85

Power factor – 0.4

Energy consumption – 4 Kwh\Kg of metal deposit.

Comparison of A.C. and D.C. arc welding.

Sl.No.	Alternating Current (From Transformer)	Direct Current (From Generator)
1.	More efficiency	Less efficiency
2.	Power consumption is less	Power consumption is more
3.	Cost of equipment is less	Cost of equipment is more
4.	Higher voltage and hence not safe	Low voltage-Safer Operation
5.	Only coated electrodes can be used	Bare electrodes can be used
6.	Not suitable for welding non ferrous metals.	Suitable for both ferrous and Non ferrous metals.
7.	Any terminal can be connected to the work or electrode.	Positive terminal connected to the work. Negative terminal connected to the electrode.
8.	Noiseless operation	It give noise.
9.	Maintenance of equipment costly and difficult.	Maintenance of equipment cheaper and simple.
10.	Not preferred for welding thin sections.	Preferred for welding thin Sections.

2.4.2. Carbon arc welding:

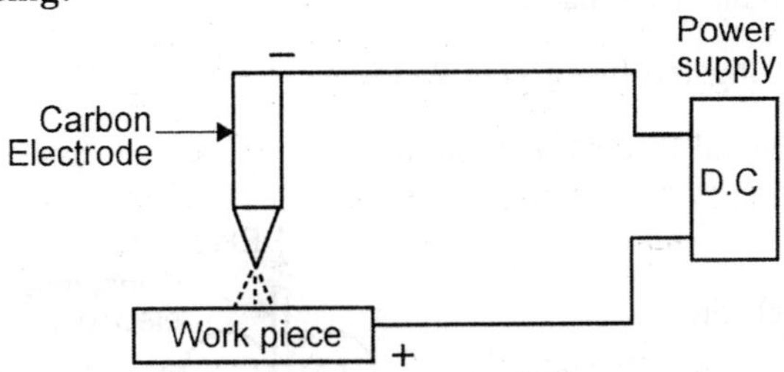

Figure.2.3 Carbon Arc Welding

In carbon arc welding, the intense heat of an electric arc between a carbon electrode and workpiece metal is used for welding. DC power supply is used. The carbon electrode is connected to negative terminal and workpiece is connected to positive terminal. This is because the positive terminal is hotter (4000°C) than the negative terminal (3000°C) when an arc is produced. So carbon from the electrode will not fuse and mix up with metal weld. If carbon mixes with the weld, the weld will become weak and brittle. Carbon arc welding is shown in figure 2.3.

To protect the molten metal from the atmosphere, the welding is done with a long arc. In this case, a carbon monoxide gas is produced which surrounds the molten metal and protects it.

Generally filler rods are not required. If needed, a flux coated filler rod may be used.

Carbon arc welding is used to weld both ferrous and non ferrous metals. Sheets of steel, copper alloys, brass and aluminium can be welded in this method.

This method can be easily done by automatic devices. Temperature control is easily done. Production of arc is very easy. But blow holes occur in the weld.

2.4.3. Inert gas welding:

In this process, an electric arc is introduced between the electrode and the work piece. An inert gas-argon surrounds the arc and prevents any atmospheric action on the molten metal. The electrode is mounted centrally in a nozzle shaped holder. When the arc is produced, argon gas is passed through the holder and it surrounds the arc. There are two types of electrodes used in this welding method.

a) Non-Consumable Tungsten Electrode:

It is used for producing the arc and heat only. Alternative current is used here.

b) Consumable Metal Electrode:

Here the electrode is consumed continuously. The electrode is used to produce the arc and also to supply filler metal. Direct current is used here.

The argon gas surrounds the arc and prevents the atmospheric defects. So very good weldings can be made without defects. This process is used for welding copper, aluminium, stainless steel, etc.

The two methods of argon arc welding are:

i) TIG Welding (Tungsten Insert Gas Welding)

ii) MIG Welding (Metal Insert Gas Welding)

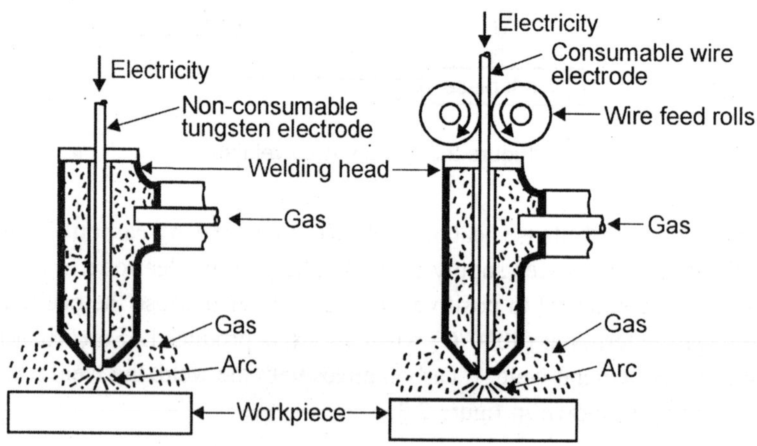

(a) TIG Welding (b) MIG Welding

Figure.2.4 Methods of Argon arc welding.

i) TIG Welding (Tungsten Insert Gas Welding):

(Refer figure 2.4a). In this welding an electric arc is produce between a non-consumable tungsten electrode holder in which the non-consumable tungsten electrode is fixed. When the arc is produce between the electrode and work, the inert gas from the cylinder passes through the welding head around the electrode. The inert gas surrounds the arc and protects the weld from atmospheric effects. So welds are made without defects. The process is used for welding steel, aluminium, cast iron, magnesium, stainless steel, etc. The inert gases used are argon or helium.

Advantages:

1. No flux is required.
2. High welding speed.
3. Can be used for both ferrous and non-ferrous metals.

ii) MIG Welding (Metal Inert Gas Welding):

(Refer figure 2.4b). In this arc welding, the electric arc is produced between a consumable metal electrode and the work metal. During welding, the arc and welding zone are surrounded by an inert gas. This inert gas cover protects the weld from atmospheric effects. Argon or helium is used as the inert gas. The electrode wire is continuously fed from a wire reel. The electrode wire from the reel passes through the electrode holder. It is melted by the arc and deposited over the joint. The arc is surrounded by the inert gas supplied from the gas cylinder. This welding can be done manually or automatically. MIG welding is used for welding aluminium, stainless steel and magnesium without weld defects.

Advantages:

1. No flux is required.
2. High welding speed.
3. Possible to weld non-ferrous metals.
4. Cheaper process.

2.4.4. Submerged Arc Welding:

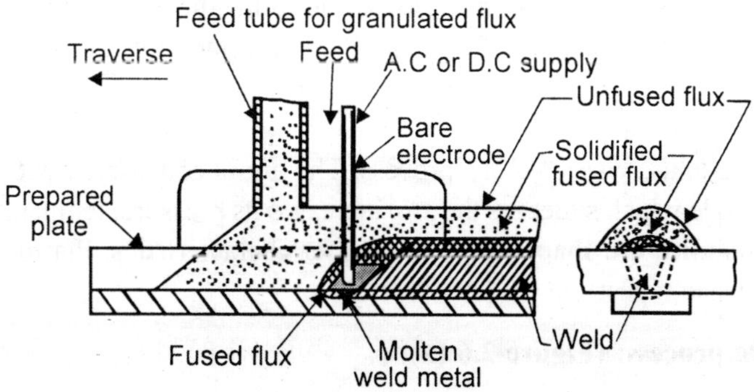

Figure.2.5 Submerged Arc Welding

In this welding, an electric arc is produced between the electrode and the workplate. But the arc is completely submerged (hidden) under the flux powder (Refer figure 2.5). The arc is not visible outside. The electrode is a continuous bare wire electrode. When the arc is produced in the welding zone, the end of electrode and the arc are completely covered by flux powder. So there will not be any defect in the weld due to atmospheric effects. The electrode is contained in a hopper. The flux powder is supplied over the work in front of the welding area continuously. The flux covers the arc and molten metal. Some of the flux melts and forms slag on the weld. The unused flux is sucked by a pipe. The electrode is fed down automatically. The welder need not wear the protective glass. Submerged arc welding process is specially used for welding carbon steels and alloy steels.

Advantages:

1. Submerged arc welding is very fast.
2. Very high quality welds are produced.
3. There will not be welding defects because the arc is submerged.
4. Very thick plates can be easily welded.
5. Long jobs can be easily welded.
6. The operator can work easily without safety device.

Limitations:

1. The weld can not be seen. So it is more difficult to guide the path.
2. Limited to flat position welding. It can be used for overhead welding.

2.4.5. Plasma Arc Welding:

Definition and Concept:

Plasma arc welding is an arc welding process wherein coalescence is produced by the heat obtained from a constricted arc set up between a tungsten / alloy tungsten electrode and the water-cooled (constricting) nozzle (non-transferred arc) or between a tungsten / alloy tungsten electrode and the job (transferred arc). The process employs two inert gases, one forms the arc plasma and the second shields the arc plasma. Filler metal may or may not be added. Pressure, normally, is not employed.

Principle of Operation:

Plasma arc welding is a *constricted arc* process. The arc is constricted with the help of a water-cooled small diameter nozzle which squeezes the arc, increases its pressure, temperature and heat intesely and thus improves are stability, arc shape and heat transfer characteristics. Plasma arc welding process can be divided into two basic types:

a) Non-transferred arc process: (Figure 2.6 (a))

The arc is formed between the electrode (–) and the water-cooled constricting nozzle (+). Arc plasma comes out of the nozzle as a flame. The arc is independent of the workpiece and the workpiece does not form a part of the electrical circuit. Just as an arc flame (as in atomic hydrogen welding), it can be moved from one place to another and can be better controlled. The non-transferred arc plasma possesses comparatively less energy density as compared to a transferred arc plasma and it is employed for welding and in applications involving ceramics or metal plating (spraying). High density metal coatings can be produced by this process. A non-transferred arc is initiated by using a high frequency unit in the circuit.

b) Transferred arc process (Figure 2.6(b)

The arc is formed between the electrode (–) and the workpiece (+). In other words, arc is transferred from the electrode to the workpiece. A transferred arc possesses high energy density and plasma jet velocity. For this reason it is employed to cut and melt metals. Besides carbon steels, this process can cut stainless steel and non-ferrous metals also, where oxy-acetylene torch does not succeed.

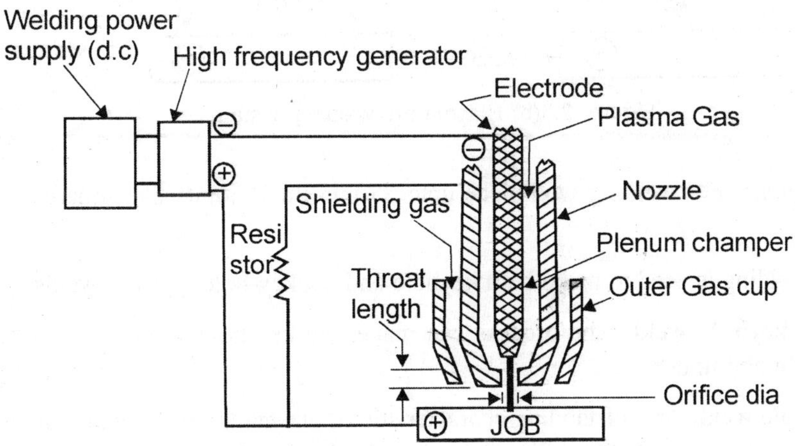

Figure. 2.6(a) Concept of transferred and non-transferred arc processes

Transferred arc can also be used for welding at high arc travel speeds. For initiating a trasferred arc, a current limiting resistor is put in the circuit (Figure 2.6(b) which permits a flow of about 50 amps between the electrode and the nozzle. As the pilot arc touches the job, main current starts flowing between electrode and job, thus igniting the transferred arc. The pilot arc initiating unit gets disconnected and pilot arc extinguishes as soon as the main arc between the electrode and the job is started.

The temperature of a considered plasma arc may be of the order of 8000-25000°C.

Figure 2.6(b) shows a plasma arc welding system.

Process description:

A brief description of the plasma arc welding (PAW) process is follows:

1. Technique of workpiece cleaning and filler metal addition is similar to that in TIG welding. Filler metal is added at the leading edge of the weld pool. Filler metal is not required in making root pass weld.

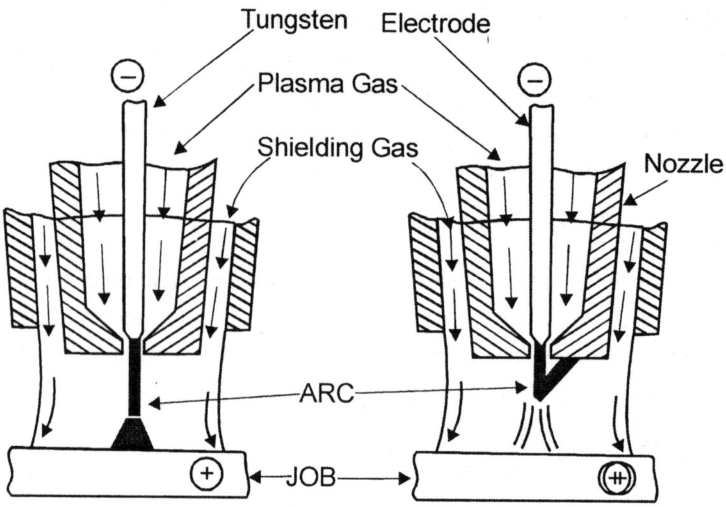

Figure. 2.6(b) Plasma arc welding system

2. *Type of joints:* For welding workpiece upto 25 mm thick, joints like square butt, J or V are employed.

3. Plasma welding is used to make both keyhole and nonkey-hole types of welds.

a) Making a non-keyhole weld: The process can make non-key-hole welds on workpieces having thickness 2.4 mm and under.

b) Making a keyhole weld: An outstanding characteristic of plasma arc welding, owing to exceptional penetrating power of plasma jet, is its ability to produce keyhole welds in workpieces having thickness from 2.5 mm to 25 mm. A keyhole effect (Figure (2.6(c)) is achieved through right selection of current, nozzle orifice diameter and travel speed, which create a forceful plasma jet to penetrate completely through the workpiece. Plasma jet in no case should expel the molten from the joint.

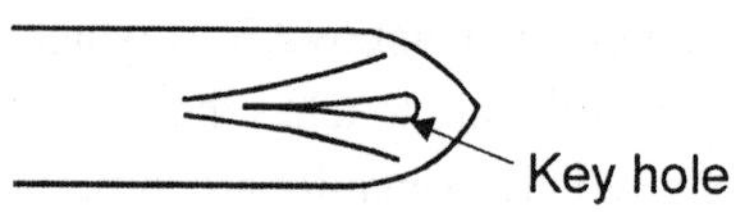

Figure. 2.6(c) Keyhole weld

The majot advantages of keyhole technique are the ability to penetrate rapidly through relatively thick root sections and to produce a uniform underbead without mechanical backing. Also, the ratio of the depth of penetration to the width of the weld is much higher, resulting in a narrower weld and heat-affected zone.

At the weld progresses, base metal ahead the keyhole melts, flows around the same, solidifies and forms the weld bead. Keyholing aids deep penetration at faster speeds and produces high quality bead.

While welding thicker pieces, in laying others than root run, and using filler metal, the force of plasma jet is reduced by suitably controlling the amount of orifice gas. The purpose is to have enough penetration but no keyhole. Filler metal is added in similar way as in TIG welding.

Base Metals Welded: Base metals welded by plasma arc welding are:

i) Carbon and low alloy steels ii) Stainless steels iii) Copper alloys

iv) Nickel and cobalt base alloys v) Titanium alloys vi) Aluminium alloys

Joint Designs: Joints made in base metals for plasma arc welding, are:

i) Single groove (for base plate thickness t = 1.6 – 6.4 mm)

ii) Single and double U groove (t = 6.4 – 25mm)

iii) Single and double V groove (t = 6.4 – 25 mm)

iv) Filler welds (v) T welds

v) T welds

vi) Edge-flange weld (t = 0.05 to 0.25 mm)

Advantages:

1. Deep penetration.
2. Welding directional stability.
3. Faster process.
4. No contamination with tungsten takes place.
5. All metals can be welded.
6. Welding can be done even without a filler rod.

Disadvantages:

★ Automatic control is necessary.

★ Welding of short length is difficult.

★ High initial cost.

★ Affect health of operator — as UV and IR rays are produced by plasma.

2.4.6. Flux - Cored Arc Welding:

As the arc-welding has been becoming popular, engineers start thinking on its automation. Stick electrode has got to be changed, once it is consumed, so it becomes necessary to make this process automatic. Continuous feeding of wire seemed to be answer, but the problem of gas-shielding must be overcome. So a feasible approach was to put the flux inside of the hollow wire.

Therefore, the equipment needed includes - a constant voltage, D.C. power source (32 volts and 375-575 amps), a wire feeder, welding gun (similar to MIG holder without gas and cooling attachment). Today, there are two processes used - a non gas shield and gas shield. Non gas shield is designed to develop sufficient gas shielding from the flux and gives satisfactory results for most of the mild-steels.

With flux-cored wire, speed of metal deposition, depth of penetration can be obtained as per requirement. Deposition rate upto 15.75 kg/hr can be maintained with 3 mm diameter electrode. Flux-cored wire has been limited to flat and horizontal welding without additional gas shielding.

2.4.7. Electro Slag Welding:

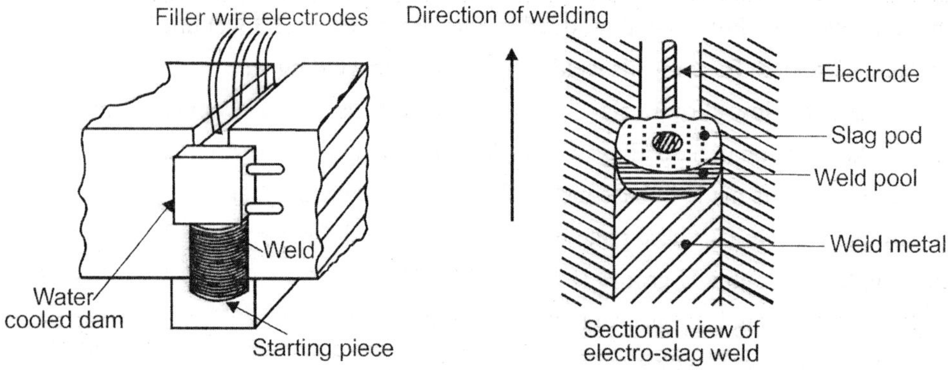

Figure. 2.7 Electro Slag Welding

In this process, the plates to be welded are put in a vertical position with a gap of about 20 to 30 mm, as shown in Figure 2.7. The filler wires and the flux are kept in this gap. The filler wires are used as electrodes. Initially, an arc is struck between the electrode and the workpiece. The heat of the arc melts the flux. The arc is stopped and the slag is maintained in the molten state by the resistance of the electric current passing between the electrode and workpiece. The inside temperature of the molten metal pool is approximately 1950°C and the outside temperature is about 1650°C. This temperature is sufficient to melt the workpiece and filler wires to form a weld after solidification.

Since the molten metal pool is large and the welding speed is slow, therefore, the cooling rate is quite low. This result in a coarse grain size and a follow up heat treatment is normally required to restore the strength.

Advantages of ESW:

★ Uniform heating reduces the risk of warpage.

★ Slower cooling reduces the risk of cold cracking.

★ Edge preparation of plate is not required.

★ Weld is free from pores.

Disadvantages of ESW:

★ Suitable only for butt welding.

★ Not preferable for thin plates.

★ Very slow cooling results in coarse, columnar grain structure which needs subsequent heat treatment to restore the strength.

Applications of ESW :

★ Welding of heavy steel forgins, large steel castings, thick steel plates.

★ Frames of Locomotive and rolling mill.

★ Frames of heavy mechanical and hydraulic press.

2.4.8. Electro Gas Welding:

The electro-gas welding is the development of electro-slag welding. In this method, either a solid electrode or a flux cored electrode is used. The electrodes are fed automatically by a slow speed motor. The main difference between the electro-slag welding and electro-gas welding is that in electro slag welding no flux is fed into the joint and the heat is created by an electric arc and an inert gas (carbon dioxide or argon) is fed into the joint to provide shield to the arc and the molten weld puddle. This process is used for welding workpieces of 20 to 80 mm thickness in a vertical position like fabrication of ships, bridges, large tanks and pressure vessels.

2.4.9. Thermit Welding:

This process is basically a fusion-welding process in which welding is effected by pouring super heated steel around the parts to be united. Thus in this process neither arc is produced to heat the parts nor flames are used. But an exothermic chemical reaction is utilised for developing high temperature.

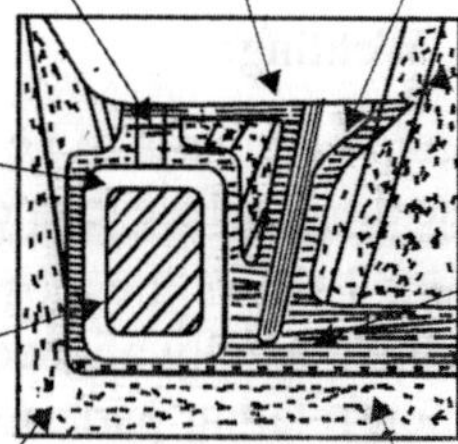

Figure. 2.8 Thermit Welding

The thermit is a mixture of finely divided aluminium and iron oxide. When this mixture in the ratio of three to one by weight (three parts of iron oxide, one part of aluminium) is ignited (with the help of highly inflammable powder composed mainly of barium peroxide). The reaction takes about 30 sec. The heat liberated, raises the temperature approximately to 2760°C which is about twice the temperature of the melting point of steel. Thus the steel produced during the chemical reaction remains in superheated condition. The following reaction takes place

$$8Al + 3Fe_3O_4 = 4Al_2O_3 + 9Fe + 179{,}000 \text{ BTU/mole of aluminum}$$

The resultant products are a highly purified iron (actually steel) and an aluminium oxide slag which floats on top and is not used. Other reactions also take place as most thermit metal is alloyed with manganese, nickel or other elements.

The figure 2.8 shows the method of preparing the material for such a weld. In thermit welding first of all a wax pattern of the weld is prepared around the break where the weld is to be made. It is placed within a sheet iron box and the empty space (i.e., around the pattern) is filled in with refractory sand, rammed properly and then gates and riser are provided. The pattern is heated with a flame so that the wax melts. The heating is continued till the joints become red hot. The flame or torch is removed and the pre-heating gate is plugged with sand. The super heated iron produced in the crucible due to the reaction of the thermit is poured into the mould surrounding the surfaces to be welded. Since the weld metal temperature is approximately twice the melting temperature of steel, it readily fuses the joint.

In thermit pressure welds, the ends of joints to be welded are pressed together to effect welding. In fact the pressure is applied only when the temperature of the joint comes to welding temperature. In this respect it is similar to resistance butt welding.

Advantages:

i) The welds are sound and free from internal residual stresses.

ii) There is no limit to the size of welds that can be made by this process. In fact this process is used to weld or repair large broken parts such as spokes of driving wheels, large parts of trucks and automobiles.

iii) It becomes quite economical to weld large broken parts by this method.

Limitation:

i) This method is limited due to the requirements of mould etc.

Non-ferrrous metals, such as copper alloys, can be thermit welded, but it is more commonly used with ferrous metals.

2.4.10. Friction Welding:

In this method the parts to be joined are not heated by any external means. But frictional heat is developed to raise the interface temperature (of the parts) until the metal is adequately plastic for welding.

The figure 2.9 shows the most common form of friction welding. The two parts (components, or work pieces) are held axially aligned. One is rotated at a predetermined speed while the other remains stationery. The non-rotating part is gradually advanced towards the rotating one till contact is made. Pressure is applied between the parts (over the non-rotating one only) for a specific period while the rotation is maintained so that frictional heat is developed to raise the interfacial temperature until metal is

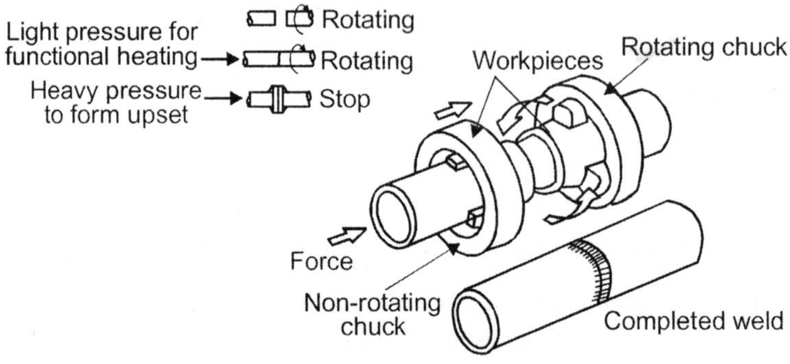

Figure. 2.9 Friction welding

adequately plastic for welding, permitting the parts to be forged together. At this stage there is no relative motion between the two parts (since they got joined together), but the original pressure is maintained to consolidate the solid-state weld.

It is a high-production process, More suitable for plastic and high alloys of comparatively smaller dimensions.

The greatest limitation is that at least one part should be of circular cross-section otherwise the relative annular position of the parts cannot be guaranteed.

2.4.11. Induction Pressure Welding:

In this process low frequency induction heating is substituted for oxy-acetylene flame heating. An important difference between flame and induction heating is that in the latter heat is generated by the circulation of induced currents within the metal. Induction heating is therefore, rapid but can be controlled effectively. Even special shielding atmosphere can be provided around the welding zone, if required.

2.5 GAS WELDING

2.5.1. Types of Gas Welding:

Gas welding process consists of joining the metals with the help of high temperature flame and filler rod. High temperature flame is generated by combusting a mixture of two gases (Oxygen + Acetylene) and the flame is used to heat the workpieces and filler rod to fills the gap. The filler rod melts along with the parent metal and fills the parent metal cavity (Figure. 2.10).

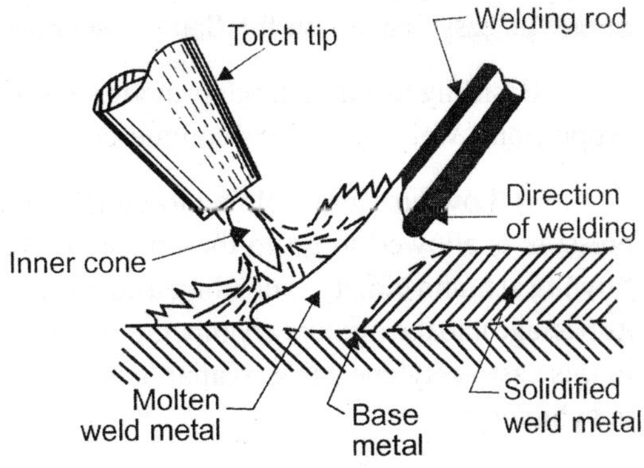

Figure. 2.10 Oxy-Acetylene Welding Process

The mixture of these two gases. (oxygen and acetylene) is commonly used for gas welding purposes. However, the mixture of other combustible gases can also be used. The production of gases in this reference is very important.

2.5.2. Gas Welding Equipments

The most commonly used equipments are listed here:

i) Welding torch
ii) Welding tip
iii) Pressure regulator
iv) Gas cylinders
v) Hose and hose fittings
vi) Goggles, gloves, spark lighter, etc.

i) Welding Torches:

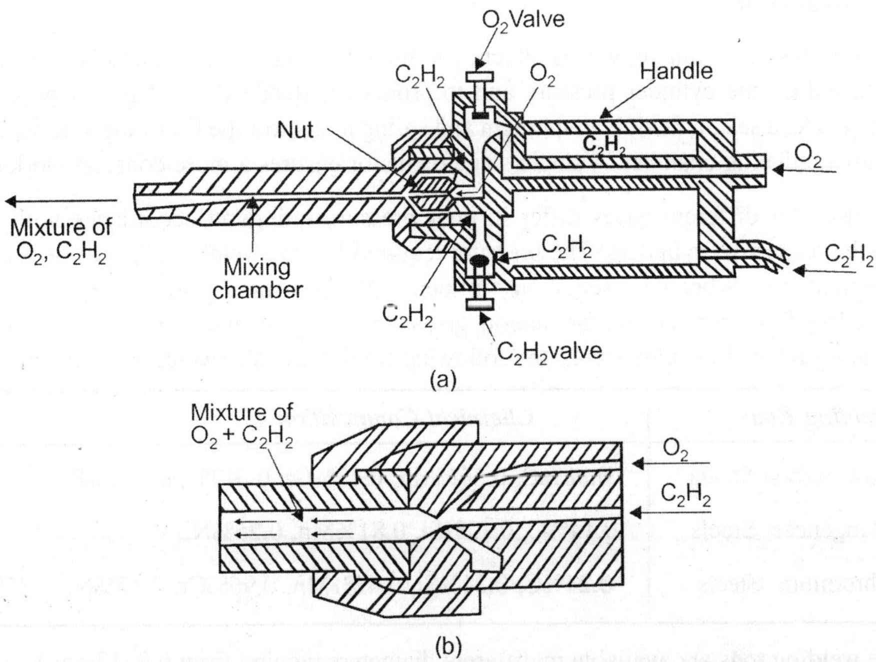

Figure. 2.11 Welding Torch (a) Gas Welding torch injector type (Low Pressure Welding Torch),

Figure.2.11 Every welding torch should be used in operation and produce a steady flame of constant composition. Since most of the torches are manually handled therefore, it must be light in weight and permit an easy control on the flame composition.

Basically, welding torch provides a mixing chamber for mixing oxygen and acetylene in the correct proportions. Welding torch is commercially available in two types:

a) Low pressure welding torch is an injector type shown in the figure 2.11(a), oxygen under high pressure is allowed to go to the nozzle in the injector through a tube at a very high velocity. Oxygen flowing at high velocity is creating a suction and thus it sucks the acetylene available at low pressure from its passage into the mixture chamber to produce the oxygen and acetylene mixture in the required proportion. In this case acetylene gas is available to use at low pressure, whereas oxygen is available at high pressure in cylinders.

b) High pressure welding torch has neither air injector nor a mixer. Instead, it uses a mixing nozzle screwed into the torch head. This type of torch utilizes both the gases under roughly equal pressures (0.5-0.7 atm). The flow rates of the gases and the composition of the mixture can be adjusted by operating the valves mounted on the torch Figure.2.11(b).

ii) Welding Tip:

It is part of the welding apparatus through which the mixed gases pass just before ignition. A great variety of interchangeable welding tips are available in the market. The tip size is governed by the diameter of the tip opening. Accordingly, a tip is selected depending upon the type of metal to be welded, the thickness of the metal to be welded. The greater the thickness, the greater the tip diameter is required.

iii) Pressure Regulator:

They are pressure reducing valves attached to the cylinders of oxygen and acetylene. Their main function is to reduce the cylinder pressure and to produce a steady flow of gas. A pressure regulator consists of a nozzle, a seat, a flexible diaphragm and spring to control the lift of the seat depending on the pressure in the regulating chamber. A single stage regulator ensures a more constant working pressure.

Regulators for different gases differ in their pressure ratings in accordance with the limits of pressure for the respective cylinders. The pressure required for gas welding depends on the thickness of the plate filler material. When it is used, it may influence the mechanical properties of gas weld. The filler metal must be free from rust, any contamination, grease, paint, non-metallic inclusions. However, for gas welding the low carbon, low alloy steels, the following filler materials (welding rods) can be used.

Welding Rods	*Chemical Composition*
Low-carbon Steels	0.08%C, 0.36%Mn, 0.13%Cr, 0.013%Ni, 0.2%P
Mangenese Steels	0.14%C, 0.12%Si, 0.81%Mn, 0.25%Ni,
Chromium Steels	0.24%C, 0.21%Si, 0.42%Mn, 0.96%Cr, 0.17%Ni, 0.35%S.

These welding rods are available in different diameters ranging from 0.3–12mm.

2.5.3 Filler and Flux materials

Filler Metal:

Filler metal or welding rod, can be added, if necessary, to all joints in all positions welded by inert-gas shielded arc welding process. Weld metal additions should be identified to the base metal, strips sheared from the plates being welded may sometimes be used as filler metal, but drawn welding rods give more satisfactory operation.

Filler metal, if not of the proper size, may project into the pool and interface the arc contaminate the electrode. For manual welding operation, if filler rod is too small, it will melt back and form a ball on the end, which oxidises and prevents control of uniform rod feeding.

The welding rod diameter "d" can be approximately determined by the empirical relation:

$$d = \frac{t}{2} + 1, \text{mm}$$

where t = thickness of base plate, mm.

Filler Materials (Welding rods) for gas welding:

Some typical compositions of welding rods for gas welding are given below:

Sl. No.	*Material to be welded*	*Welding rod chemical composition*
1.	L - C Steels	0.08%C, 0.36%Mn, 0.13%Cr, 0.013% Ni, 0.20%P
2.	Mn - Steels	0.14%C, 0.12%Si, 0.81%Mn, 0.25% Ni
3.	Cr - Steels	0.24%C, 0.21% Si, 0.42%Mn, 0.96%Cr, 0.17%Ni, 0.35%S

These welding rods are available in different diameters ranging from 0.3 to 12 mm.

Fluxes:

One of the principal advantage of this process is the ability to produce sound welds in most metals without employing fluxes for satisfactory results, e.g., with some large weldments composed of rather heavy sections, it is often desirable to use a flux along the botton edges of the seams for preventing oxidation of the undersurface and promoting root fusion.

Except for lead, Zinc and some previous metals, gas welding of non-ferrous metals generally requires fluxing. Fluxes are also needed for cast iron and stainless steel. In welding mild steel, the gas flame adequently shields the weld pool and no flux is needed. For cast iron fluxes are required to increase the fluidity of the fusible iron silicate slag, as well as to aid in the removal of slag.

Fluxes compose of borates or boric acid, soda ash and small amounts of other compounds such as sodium chloride, ammonium sulphate and φ iron oxide. Mixture of equal amounts of boric acid and soda ash, 2% aluminium sulphate and 15% powdered iron makes a φ-satisfactory flux.

Gas welding fluxed must melt at a lower temperature than the metals being welded so that surface oxides will be dissolved before the metal melts.

2.6 GAS WELDING DESIGN

All the welds can be classified on the basis of welding position into three broad classes:

i) Horizontal Welding:

Sometimes, it is also known as downhand or flat welding where the weld is deposited on the parent metal in any direction on a horizontal plane, and the metal to be welded is below the flame.

ii) Vertical Welding:

The weld is deposited on the parent metal in any direction (upward, downward and horizontal) on a vertical plane. This is very useful and economical method of welding when the metallic plates of any thickness are to be placed in a vertical plane. If the plate thickness is less than 6mm, welding from one side is enough. But for thicker plates (>6mm), the welding is to be performed from both the sides. This method does not need any edge preparation upto 16 mm thick plates. In this type of welding, the welding should start from the bottom of the plates and proceeds upward along the seam, and the filler-rod should move ahead of the torch i.e. the filler rod should be followed by blow-pipe.

iii) Overhead Welding:

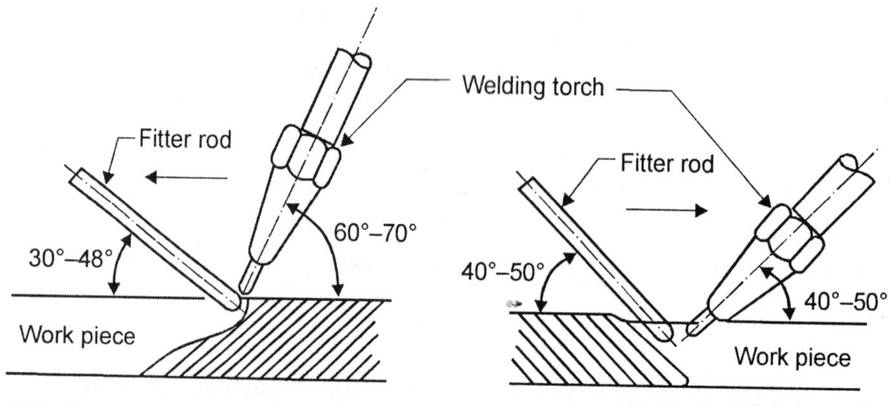

(a) Leftward technique of gas welding (b) Rightward technique of gas welding

Figure.2.12

The weld is deposited on the parent material in the horizontal direction but the metal plate is above the flame.

Apart from these three, sometimes the welds are also deposited when the parent metal is inclined at any angle with the horizontal plane.

In the horizontal welding, the direction of travel and inclination of the torch have got a significant effect on the speed and quality of welding. Hence one can have Leftward Technique Figure.2.12(a) and Rightward Technique Figure.2.12(b).

iv) Leftward Technique:

Figure.2.12(a). In this technique, the weld is started from the right hand side of the seam and moves towards the left hand side of the seam. The blowpipe is given small sideways movement while the filler-rod is moved steadily across the seam. The filler-rod is given backward and forward movement along the seam allowing the flame to melt the bottom edges of the plate just ahead of the weld pool. The welder holds the torch in the right hand and the filler rod is in left hand and the welding flame is directed away from the welded portion and towards the unwelded portion, so automatically preheat the edges of the joint.

This method is widely used for butt-welding of steel plates upto 5 mm thick. Above 5 mm, this process becomes uneconomical. When welding the plates (>5 mm), it is difficult to obtain even penetration and therefore quality of weld decreases. So, make a V-groove of 80°C on the plates to be joined to make it success. Experience has indicated that the operator can see the solidifying front of the puddle and make the bead more uniform. It is more efficient (higher rate of welding) on light gauges of plates.

v) Rightward Technique:

Figure.2.12(b). It is also referred to as backhand welding. The welding starts at the left hand end of the joint and proceeds towards the right hence named as Rightward Technique. The flame is directed towards the completed weld section. During welding, the filler rod may be moved in circles or semicircles within the puddle. The weld puddle formed is less fluid and this appears from the weld surface since the ripples are heavier and spaced significantly apart. It is usually used on heavier (>5mm) base plates since the heat concentration is high and the penetration which can be achieved is of the order of 12 mm in a single pass.

Since the flame is directed towards the completed weld, so the welded portion is better shielded from the atmospheric oxidation and also slows down the cooling rate. The heat-affected zone is also smaller, weld metal is denser, stronger and tougher with less metal deposit. The economic efficiency is better on heavier gauges owing to reduction in welding time and gas consumption.

2.7 FUEL GASES (ACETYLENE AND OXYGEN)

Acetylene is the most important hydrocarbon in the gas welding industry since it produces higher temperature than other gases. It is generally produced in industries through a controlled reaction of calcium carbide in water.

$$CaC_2 + 2H_2O \rightarrow C_2H_2 \,(\text{acetylene}) + Ca(OH)_2$$

Calcium carbide (CaC_2) is a stone-like substance which is made by fusing limestone and coke together in an electric furnace. Calcium carbide so obtained cannot be stored in open since it reacts with atmospheric moisture and generated acetylene gas. So it is stored either in moistureless atmosphere (sealed drums) or in kerosene. The gas may be generated in a centrally located plant and is compressed at less then 0.17 MPa pressure into the cylinders. Since it explodes at pressure more than 0.17 MPa (25 psi). But, if it is dissolved in acetone, then it can be safely stored upto 1.4 MPa (200 psi). Therefore, the acetylene storing cylinders are packed with 80% inert porous materials (asbestos, wood, charcoal, silk fibre, etc.) which are saturated throughly with acetone. Thus the saturated porous materials are capable to absorb acetylene to the extent of 25 times its volume per atmosphere of pressure. The quality of dissolved acetylene is reasonably good and it does not contain more than 2% impurities. Gas is forced into charge the cylinder and withdrawn for use through a valve mounted on the top of the cylinder. Safety fuse plugs are provided to relieve pressure in emergency.

The advantage of dissolved acetylene is not only a large amount of acetylene is available for use, but also a reasonably pure and easy to transport gas cylinders is available. The need of high pressure inside the cylinders is essential for steady operation of torches, a constant composition of combustible mixture under the most adverse condition (heavy heating at the tip).

Oxygen is also an important constituent for gas welding. Commercially, oxygen is produced either by electrolysis of water or by air liquefaction. Any amount of oxygen can easily be produced commercially from air with least possible energy consumption. Though oxygen is not a combustible or explosive gas but supports explosive, so all cares have got to be taken. After removing all mechanical impurities, moisture and carbon dioxide, the oxygen is compressed to 30 atmosphere in the gas cylinders which can be transported to weld shops.

2.8 TYPES OF FLAMES

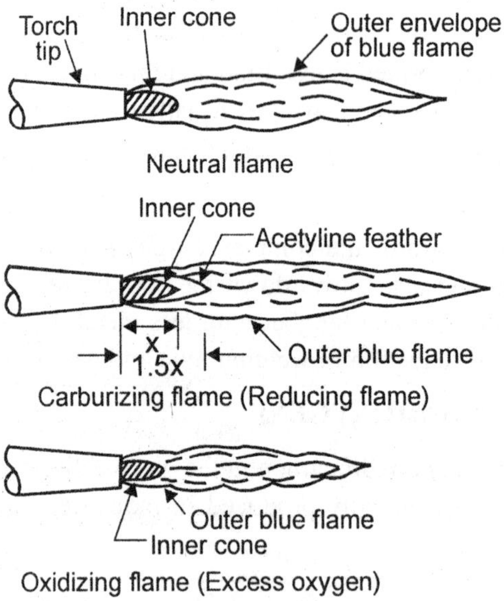

Figure.2.13 Various Types of Oxy-acetylene Flames

Oxy-Acetylene Flame (Ref Figure. 2.13). Flame temperature is a very important characterisitic for welding. The flame temperature is not constant throughout the entire length of flame and it varies with the flame composition. Obviously, the heat for welding is to be taken from the zone of highest flame temperature. The ratio of these two gases is to be kept such that the highest flame temperature should be obtained, i.e. the ratio varies from 1.2 to 1.9 for oxygen to acetylene ratio and the flame temperature varies from 3100°C to 3600°C.

Fuel – gas properties include the flame temperature, flame propagation rate and gross heat of combustion. Flame temperature and gross heat are connected to each other and under ideal conditions, these are made available. The oxy-gas welding flame consists of three regions of zones, viz the inner luminous one; the intermediate (reducing) zone; and the flame envelop (oxidizing) zone.

However, one can classify the oxy-acetylene flame into three categories:

i) Neutral or balance flame has a well defined luminous cone gradually rounding off towards the tip. It is produced by burning an approximately one to one mixture of oxygen and acetylene.

ii) The carburizing or reducing flame is obtained by burning an excess of acetylene i.e. the rate of oxidation is slowed down owing to more acetylene is burnt with the oxygen from the surrounding air, so the flame grows its size. The inner cone is not so well defined as in the case of neutral flame. There is no sharp demarcation between the outer flame envelope and the innercone. (intermediate excess of acetylene zone). The outer flame takes on a reddish color.

iii) The oxidised flame is produced by turning on excess of oxygen. The inner white cone is very short and not so well defined. The intermediate cone and the envelope are also short due to intensive oxidation. The flame is blue and produces a specific sound.

The complete combustion of oxygen and acetylene takes place and thus the flame produced at the tip of torch, is due to the reactions taking place in two stages:

(a) Combustion due to the oxygen supplied by torch (primary combustion)

$$C_2H_2 + O_2 \rightarrow 2CO + H_2 + \text{Heat energy}$$

(b) Combustion due to the oxygen of air (secondary combustion)

$$4CO + 2H_2 + 3O_2 \rightarrow 4CO_2 + 2H_2O + \text{Heat energy}$$

2.9 FLAME CUTTING

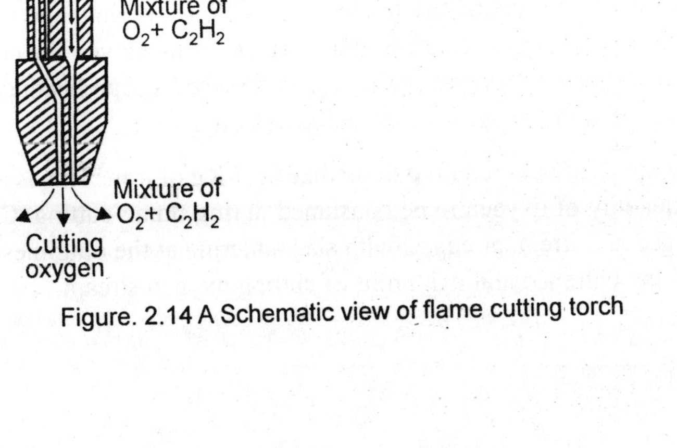

Figure. 2.14 A Schematic view of flame cutting torch

Every gas welder can cut the materials for various fabrication purposes. A cutting torch, (Refer Figure.2.14) a universal tool is extensively used in welding and fabrication. Oxy-acetylene cutting process has made it economically feasible to fabricate heavy gauges of rolled steels.

The procedure for torch gas-cutting is based on the fact that the steel will burn in an atmosphere of oxygen after having brought to pool temperature. After burning, iron oxide which is solid at ordinary temperature, is formed and has lower melting temperature than that of steels. So the heat generated by burning iron is enough to melt the iron oxide and run off as molten slag and fresh surface of steel will be exposed to oxygen jet. Jet of oxygen is responsible to oxidize the steel rapidly, from oxide goes down as molten slag owing to the pressure of jet and so on.

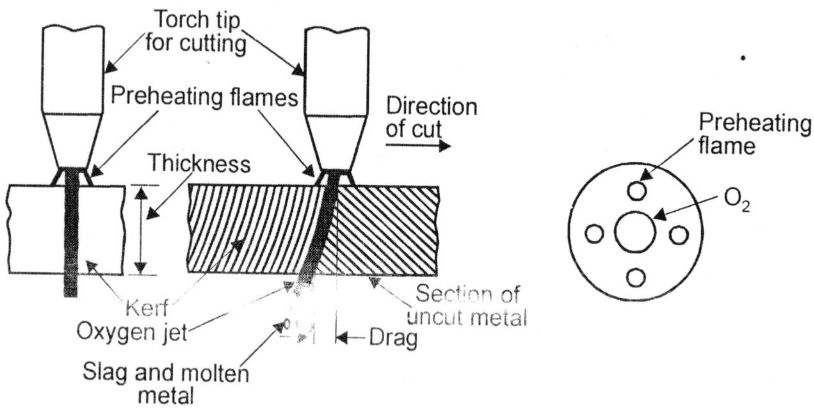

Figure.2.15 Principles of Flame Cutting

The equipment consists of oxygen cylinder, acetylene cylinder, cutting torch along with tips and various controlling assessories.

Cutting torch is an apparatus for mixing oxygen and acetylene in the required proportion. The cutting torch is very similar to welding torch but it has a provision of several neutral flames and a separate jet of high pressure oxygen in the centre. The cutting tip has a central hole for high pressure oxygen and around this centre hole, number of holes are there through which flames can come out which are used for preheating. Refer figure 2.15 for flame cutting.

These neutral flames are brought close to metal as we do in welding and when the metal becomes red hot (>820°C), the jet of pure oxygen is released by a quick acting valve. The cutting action takes place and its quality depends on the tip size, oxygen pressure, cutting speed and preheating flames and temperature of metal. The tip size controls the oxygen pressure and velocity.

Theoretically $0.085m^3$ of 0_2 is required to oxidize 0.45 Kg of iron. This is an indication for having the knowledge of the quantity of oxygen to be consumed during flame-cutting. Cuts that are improperly made will produce ragged and irregular edges, with slag adhering at the draglines. Drag is the horizontal lag that exists between the entrance and exit point of cutting oxygen stream.

It is specified as a ratio or as a percentage:

Drag = d/t (Both in mm)

Drag (%) = d/t x 100

Where d – amount of horizontal lag.

t – thickness of metal plate to be cut.

Only 10% drag is permissible during cutting. Excess value of drag may result into many defects in cutting.

Basically, flame cutting is carried out manually for rough work since manual manipulation of cutting torch cannot give faster, straight, fine and economical results. However, steadier, faster, economical result are obtained from mechanized cutting where there is enough use to justify the investment. A flame-cutting machine consists of a table to hold the work and a framework to carry and move the torches. The work rests on series of disposable bars on an open top-table. A production machine with four/five torches on an arm controlled from a master template to cut out four/five pieces at a time (multiple-torch cutting). A high degree of skill is needed to obtain best results.

2.10 ARC – WELDING ELECTRODES

As noted above, the electrodes used for arc welding of two types:

i) Non-consumable Electrodes

ii) Consumable Electrodes

i) Non-Consumable Electrodes:

These electrodes are made of carbon, Graphite or Tungsten. Carbon and Graphite electrodes are used for D.C. welding only. Tungsten electrodes are used for D.C. as well as A.C. welding. When non-consumable electrodes are used, the filler material is added separately. Since the electrode is virtually not consumed, the arc length remains constant, so that it is stable and easy to maintain.

ii) Consumable Electrodes:

The consumable electrode provides the filler material also. These may be made of various metals, but should have the same composition as the material to be welded. During welding, they melt and supply the filler material. Therefore the arc length and the resistance of the arc path vary, as a result of the change in arc length and the presence or absence of metal particles in the path. This requires that the electrode be moved toward or away from the work to maintain the arc and satisfactory welding conditions.

The consumable electrodes are of three kinds:

a) Bare electrodes

b) Fluxed or lightly coated electrodes

c) Coated or extruded/shielded electrodes.

a) Bare electrodes may be used to weld wrought iron and mild steel. They must be used only with straight polarity. Bare electrodes in the form of sticks or rods are used for hand arc welding. For automatic welding, bare electrode is in the form of continuous wire (coil).

b) Lightly coated electrodes have a coating layer several tenths of a millimeter thick, the weight of the coating is from 1 to 5 percent of the electrode weight. The primary purpose of a light coating is to increase are stability and are also called as ionizing coatings. This type of coating contains chalk (80 to 85 parts by weight) and water glass (15 to 20 parts by weight). These electrodes are used in welding only non critical structures. The welds produced have poor mechanical properties due to the lack of the protection of the molten metal.

c) Coated electrodes are covered with a relatively high quality covering applied in a layer of from 1 to 2 mm. The weight of such a coating is from 15 to 30% of the electrode rod. The greatest amount of welding is done with coated electrodes. Since the electrode coating is brittle, only straight stick electrodes can be used. This restricts the process to a slow manual operation. If the flux coating is placed inside a long tube, the electrode can be in the form of a bare wire in the form of a coil and then the shielded arc process can be made continuous and automatic.

The stick electrodes are available in the following sizes: 3.2,4,5,6,8,9 and 12 mm diameter and lengths of 350 or 450 mm.

2.11 FUNCTIONS OF COATINGS

1. Improves arc stability by providing certain chemicals which have this ability by ionizing the path of arc.
2. Provides a protective gaseous atmosphere to prevent O_2, H_2 and N_2 pick-up by the molten metal.
3. Provides a protective slag over hot metal.
4. Provides flux, which helps to remove oxides and other impurities from the molten weld metal.
5. Reduces spatter of weld metal – when coating burns off slower than core.
6. Acts as deoxidizer.
7. Adds alloying elements.
8. Increases deposition efficiency.
9. Slows down the cooling rate of weld (due to protective layer slag) to prevent hardening.
10. Coatings are normally insulators of electricity and so permit the use of electrode in narrow grooves etc.

2.11.1. Electrode Coatings:

1. *Slag forming ingredients:* Asbestos, mica, silica, fluorspar, titanium dioxide, iron oxide, magnesium carbonate, calcium carbonate and aluminium oxide.
2. *Arc stabilising ingredients:* Or ionizing agents: Potassium silicate, Ti O_2 + ZrO_2 (Rutile), Mica, Calcium oxide, sodium oxide, magnesium oxide, feldspar ($KA\ell\ Si_3O_8$).
3. *Deoxidizing ingredients:* Cellulose, Calcium carbonate, dolo-mite, starch, dextrin, wood flour, graphite, aluminium, ferromanganese.
4. *Binding materials:* Sodium silicate, potassium silicate, asbestos.
5. Alloying constituents to improve strength of weld.

6. Ti O_2 and potassium compounds increase the melting rate of the base metal for better penetration.
7. Iron powder provides higher deposition rate.

2.11.2. Coding of Electrodes:

There is a no standard code designation used to classify the bare electrodes. Although, the composition of the electrode is basically the same as the base metal, small amounts of special additives, such as deoxidizers, are included by many manufacturers to improve the weld quality.

There is a code designation for coated stick electrodes for metal arc welding, for selecting a correct electrode for a specific application IS: 815 – 1974.

The coding for an electrode shall consist of a prefix letter, a code number of six digits in numerals and in some cases one or more suffix letters, each indicating specific property or characteristic or both of the electrode.

1. Prefix letter indicates the method of manufacturing.

Prefix letter	***Method of manufacturing***
E	Soild extrusion
R	Extruded with reinforcement

2. First digit: The first digit of the code indicates the type of covering.

Digit	Type of covering
1.	Having a high cellulose content
2.	Having a high content of titaniam and producing a fairly viscous slag.
3.	Contains appreciable amount of titaniam and producing a fluid slag.
4.	Contains a high content of iron oxides and / or silicates or iron and manganese and producing an inflated slag.
5.	Contains a high content of iron oxides and / or silicates and produce a heavy solid slag.
6.	Having a high content of calcium carbonate and fluoride.
7.	A covering of any other type not classified above.

3. Second digit: The second digit of code number indicates the welding position or positions in which the electrode may be used.

Digit	**Welding Positions**
0	F, H, V, D, O
1	F, H, V, O
2	F, H
3	F
4	F, H_f (H_f - horizontal fillet)
9	Any other welding position not classified above.

Where F – stands for flat

H – stands for horizontal–vertical

V – stands for vertical–up

D – stands for vertical–down

O – stands for over–head

4. Third Digit: The third digit of the code number indicates the welding current conditions recommended by the manufacturer of the electrode, as follows:

Digit	Welding Current condition
0	D+
1	D+, A90
2	D-, A70
3	D-, A50
4	D+, A70
5	D±, A90
6	D±, A70
7	D±, A50
9	Any other current conditions not classified above

The notations are as follows:

D+ DC with electrode positive

D– DC with electrode negative

D± DC with electrode positive or negative

A90 AC with an open circuit voltage not less than 90V

A70 AC with an open circuit voltage not less than 70V

A50 AC with an open circuit voltage not less than 50V

Note 1: The electrodes may not function satisfactorily at a lower voltage than that for which it is classified but a higher voltage may be used in service with advantage.

2: The open circuit voltage necessary for striking the arc varies according to the core diameter of the electrode. The reference core wire diameter for coding is 4 or 5 mm. Within the range of sizes 2.5 mm to 8 mm, the open-circuit voltage necessary may be expected to vary approximately as follows:

Code voltage	Variation in voltage
90	100 to 70
70	80 to 55
50	60 to 40

The higher voltages are associated with 2.5 mm electrodes and the lower voltage with 8 mm electrodes.

5. Fourth and Fifth Digits: Fourth and fifth digits indicate mechanical properties such as tensile strength and the appropriate yield stress. The range of tensile strength 410 to 510 N/ mm^2 is indicated by the digit 41 and the digit 51 indicates the range 510 to 610 N /mm^2. The minimum yield stress for the strength range of 410 to 510 N/ mm^2 is 330 N/mm^2 and for the strength rate of 510 to 610 N/ mm^2, is 360 N/mm^2.

6. Sixth digits: The sixth digit of code number indicates the percentage elongation in combination with impact value of the deposited metal.

Digit	Strength code number	Minimum elongation in % of guage value	Temp. for min. impact value of 47 J, °C
0	41	–	–
1	41	20	+27
2	41	22	0
3	41	24	–20
4	41	24	–30
5	41	24	–40
0	51	–	0
1	51	18	+27
2	51	18	0
3	51	20	–20
4	51	20	–30
5	51	20	–40

7. Suffix Letters: The Suffix letters indicate the special properties / characteristics or both of the electrode, as following:

Suffix Letters	Special properties / Characteristics
H	Hydrogen controlled electrodes
J	Iron powder covering, giving a metal recovery of 110 to 130%
K	Iron Powder covering, giving a metal recovery of 130 to 150%
L	Iron powder covering, giving a metal recovery of over 150%
P	Deep penetration

A typical electrode coding will be written as follows:

D 426 – 513 H

2.12 SPECIAL WELDING PROCESSES

2.12.1. Electron Beam Welding:

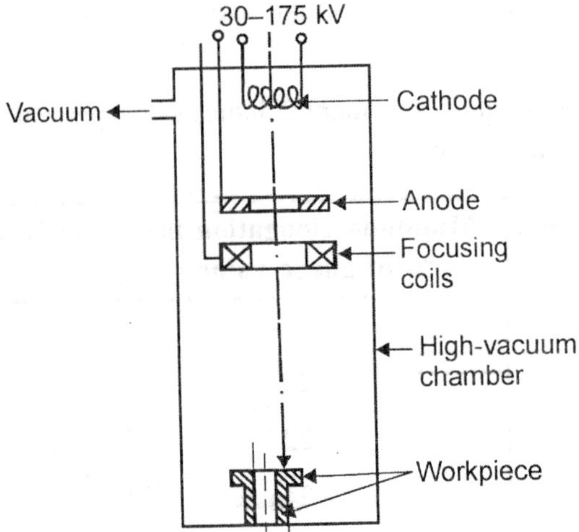

Figure.2.16 Electron Beam Welding

Electron beam welding utilizes the energy from a fast moving beam of electrons focussed on the workpiece. The electrons strike the metal surface, which gives up kinetic energy almost completely into heat. The beam is created in a high vacuum (10^{-3} to 10^{-5} mm Hg). If the work is done in such vacuum, no electrodes, gases, or filler metals can contaminate it, and pure welds can be made. Moreover, high vacuum is necessary around the filament so that it will not burn up and will also produce and focus a stable beam (figure.2.16).

In all types of electron beam machines, a tungsten filament which serves as a cathode emits a mass of electron that are accelerated and focussed to a 0.25 - 1 mm diameter beam of high energy density upto 0.5 to 10 kW/mm2. The heat generated is about 2500°C. This is sufficient to melt and vaporize the work piece material and thus fills a narrow weld gap even without a filler rod (although filler rods may be used).

The speed of the beam is stepped up to one-half to two-thirds of the speed of light by passing it through a high-voltage electrostatic field. An electromagnetic lens is employed to obtain correct focusing of the beam.

Near zero pressure is necessary for the formation of the beam. The chamber is evacuated to a pressure to 10 mm Hg. Welding begins at this point. A similar process, known as nonvacuum electron beam welding is rapidly without elaborate arrangement required for vacuum electron beam welding. In that case, the vacuum chamber in which the beam is created is evacuated to a lower pressure. In situation where contamination of the workpiece must be held to a minimum, the beam should be passed through argon or helium. To make the chamber high vacuum, it takes about 5 to 30 min to evacuate the air, depending on the size of the chamber. A medium size electron beam welder operates below 60 kV. The welding head or the work is moved by numerical control or by hand.

Today automobile, airplane, aerospace, farm and other types of equipment including ball-bearing over 100 mm are being welded by the electron beam process.

Advantages:

The advantages of EBW are that the welds are clean, with no porosity since there is no air; so shielding gas is needed; and as the energy input is in a narrow, concentrated beam, distortion is almost eliminated. The speed may be as fast as 2500 mm/min, and it will weld or cut any metal or ceramic, diamond, sometimes as thick as 150 mm.

Disadvantages:

★ The equipment is expensive.

★ High operating cost.

★ Limitation of the vacuum chamber. Work size is limited by the size of the chamber.

2.12.2. Laser Beam Welding:

Lasers are devices which are capable of generating a very intense beam of optical radiation. The word "LASER" is an acronym of Light Amplification by the Stimulated Emission of Radiation. Laser beam welding process is shown in figure 2.17.

An even more concentrated beam is produced, but at a lower overall efficiency, with the laser beam a CO_2 laser pumped with 500 W emits far-infrared light (10.6 μm wavelength) and develops a peak energy density of 80 kW/mm^2, yet the heat affected zone is only 0.05 to 0.1 mm wide. Oxygen blown on the surface of the metals reduces the heat reflection and increases material removal rates by oxidation; inert gas increases heat transfer for non-metals.

The laser has the advantage that vacuum is not necessary and it is finding limited but growing application, particularly for thin gauge metals. Lasers using for work such as welding very small wires to electronic devices and similar work is called micro welding. Welding speed of about 2500 mm/min is achieved on steel sheet 1.5 mm thick.

In practice, numerical control is used to move the workpiece; and laser's use in heavy production work is still limited.

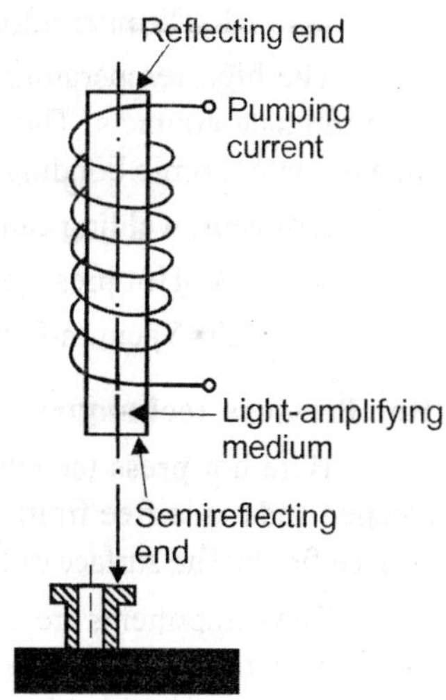

Figure.2.17 Laser Beam Welding

Advantages of LBW:

★ Excellent heat control.

★ Welding is accurate.

★ No thermal distortion.

★ Welding can be done at any position of weld.

★ As laser is a ray of light, electrode is not necessary.

Disadvantages of LBW:

- ★ Materials upto 0.5mm thick alone can be welded.
- ★ High energy consumption.
- ★ High cost of equipment.

Application of LBW:

- ★ Dissimilar metals can be joined.
- ★ Objects which are sealed in a glass container can be welded without breaking the seal.
- ★ The process is used for making light quality welding in aero space and electrical industries, automobiles etc.

2.12.3. Plasma Arc Welding:

Already discussed in 2.4.5

2.12.4. Diffusion Welding:

In this process, both heat and pressure are used to get the weld by interatomic bonding. Diffusion welding process is controlled by the following factors:

1. Temperature, 2. Pressure and 3. Time.

The high temperature makes the metal surfaces soft. The pressure brings the metal surfaces to have intimate contacts. The parts to be welded are kept at specified temperature and pressure till such time the interatomic bonding is complete.

Diffusion welding can be done in three basic techniques,

i) Hot press technique. ii) Gas pressure technique.

iii) Vacuum-furnace technique.

Hot Pressure technique:

Here hot press technique is explained. First the surface of the parts to be welded are throughly cleaned and made free from dust and contaminants. The surfaces to be joined are machined to have good surface finish. The surface of the parts to be joined are clamped together. This is done using a hydraulic press.

The components are kept in a hot chamber at a temperature of about 50% of the melting temperature of component material. The clamped components are kept inside a hot chamber for about 20 minutes. The surfaces get welded by interatomic bonding due to temperature and pressure.

There are three kinds of joints made in diffusion bond welding.

1) Metal to metal without intermediate material.
2) Metal to metal with intermediate material.
3) Metal to ceramic material.

2.12.5. Explosive Welding:

In this process, two metal surfaces are brought together with high relative velocity are developed by igniting an explosive. The figure 2.18 shows the arrangement for explosive welding. One plate is kept on the anvil. The other plate is placed at an angle to the first plate maintaining a gap. This angle is called contact angle. This angle varies according to material. For example, for Aluminium it is 15° and for steel it is 10° to 50°.

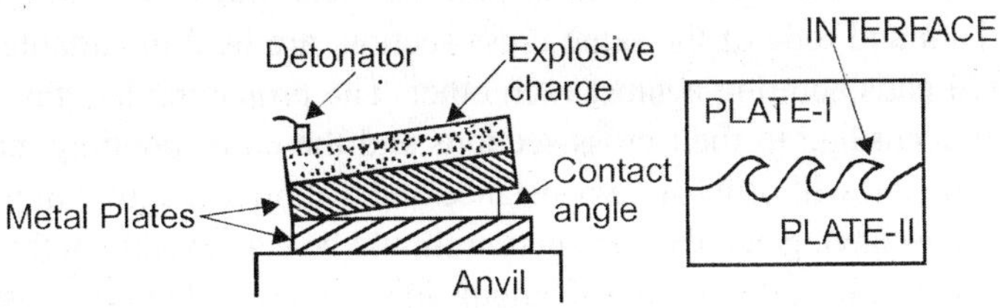

Figure.2.18 Explosive Welding

The explosive charge is placed on the top plate. It is ignited. The explosion causes very high impact and forces the top plate to join with the bottom plate at a high velocity. Metallurgical bond takes place between the surfaces. Microscopic swirls are produced on the interface of the plates as shown in the figure. This increases the strength of the weld because of interlooking action.

Application:

1) Aluminium and mild steel are welded using explosive welding.

2) Tantalum can be welded to steel only by this method, as the melting point of tantalum is higher than the vapourisation temperature of steel.

Advantages:

1) Process is simple.

2) Larger surfaces can be welded.

3) Fast process.

4) Excellent strength.

5) Heat treated sheets can be easily welded without spoiling the surface.

Limitations:

1) Metals with less impact strength can not be welded.

2.13 RESISTANCE WELDING:

2.13.1. Resistance Butt Welding:

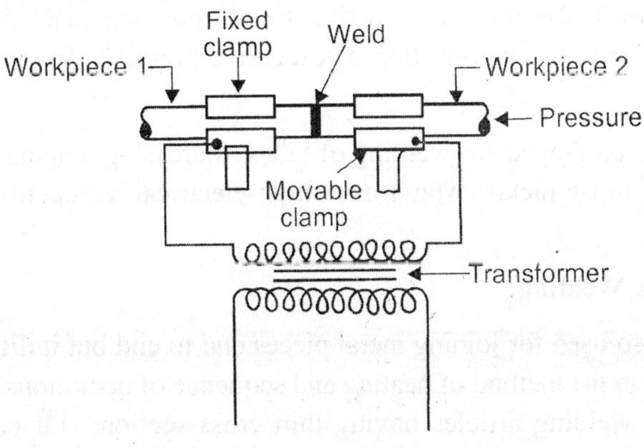

Figure.2.19 Principle of Butt Welding

Also known as upset welding. It is used to join the metal pieces end to end. In this process the metal pieces, usually bars and rods of the same cross-section, are held in suitable clamps or vices with their previously squared ends abutting against each other. The projecting lengths of the pieces between the clamp are adjusted according to their cross-sections and the corresponding materials of which they are made. A longer projection will offer more resistance and a shorter length will provide less resistance. A butting ends of the pieces having equal cross-section should be kept exactly in the middle of the clamps. The endeavour should be to adjust the projected lengths in such a way that the resistance offered by them to the flow of electric current is sufficient enough to generate the desired amount of heat at the respective ends. The clamps holding the pieces either form the electrodes themselves or are fitted with separate electrodes in them. One of these clamps is rigidly fixed to the frame of welding machine and the other is mounted on a movable slide operated by a hand lever in case of large machines and a spring in case of small machines having welding capacity upto 12.7 mm.

After abutting the ends together the current is switched on and the contacting surfaces heated to the fusion point. At this moment additional mechanical pressure is applied by means of the hand lever or the spring attachment and this completes the weld. This pressure should be maintained for a few seconds, actual time depending upon the cross-section of the pieces, to allow the metals to join together. In hydraulic type machines the hand lever is replaced by a hydraulic plunger. The butt-welding method is very suitable for joining end to end the items like bars, rods, tubes and wires etc. It is shown in Figure.2.19.

Application of correct pressure at right moment plays a vital role in making of a good quality weld. For welding of wires and rods upto 12.7 mm diameter the machine may be spring operated. After clamping the parts in position the spring is released to force the end forces against each other, current is switched enough to squeeze the ends and form the joint. A trip switch automatically breaks the circuit as soon as the upsetting is over.

For larger sizes of stock a better practice is to start the weld with a lower pressure to localise the heat at the joint. This can be accomplished by employing a hand wheel or lever in place of the spring. When molten metal is seen flowing from the outer surface the weld is consolidated by application of higher pressure, preferably hydraulically or pneumatically. Enough care should be taken in gripping very thin wires to ensure that the grip is at the edge of the clamp. Failing this, the wire may bend, instead of being upset, when the pressure is applied. Also, in such cases, it is a usual practice to mount the movable clamp on a swinging arm instead of the slide, which in turn is mounted on a ball bearing in order to reduce friction and prevent the chances of application of excessive pressure. Too much pressure may damage the wire by over heating.

But welding can be employed for welding of tubes, increasing lengths of wires and rods, making chains and in welding of those metals which have high electrical conductiivity such as copper, brass, aluminium, etc.

2.13.2. Resistance Flash Welding:

Flash welding is also used for joining metal pieces end to end but it differs from the butt-welding process, described above, in the method of heating and sequence of operations. It has largely replaced the butt-welding method for welding articles having thin cross-sections. Of course, it is used for thick sections also with equal advantage.

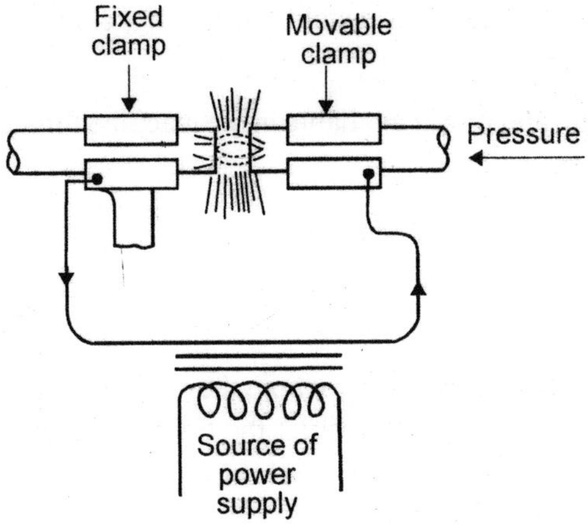

Figure.2.20 Principle of Flash Welding

In this method, first the current is switched on and then the ends of the pieces to be welded are slowly brought closer until they finally come in contact with each other. This forces the heat generated to localise at the ends and thus raises the temperature of the ends quickly to the welding heat. The ends, after they have acquired the contact with each other, are then pressed against one another by applying mechanical pressure. This forces the molten metal and slag to be squeezed out in the form of sparks enabling the pure metal to form the joint and disallowing the heat to spread back. (See figure. 2.20). Single phase A.C. machines are most commonly used for flash welding. The main advantages and disadvantages of flash welding over simple butt-welding are as follows:

Advantages:

1. It is comparatively much quicker than butt-welding.
2. On account of only a small portion of the metal being heated the current consumption is less as compared to butt-welding.
3. A flash-welding joint is stronger than the butt-welding joint. It is also known that the strength of the weld produced is comparable to or even more than that of the base metal.
4. The end faces of the metal pieces need not be squared which is a primary requirement in butt-welding.
5. Lengths and alignment of workpieces is maintained to a high degree of accuracy.

Disadvantages:

1. During flashing particles of molten metal are thrown out, which may enter into the slideways and insulation, etc. This needs periodic maintenance of machine and replacement of insulation.
2. Operator has to take enough care against possible fire hazard due to flashings.
3. Additional stock has to be provided to compensate the loss of metal during flashing and upsetting. This adds to the cost of product.

4. Cost of removal of flash and upset metal by trimming, chipping, grinding, etc. further adds to the product cost.
5. The upsetting pressure and power available in the machine limit the size and cross sectional area of the workpiece to be welded.
6. Opening between the jaws of the gripping clamps also limits the size of the workpiece.
7. For proper heat balance between workpieces it is necessary that their abutting ends should have same shape and size.
8. Surface of the workpieces, particularly where they come in contact with the gripping surfaces, should be clean otherwise they will restrict the flow of electric current.

Metals welded by this process:

As a general rule it can be said that any metal which can be forged can also be flash welded. Also it is possible to weld a number of dissimilar metals by controlling the welding conditions carefully. Metals commonly welded are low carbon steels, low-alloy steels, tool steels, stainless steel, copper alloys, aluminium alloys, nickel alloys, molybdenum alloys, magnesium alloys and titanium alloys. This process is unsuitable for welding of lead, tin, antimony, zinc, bismuth and their alloys.

Some applications:

This process is widely used in automobile industy, welding of tubular and solid structural assemblies, etc. in air-craft industry, welding of band saw blades, welding of tool steel drills, reamers and taps etc. to mild steel or alloy steel shanks, welding of pipes and tubes to increase their lengths, in joining wire ends for producing coils of wires, and many other similar jobs.

2.13.3. Resistance Seam Welding:

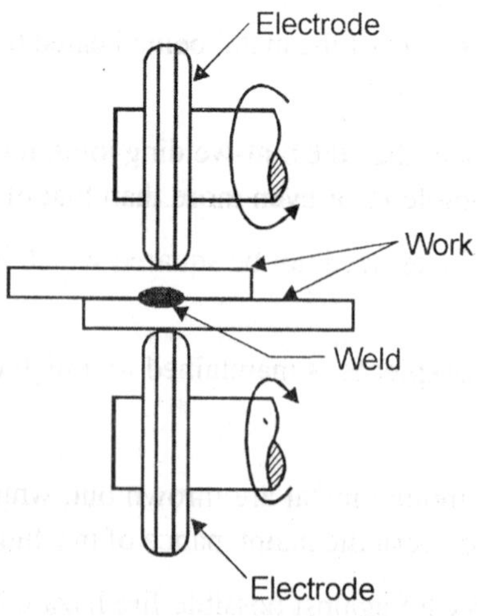

Figure.2.21 Principle of Seam Welding

In principle it is very similar to spot welding except that in this process the spot welding tips are replaced by continuously rotating wheel type electrodes (see figure. 2.21). With the result, the weld produced is continuous instead of being intermittent. A seam weld can best be described as consisting of a continuous series of spot welds produced by passing the workpieces between the revolving electrodes. In operations, the current switched on and the metal pieces pushed together to travel between the revolving electrodes. The metal between the electrodes gets heated to welding heat and welded continuously under the constant pressure of rotating electrodes as it passes between them. This is a quicker operation than spot welding and gives a stronger joint than that. The surface to be joined should be cleaned before being subjected to this process.

This process is employed with equal advantage for making lap welds as well as butt welds. In welding thick sections the use of an "Interrupter" is necessary. It is for the reason that if a continuous current is allowed to flow, the amount of heat generated on account of the high resistance offered to its flow by the thick section, will be too much and the metal will get melted on its surface which will stick to the contacting surface of the electrodes. For thin sections, say upto 20 W.G. there is no need of using any interrupting devices.

Process Limitations of Seam Welding:

1. It cannot be applied to those portions where abrupt change in contour occurs along the path of electrode wheels, such as on sharp corners.
2. In longitudinal seam welding machines the maximum length of the seam joint that can be made equals the throat depth of the machine.
3. It is necessary to avoid obstructions in the path of electrode wheel or else a corresponding recess should be provided on the wheel periphery to accommodate these.
4. It is necessary that the weld should proceed along a straight line or a uniform curve.
5. Stock thickness above 3 mm cannot be welded with normal ease.
6. For successful welding and production of defect free welds it is essential that the work surfaces should be perfectly clean and free from grease, paint, oil, rust and scale.

2.13.4. Resistance Projection Welding:

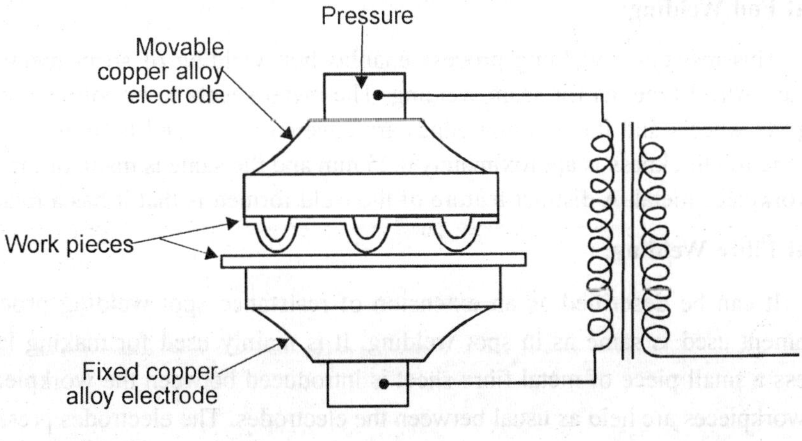

Figure.2.22 Projection Welding

This process is similar to spot welding, but differs from the latter in that the spots at which welding takes place are previously located by providing projections at the desired locations on the surfaces of one of the work-pieces, as shown in Figure. 2.22. Thus, the surfaces of the workpieces are in contact with each other only at the projections. As current is switched on the projections are melted and the workpiece pressed together to complete the weld, by pressing the upper electrode downwards. The melted projections forms the welds. This method enables production of several spot welds simultaneously. The electrodes, if required, may be designed and shaped to work as holding fixtures for workpieces and assemble them in proper relative location through welding. Closer welds than the rough spot welding can be obtained by this process. However, this process is economical only for large-scale production.

2.13.5. Percussion Welding:

It is a very fast method of welding. It consists of holding the parts at a small distance with their end faces opposite to each other, bringing them closer at a fast speed after switching on the current, thus creating an arc between their end faces just before they come in contact and completing the weld under impact. Some of the metal may squeeze out of joint, but it is very small. The use of this process is limited to very thin wires, with their diameters ranging between 0.05 mm and 0.38 mm. It can also be used for joining wires of dissimilar metals, such as copper to nichrome and copper to stainless steel.

2.13.6. Other Resistance Welding Processes:

i) Spike Welding:

It is a resistance welding method in which large amount of electricity is stored up in capacitors and then the same is released rapidly through the electrodes and the metal pieces to be joined. This current flow is allowed for a very small interval and its timing is controlled electronically. The welding is, therefore, very rapid and consequently the chances of warpage of workpieces and of contamination are altogether eliminated. This process can be successfully used for welding of almost all metals and alloys, metals of different thicknesses together and dissimilar metals also.

ii) Metal Foil Welding:

This resistance welding process enables butt welding of sheet metal. The electrodes used are wheel type, as for seam welding. The metal sheets to be joined are placed between electrode wheels and the abutting edges are covered at top and bottom with very thin metal foils, the foil thickness is approximately 0.25 mm and the same is made of the same material as the workpiece metal. A distinct feature of the weld formed is that it has a raised bead.

iii) Metal Fibre Welding:

It can be described as an extension of resistance spot welding process, In this, the equipment used is same as in spot welding. It is mainly used for making lap joints. In this process a small piece of metal fibre sheet is introduced between the workpieces to be joined. The workpieces are held as usual between the electrodes. The electrodes pressure employed is very low and, therefore, the indentation on the work surface is also very small.

iv) Gas Welding:

It is a fusion or non-pressure welding method in which a strong gas flame is used to raise the temperature of the ends of the pieces to be joined to a heat sufficient to melt them. The metal thus melted starts flowing along a strong weld. A filler metal may be added to the flowing molten metal to fill up the cavity made during the end preparation. So many different combinations of gases can be used to obtain a heating flame, but the most common of these are oxygen and acetylene, oxygen and hydrogen and oxygen with coal gas, of which the first one is very extensively used. The filler rod or welding rod which provides the additional metal required is of the same or nearly same composition as that of the metal of which the pieces to be joined are made.

Oxy-hydrogen mixture can be used for welding of only such metals of which the melting point is low, such as aluminium, magnesium, lead, etc. It is for the reason that the temperature of the flame produced is too low to produce effective fusion of high melting point metals. For similar reason the other gas mixtures, like oxygen—natural gas, oxygen—propane, oxygen—coal gas, etc., can not be used for welding of high melting point metals. If, however, effort is made to raise the flame temperature of these mixtures, by increasing the proportion of oxygen, the flame atmosphere becomes highly oxidizing, which is an abvious disadvantage. It is only for these reasons that the use of these gases is largely confined to brazing and to some extent to braze welding only.

The oxygen used should be highly pure since even a small proportion of impurities has a considerable effect on the combustion effect of oxygen. Its purity should be above 99.5%. The common impurities associated may be nitrogen, argon and water vapour.

2.14 BRAZING - INTRODUCTION

Brazing is a process of joining metals with a nonferrous filler metal that has a melting point below that of the metals being joined. By AWS definition, the melting point of the filler metal will be above 800°F (427°C). Below this temperature are the solders. The filler metal must wet the surfaces to be joined, that is, there must be a molecular attraction between the molten filler material and the materials being joined. The brazing alloy, when heated to the proper temperature, flows into the small joint clearances by capillary action. A limited amount of alloying occurs between the filler metal and the base metal at elevated temperature. As a result, the strength of the joint when properly made may exceed that of the base material. The strength is attributed to three sources, atomic forces between the metals at the interface; alloying, which comes from diffusion of the metals at the interface; and intergranular penetration.

The heat for brazing may be provided in many different ways, the most common of which are by torch, induction furnace, and hot dipping.

2.14.1. Types of Brazing:

i) Torch Brazing:

Torch brazing is one of the oldest and most widely used methods of heating for brazing; it is versatile and adaptble to most jobs. It is especially good for repair work in the field and for small-lot jobs in the shop. As it is a manual operation, torch brazing may have high labor costs, and the skill and judgement of the torch operator will determine the quality of the work.

Torch brazing may be done with the use of natural gas and oxygen, air and acetylene, butane, propane, MAPP, and the regular oxyacetylene. If the latter is used, a neutral flame is recommended, which is a balanced condition of acetylene and oxygen. A slightly reducing flame (excess acetylene) may be used to avoid oxidation of the metal. The filler metal is touched to the joint when the flux becomes a clear liquid. It is the parent metal, not the flame, that transfers the heat to the brazing alloy.

ii) Induction Brazing:

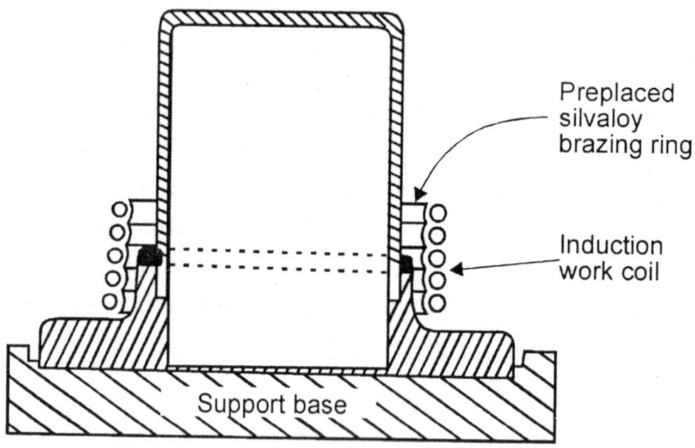

Figure.2.23. Example of External Induction Brazing

The heat of induction brazing is furnished by an a AC coil placed in close proximity to the joint (Figure. 2.23). The high-frequency current is usually provided by a solid-state oscillator that produces a frequency of 200,000 to 5,000,000 Hz. These alternating high-frequency currents induce opposing currents in the work, which, by electrical resistance, develop the heat. This method has the advantage of providing (1) good heat distribution, (2) accurate heat control, (3) uniformity of results, and (4) speed. It is especially good for certain types of repetitive work that require close control.

iii) Furnace Brazing:

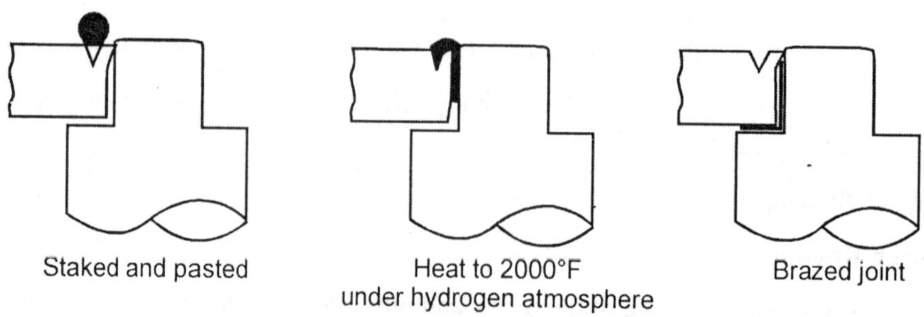

Figure.2.24. Three Phases of Furnace Brazing.

Parts are to be furnace brazed have the flux and brazing material preplaced in the joint. If the furnace has a neutral or shielding atmosphere, the flux may not be necessary. The furnace may be batch or conveyor type. Automatic controls regulate both time and temperature and, where applicable, atmosphere. The three steps used in copper-hydrogen funace brazing are shown in figure 2.24.

iv) Dip Brazing:

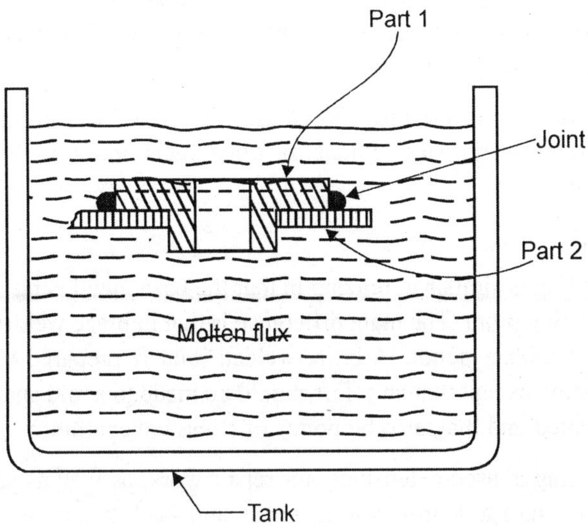

Figure.2.25 Parts to be dip brazed are first placed in Jigs or, as shown here, the parts may be self-jigging.

Dip brazing gets its name from the fact that parts are jigged (or in some cases the parts are self-jigging, as shown in Figure 2.25) and are placed in a chemical or molten-metal bath maintained at the correct brazing temperature. In a chemical bath, the parts are first thoroughly cleaned and then assembled with a filler-metal placed on or in the joint. The bath, usually molten salts, is maintained at a higher temperature than the filler metal being used. After dipping, the parts are removed and immediately cleaned to remove the flux. The number and size of the parts to be brazed are limited only by the capacity of the bath and the handling facilities.

v) Paste Brazing:

A newer development in brazing is paste brazing. Anyone who has done any brazing knows there is some skill involved. After the appropriate type and amount of flux and filler metal are selected, the appropriate temperature must be chosen. Too little flux means the base metal will not be sufficiently cleaned, resulting in a weak joint. To ensure strong joints, excessive flux is often used; this wastes material, requires more cleaning, and can errode parts and fixtures.

vi) Paste Alloy Process:

Paste alloys are designed to overcome the problems given in the previous paragraph. Paste alloys are blends of filler metal in powdered form and a flux that are held in suspension by a paste-type binder. Premixing of flux and filler ensures the proper proportions of each and keeps the alloy localized to the joint area. Pastes rely on capillary action to spread the metal and flux over the joint area. Paste brazing filler materials are available in silver, nickel, copper, and gold based formulations.

For automatic past brazing, rotary index tables are used that convey fixture held parts through a rimed sequence of operations at production rates that typically average from 500 to 1000 parts per hour. In most systems, paste is applied at only one station, where one or more pneumatically controlled guns descend to the indexed parts. Paste deposition occurs when a limit switch contacts the assembly.

A natural gas and air mixture is used as a source of heat on most operations; however, alternative methods use propane, butane, or other fuels, along with pure oxygen for higher heating temperature.

vii) Braze Welding:

Braze welding is similar to brazing in that the base metal is not melted but joined by an alloy of lower melting point. The main difference is that in braze welding the alloy is not drawn into the joint by capillary action. A braze-welded joint is prepared very much like a joint is prepared for welding except that an effort should be made to avoid sharp corners, because they are easily overheated and may also be points of stress concentration.

Braze welding is used extensively for repair work, as well as some fabrication on such metals as cast iron, malleable iron, wrought iron, and steel. It is used, but to a lesser extent, on copper, nickel, and high-melting point brasses.

2.15 BRAZING FLUXES

The primary purpose of brazing fluxes is to dissolve and absorb oxides that heating tends to form. Therefore, the parts to be brazed must first be cleaned of all existing oxides, oil, grease, and other foreign matter either by degreasing, grinding, pickling, or a combination of these cleaning processes. When the metal is heated to the brazing temperature, the flux becomes a clear liquid that wets the surface and aids in the flow of the filler metal. Some of the most common ingredients of fluxes are sodium, potassium, and lithium. They are used in making up chemical compounds such as borates, fluorides, chlorides, boric acids, akalies, and wetting agents.

Generally, fluxes are available in three forms; powder, paste, or liquid. Of the three, paste is most commonly used, although powdered flux is frequently mixed with water or alcohol to give a paste like consistency.

2.15.1. Brazing Filler Metals:

Brazing filler metals are divided into seven classifications. In order of popularity, these are silver, copper, copper-zinc, copper-phosphorus, aluminium-silicon, copper-gold, and magnesium. These alloys are produced in many forms, such as wire, rods, coated sheets, and powder. Frequently, brazing alloys are preformed into rings or special shapes to simplify placement of the correct amount at the joint.

The service temperature of brazing varies widely. Most are for room temperature, but some have been developed for stainless steel and have a service temperature of upto 2200°F (1204°C).

2.15.2. Advantages of Brazing:

Brazing is well suited to mass-production techniques for joining both ferrous and non-ferrous metals. Some of the principal advantages are:

1. Dissimilar metals can be joined easily.
2. Assemblies can be joined in a virtually stree-free condition.
3. Complex assemblies can be joined in several steps by using filler metals with progressively lower melting temperatures.
4. Materials of different thickness can be joined easily.
5. Brazed joints require little or no finishing other than flux removal.

Limitations of Brazing:

1. Joint design is somewhat limited if strength is a factor.
2. Joining is generally limited to sheet-metal thicknesses and relatively small assemblies.
3. Cost of joint preparation can be high.

2.16 TESTING OF WELDED JOINTS

Welded joints are tested to determine the strength and to find out the defects if any. Welded joints are tested by using:

a) Non Destructive Tests.

b) Destructive Tests.

Non destructive tests are tests during which the welded joints are not destroyed. The welded joints will be destroyed during destructive tests. The various types on non destructive tests and destructive tests are explained below.

2.16.1. Non Destructive Tests:

i) Magnetic particle test:

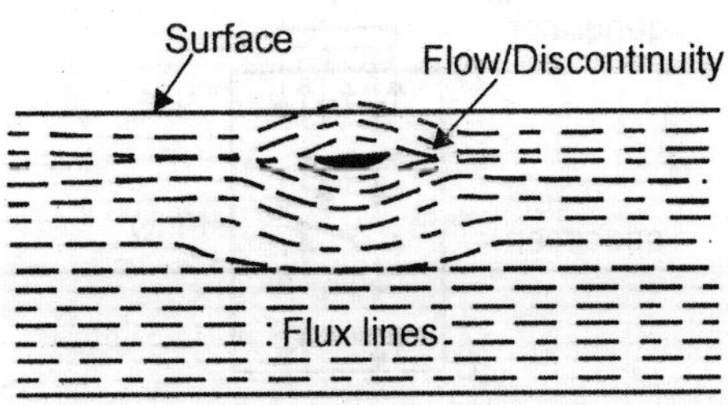

Figure.2.26 Magnetic particle test

Cracks and slag inclusions can be found by magnetic tests. The workpiece is magnetised. Fine iron powder is sprayed on the surface of job. If there is a crack, magnetic poles will be formed at the crack. The magnetic particle test is shown in figure 2.26.

The iron powder will be attracted to the crack and the crack will appear as a line. This method is called magnetic crack detection. This method is applicable only to ferrous metals.

ii) X-ray test:

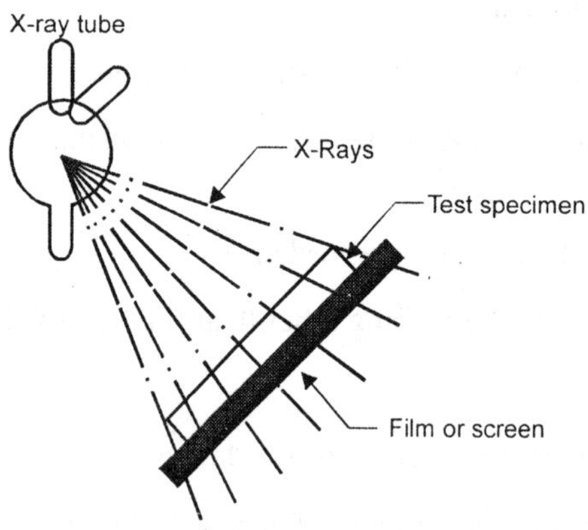

Figure.2.27 X-ray test

From an X-ray tube X-rays are passed through the welded metal. Refer figure 2.27. The X-ray films will show weld defects like blow holes, cracks, inclusions, cavities, porosity and imperfect fusion. The cost of the test is high. This is a quicker method. It gives a permanent record of the result. We have to take special precautions to avoid radio active effects on human beings.

iii) Ultrasonic test:

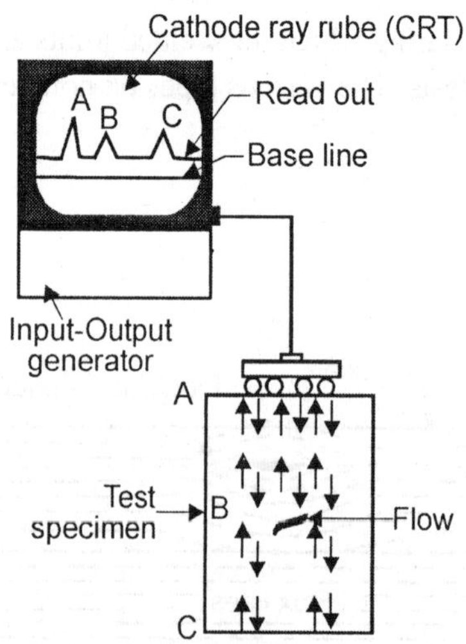

Figure.2.28 Ultrasonic Test

In ultrasonic test, a beam of ultrasonic waves (high frequency vibrations) produced by a transducer (input-output generator) is passed into the part being tested. If there are defects, the echoes will be returned and they will be picked up by a receiver. The receiver converts the ultrasonic waves into a electric signal. The signals are amplified and is projected on the screen (CRT) as shown in figure 2.28. The top surface A and bottom surface C of the specimen are shown in the screen by signals A and C. The internal flaw at B in the specimen is indicated in the screen by the signal B. All the surface defects and internal defects can be located by this method of testing.

2.16.2. Destructive Tests:

Nick Break Test:

In this test the grain size, blow holes, flash eye and foreign material inclusion are inspected. Two plates are welded together. A test specimen of 200 mm long is selected. 6 mm deep slots are cut with a saw at the opposite sides of the welded portions. This is called nicking. The specimen is supported on two rollers. At the nicked portion, a hammer blow is applied to break the weldment. The fractured surface of the welded portion is inspected for defects. (Refer figures 2.29).

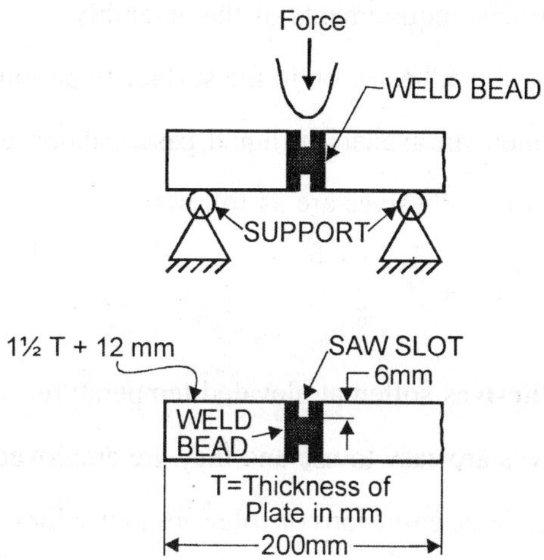

Figure.2.29 Nick break test

2.17 ADHESIVE BONDING

2.17.1 Definition:

Adhesive bonding is the process of joining materials with the help of adhesives. The term adhesives such substances as flues, cements and other bonding agents. As related to a bonding or joining process, ther term adhesion is probably best defines simply as the interaction between two surfaces.

2.17.2 Concept:

Figure 2.30(a) shows two flat, solid surfaces brought into contact. It can be seen that only a small fraction of the total surface areas actually are in contact.

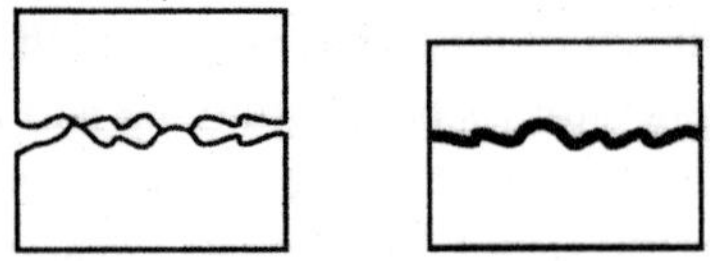

Figure.2.30 Two solid surfaces in contact

If these microscopic voids between the two solid surfaces (in contact) can be filled with an adhesive fluid (Figure 2.30 (b)) which, then curves or hardens (on cooling or chemical reaction) to form a solid adhesive, a strong adhesive bond will be developed and considerable energy will be required to separate the two solid pieces.

2.17.3 Types of Adhesive bonding

An adhesive for a particular application is selected considering:

i) The type of materials required to be joined.

ii) Service (performed) requirements of the assembly.

iii) Method to apply the adhesive on to the surface to be joined.

Many adhesive formulations are available in liquid, paste solution, emulsion tape films and powder form.

The most commonly used adhesives are as follows

a) Thermo plastic

b) Thermosetting

a) Thermo plastic type adhesives soften at elevated temperature.

Thermoplastic adhesives are easy to use and they are employed as:

i) Air dying dispersions, emulsions or solutions that achieve their strength through evaporation of the solvent.

ii) Fusible solids that liquify when heated, flow and fill the bond area, Bond strength is achieved after the adhesive solidfies by cooling.

iii) Pressure-sensitive tapes, which are highly ciscosity plastics on cloth or paper backings.

b) Thermosetting adhesives cure or harden by chemical reactions, such as polymerization, condensation, cross linking, vulcanization or oxidation caused by the addition of a catalyst; heat, pressure, radiations or other energy can also affect hardening or curing. Thermosetting adhesives, once hardened, cannot be remelted and a broken joint cannot be rebonded by heating.

2.17.4 Applications of Adhesive bonding

Adhesives bonding finds appllication is following industries:

i) Automotive

ii) Aircraft

iii) Building products

iv) Packaging

v) Shape and apparel

vi) Furniture

vii) Book binding

viii) Electrical

ix) Railroad

x) Ship-building

xi) Medical and dental

xii) Tape-industries etc.,

2.18 SOLDERING

Soldering is a way of joining two or more pieces of metal by means of a fusible alloy or metal, called solder, applied in the molten state. Soldering can be distinguished from brazing in that it is done at a lower temperature (below 800°F, 427°C) and there is less alloying of the filler metal with the base metal.

2.18.1. Types of Soldering:

1. Soft soldering:

If the melting point of the filler metal is very much lower than the melting point of metals to be joined it is called soft soldering.

2. Hard soldering:

The hard solders are alloys of copper and zinc to which silver is also added. The hard solder melt above 90°C. Hard soldering produces stronger joints than soft soldering.

2.18.2. Solder Alloys:

The four metals used as the principal alloying elements of all nonproprietary solder formulations are tin, lead, cadmium, and zinc. These alloying elements fall into three thermal groups. Those that melt below 596°F (313°C) are called low-melting point solders. Those that melt between 596° and 700°F (313° and 371°C) are called intermediate solders. Those that melt between 700° and 800°F (371° and 427°C) are termed high-temperature solders. The strongest and most corrosion resistant are the high-temperature solders. The strength of low-temperature solders in shear is a bit in excess of 5000 psi (35 MPa).

The low-melting point solders 378° to 596°F, (190° to 313°C) are the most widely used and are of the tin-lead or tin-antimony type. The tin-lead alloys provide good strength at low temperatures. A combination of 62% lead and 38% tin produces the lowest melting point and in practise is referred to as 60-40 solder. Because it is eutectic alloy, it behaves like a pure metal and has only one melting temperature, 362°F (183°C), at which it is completely liquid. Increasing the tin content produces better wetting. Solders are generally used for the same types of applications are the tin-lead alloys except that they are not rècommended for use on aluminium, zinc, or galanized steel.

The tin-zinc group of alloys is used mainly for soldering aluminium, primarily where a lower soldering temperature than that of a zinc-aluminum solder is required. Zinc-aluminum solders are designed specifically for soldering aluminum.

Indium solder, 50% indium and 50% tin, is particularly suitable for products subjected to cryogenic temperatures and for glass-to-metal bonds.

2.18.3. Aluminum Soldering:

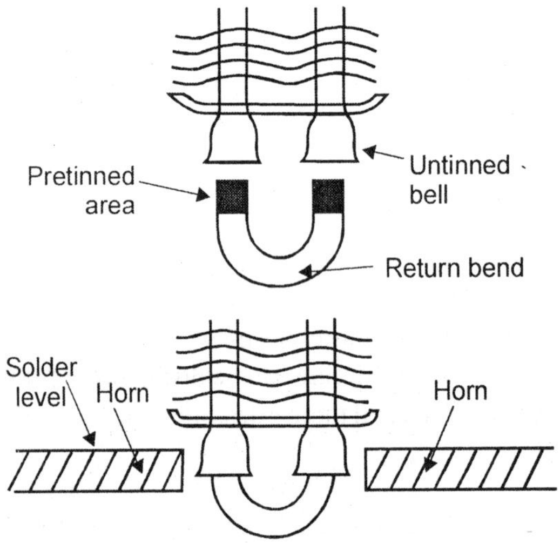

Figure.2.31 Ultrasonic energy is used to remove the oxide as the aluminum tubing is immersed in the liquid solder.

Aluminum is often thought of as being very difficult to solder. One reason is that an oxide film forms quickly over the surface after cleaning. Now, however, there are several methods of removing the oxide as the soldering is being done. These methods are flux soldering, friction soldering, and ultrasonic soldering. Ultrasonic soldering is now being used to solder aluminum tubing in heat exchanger coils without flux. The return bends, as shown in figure 2.31, are pretinned (coated with solder) and placed in the untinned bell. The assembled joint is then immersed in the ultrasonically activated solder. The sonic energy removes the oxides. The process lends itself to automation.

Of the low-temperature solders used for aluminum, a zinc-lead eutectic, which melts at 400°F (204°C), has an advantage over brazing because less heat is required, which results in a minimum loss of strength and hardness and the least distortion. This is particularly important in complex structures. If greater strength is required, the intermediate - or- high-temperature solders may be used. The high-temperature solders contain 90% to 100% zinc and are the least expensive and the strongest of all aluminum solders.

Corrosion in aluminum soldered joints can be particularly troublesome. It may be either chemical or galvanic in nature. In any case, moisture must be present. By coating the joint with a waterproof coating, the problem can be eliminated.

2.19 DEFECTS IN WELDING

A weld may have the following defects.

Porosity.

Cracking

Lack of fusion

Over penetration

Slag inclusion

Excessive grain growth

Bending distortion

Longitudinal shrinkage

Transverse shrinkage

Some of these such as slag inclusion, inadequate penetration lack of fusion are basically the faults of the welder or choice of process. Whereas, porosity, cracking are due to the fundamental difficulties in a welding operation, although faulty technique adopted by the operator can also cause these defects.

Porosity:

If gas is entrapped by the cooling weld metal, porosity results. The porosity may be isolated in small place or may be scattered throughout the entire weld.

Gas may evolve due to chemical reaction such as

Due to supersaturation of hydrogen and nitrogen in the melt, the gas may evolve. In bare wire welds nitrogen is picked up from the atmosphere. In atomic hydrogen welding, in which excessive hydrogen is evolved, the weld pool may absorb hydrogen causing porosity.

Lack of fusion:

If the temperature of the base metal is not raised to its melting point or scales and other foreign materials are removed before welding, the layers of weld metal cannot fuse with the adjacent layers of the base metal.

Over Penetration:

If the base metal melts away or burns out over penetration of the weldment results.

In a fillet weld, undercutting reduces the cross-section of the member.

SUMMARY

❖ ***Welding:*** Welding is a process of joining two similar or dissimilar metals by fusion, with or without the application of pressure and with or without the application of filler metal. The fusion of metal takes place by heat.

❖ ***Types of welding:*** The modern methods of welding can be classified into two broad headings: i) fusion or non-pressure welding, ii) Plastic or pressure welding.

❖ ***Fusion welding:*** In fusion welding the metal at the joint is heated to a molten state and then allowed to solidify.

❖ ***Plastic welding:*** In plastic welding the parts to be joined arc heated till they become plastic, then they are pressed together to form the joint.

❖ ***Some types of fusion welding:*** Gas welding, Arc welding, Thermit welding.

❖ ***Examples of plastic welding:*** Resistance welding, Thermit welding, Gas welding.

❖ ***Electric arc welding:*** Electric arc welding is used to join two metal parts. In this an electric transformer or an electric generator is used to produce an arc between the workpiece and an electrode.

❖ ***Types of cosumable electrodes:*** Bare electrodes, Lightly coated electrodes, Heavily coated electrodes.

❖ ***Filler rod in welding:*** Filler rod or welding rod is a metal added to the weld when the thickness of the pieces to be joined is more than 15 mm.

❖ ***Flux in welding:*** A flux is a substance used in welding to remove impurities and prevent oxide formation. It also helps in proper bonding of metals being welded.

❖ ***GTAW* or *IGTAW* or *TIG:*** Gas tungsten arc welding or insert gas tungsten arc welding or tungsten arc welding is similar to MIG. In this process the intense heat of an electric arc a non-consumable tungsten electrode and metal to be welded. The shielding is obtained from helium, argon or mixture of the two gases.

❖ ***MIG* or *GMAW:*** Metal inert gas welding or Gas metal arc welding. This process uses a consumable bare electrode (made of Cu, Steel, Al, etc) which is usually in the form of a coil, fed continuously by guide rollers driven by a motor.

❖ ***Major components in GMAW:*** Welding gun, Wire drive, Gas supply and suitable valve, Welding machine (Power source), Water supply system (if necessary), Control unit.

❖ ***Different models of metal transfer:*** Short circuiting transfer, Drop transfer (grobular type, buried etc), Spray transfer, Pulsed spray transfer, Rotating spray transfer.

❖ ***Advantages of GMAW*** **or** ***MIG:*** No flux is required, High welding speed, High corrosion resistance, Possible to weld non-ferrous metals.

❖ ***Submerged arc welding (SAW):*** Submerged arc welding is used for faster and high quality but twelds for thicker plates. Similar to gas metal arc welding, the arc is produced between electrode and work piece is completely submerged (hidden) under the flux powder. The arc is not visible outside.

❖ ***Applications of SAW:*** i) Fabrication of penstocks, pressure vessels, pipes, boilers, rotary kilns, earth moving machineries, cranes, bridges, rail road, chassis of railway coaches and locomotives. ii) Rebuilding of worm out parts and depositing wear resisting alloys and hard facing. iii) Used in automotives, aviation, ship building and nuclear power industry.

❖ ***Resistance welding:*** Resistance welding is a fusion welding process in which both heat and pressure is applied to the joint. No filler metal or flux is added. The heat necessary for the melting of the joint obtained by the heating effect of the electrical resistance of the joint.

❖ ***Types of resistance welding:*** Spot welding, Butt welding, Flash welding, Seam welding, Projection welding, Precussion welding.

❖ ***Flash welding:*** Flash welding process is two parts to be welded are clamped in jaws of welding machine. Two clamps are brought close enough for a small air gap between them. When power supply is on, an electric arc or flash is produced due to the jumping of current in the gap. This jumping of current produce enormous heat at this moment mechanical pressure is appled forcing the end together and the welding is completed.

❖ ***Gas welding:*** Gas welding processes are technically called oxygen-fuel gas welding, in which the material to be joined is completely melted to obtain the fusion. Gas flame is used to melt the material to be joined. Which is obtained from the combusion of fuel gas acetylene in combination with oxygen.

❖ ***Laser beam welding:*** The laser is a concentrated beam of coherent monochromatic radiation. Because of the coherency, it is possible to concentrate the laser beam by means of an optical lens to required spot of required size without losing its intensity. The laser beam is a high energy source of heat to melt the joint for fusion welding in laser beam welding.

❖ ***Basic component of the welding laser:*** Electrical storage unit, capacitor bank, triggering device, a flash tube wrapped with a wire lasing material, a focussing lens mechanism, the work table and workpiece.

- ***EBW Process:*** This is a fusion welding process. High velocity electrons is used as heat source in this process for melting the metals. The electron beam upon impinging the workpiece releases the necessary heat by converting its kinetic energy. Due to penetration of electrons, a key hole is formed melting the work around the beam. As beam traverse the key hole would also travel along with molten metal being pushed back which when solidified forms the joint.

- ***Plasma arc welding:*** In this process a gas argon, hydrogen or helium is passed through an electric arc, where it is heated up and ultimately get ionised. The stream of these hot ionised particles (plasma) is passed through a small orifice of a welding torch. This is termed as plasma arc is an electric arc combined with an energised plasma or stream of ionised gas particles. This stream is made to impinge on the area to be welded and its energy is utilised in heating the base metal.

- ***Applications of ultrasonic welding:*** Joining of fine wire and their metal assemblies, Joining thin sheets to thick sheets, Joining electrical and electronic components, Joining nuclear fuel elements, Joining of plastics.

- ***Applications of electro slag welding:*** i) Welding of heavy steel forgings, large steel castings, thick steel plates. ii) Frames of heavy mechanical and hydraulic press, iii) Used in the construction of boiler components for thermal power plant and heavy construction jobs, iv) Frames of locomotive and in rolling mill.

- ***Friction welding:*** Friction welding is a solid state pressure welding process in which coalescence of materials is produced from the hear generated (due to friction) by rotating one piece against another under controlled axial pressure. In this one of the workpieces to be joined is rotated at a high speed around 300 rpm and the other is aligned axially and pressed tightly against it. The friction between the two parts raises the temperature of the both ends.

- ***Thermit Welding (TW):*** Thermit welding (TW) gets its name from the compound named thermite, a name is based on the word them, meaning heat; the work thermit is a registered trade mark. The process of thermit welding is similar to an investment casting process. *[The process in which superheated liquid called thermit steel is spread around the parts to be joined, thermit steel is a mixture of fine aluminium powder and iron oxide. (1.3 by weight)]*.

- ***Soldering:*** Soldering is the process of joining two similar or dissimilar metals by adding a low melting alloy called solder. The work metals are not melted in soldering.

- ***Types of solders:*** The types of solders are i) Soft solder, ii) Hard solder and other types, torch soldering, furnace soldering, iron soldering, induction, resistance soldering, dip soldering, infrared soldering.

REVIEWS

1. Name the classification of welding processes.
2. Name the types of welding.
3. Name the equipments used un arc welding.
4. Listout the specification of welding generator and transformer.
5. Differentiate between AC and DC arc welding.
6. Explain the working of TIG and MIG welding.
7. Explain the working of Plasma arc welding.
8. Explain the working or electro slag welding with neat sketch.
9. Explain thermit welding process.
10. Explain friction welding process.
11. Name the types of gas welding process.
12. Name some gas welding equipments.
13. Name the gases used for gas welding.
14. What are the different types of flames used in gas welding proces?
15. Explain the flame cutting process.
16. Name the types of welding electrodes.
17. Differentiate between Bare and coated electrodes.
18. State the functions of coating in electrodes.
19. Explain electron welding process.
20. Explain laser beam welding process.
21. What do you mean by diffusion welding process?
22. Explain explosive welding process with neat diagram.
23. What do you mean by resistance welding process?
24. Explain resistance butt welding process.
25. Explain resistance flash welding process.

26. Explain resistance seam welding process.
27. Explain resistance projection welding process.
28. What do you mean by percussion welding process?
29. Name some other welding process.
30. Explain brazing process.
31. Name the various types of brazing process.
32. Explain induction brazing.
33. Explain braze welding.
34. Name the brazing fluxes used for brazing process.
35. State the different types of testing used for welded joints.
36. Explain X-ray test.
37. Explain soldering process.
38. Name some solder alloys used for soldering.
39. Explain aluminium soldering process.
40. Name some defects in welding process.

CHAPTER - 3

BULK DEFORMING PROCESSES

3.1 INTRODUCTION

A wide variety of processes have been developed to mechanically shape material, and a number of methods have been proposed to provide classification. One approach divides the processes into primary and secondary. Primary processes reduce a cast material into intermediate shapes, such as slabs, plates, or billets. Secondary processes further convert these shapes into finished or semifinished products. Unfortunately, some processes clearly fit both categories, depending on the particular product being made.

A more useful division focuses on the size and shape of the workpiece and how that size and shape is changed. Bulk deformation processes are those where the surface area of the workpiece changes significantly. Thicknesses or cross sections are reduced or shapes are changed. Since the volume of the material remains constant, other dimensions must change in proportion. Thus the enveloping surface area is altered, usually increasing as the product lengthens or the shape becomes more complex. In contrast, sheetforming operations involve the deformation of a material where the thickness and surface area remain relatively constant. Ever here, the division is not without confusion. Coining, for example, begins with sheet material but alters the thickness in a complex manner that is essentially bulk deformation. Where the division of processes is based on the workpiece temperature. The division is somewhat blurred, especially in view of the increased emphasis on energy conservation, the growth of "warm working", and new advances in technology. Processes that were traditionally performed hot are now being performed cold, and cold-forming processes can often be aided by some degree of heating. As a result, most sheet-forming operations are performed cold.

3.1.1 Hot Working Processes:

The hot working processes possess features that often make them an attractive means of producing a desired shape. At elevated temperatures, metals weaken and become more ductile. With continual recrystallization, massive deformation can take place without fear of fracture due to diminished ductility. In steels, hot forming involved the deformation of the weaker, austenite structure, which them cools and transforms to the stronger, room-temperature ferrite, or much stronger nonequilibrium structures, such as martensite.

Some of the hot-working processes that are of major importance in modern manufacturing are:

1. Rolling 2. Forging 3. Extrusion 4. Hot drawing 5. Pipe welding 6. Piercing

Advantages of Hot Working:

- ★ The strength of the metals is low at high temperature. Hence low tonnage equipments are adequate for hot working.
- ★ Very large workpieces can be deformed with equipment of reasonable size.
- ★ Inter stage annealing and stress relieving are not required.
- ★ Because of high ductility at high temperature and absence of work hardening, large deformations can be undertaken in a single stage and complex parts can be fabricated.

Disadvantages of Hot Working:

- ★ Due to oxidation and scaling, there is heavy material loss. This also results in poor dimensional accuracy and surface finish.
- ★ High energy costs to heat the metals to high temperature.
- ★ Automation is difficult due to high working temperature.
- ★ Thin parts (sheets, wires etc.) cannot be produced due to loss of ductility because of high rate of loss of heat.

Typical hot working temperatures are :

Steels	:	1100 to 1260 °C
Cu and its alloys	:	760 to 925 °C
Mg	:	315 °C
Al and its alloys	:	370 to 455 °C.

3.1.2 Cold Working Processes:

Cold working may be defined as the process of plastically deforming a metal piece below its recrystallisation temperature. In cold working, more force is required to deform the metal as the metal is in cold condition and its structure is rigid. The grains of the metal are elongated while cold working Elongation takes place in the direction of flow of metal. Internal stresses are developed. To remove these stresses, cold worked components are often annealed. The metal gets hardened, this hardness produced due to cold working is known as work hardening.

Cold working is often used as a finishing operation after hot working of metal.

Advantages of Cold Working:

- ★ Surface defects are removed.
- ★ High dimensional accuracy.
- ★ Thin guage sheets can be produced.
- ★ Physical properties of metals that do not respond to heat treatment can be unproved by cold working.
- ★ Since cold working is done at room or low temperature no oxidation and scaling of the work material occurs. This results is reduced material loss.

3.2. EFFECTS OF COLD WORKING

The cold working process produces the following effects on the metal:

1. Internal stresses are produced in the metal.
2. Grains are distorted.
3. Strength and hardness are increased.
4. Ductility of the metal decreased.

3.3 COLD WORKING OPERATIONS

Cold working operations are classified as follows:

I. Drawing Process

1. Blank drawing 2. Tube drawing 3. Wire drawing 4.Embossing
5. Metal spinning 6. Stretch forming 7. Stretch draw forming.

II. Squeezing process

1. Coining 2. Cold rolling 3. Sizing 4. Swaging or cold forging
5. Thread rolling 6. Knurling 7. Riveting

III. Bending Process

1. Angle bending 2. Roll forming 3. Plate bending
4. Seaming 5. Curling

IV. Shearing Process

1. Punching 2. Blanking 3. Cutting Off 4. Trimming
5. Perforating 6. Notching 7. Slitting 8. Lancing
9. Shaving

V. Special Forming Processes

1. Explosive forming 2. Electro hydraulic forming
3. Magnetic forming 4. Impact extrusion
5. Shot peening

3.4. DRAWING PROCESSES

Figure 3.1 illustrates the principle of cupping or drawing. Drawing is making cup-shaped products from heated sheet metal blanks. The blank is placed over a die and pressed with a punch of suitable size (diameter). The punch enters the die and draws the sheet to its own form. If the product is quite long and thin walled repeated heating and drawing through successively decreasing size of dies may bc become essential. After drawing through two dies Figure (a) and (d), the drawn cup is further drawn through a horizontal drawbench which consists of several dies set in order of decreasing diameters. So that decrease in thickness of the wall is gradual.

If the final product is to be a tube the closed end is cut-off. If necking is to be formed on the drawn cylinder, it is done by hot spinning or swaging.

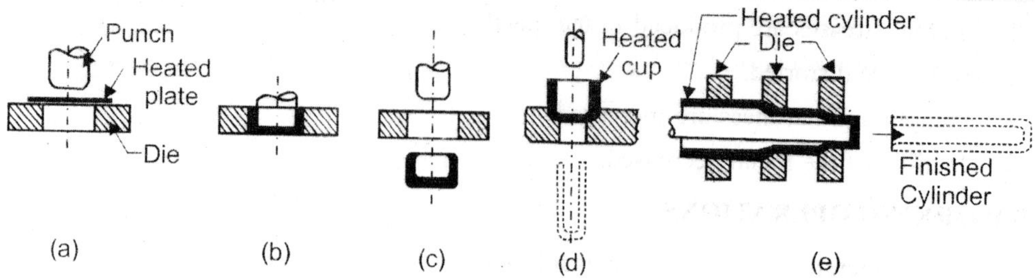

(a) Heat plate and punch set at proper position (1st Stage), (b) Punch forces the plate in the die (2nd stage), (c) Punch goes to initial position, the cup is knocked out (3rd Stage), (d) Heated cup is placed over smaller diameter die for further reduction of cup diamete, (e) The cup is finally passed through a horizontal draw bench

Figure 3.1 Hot drawing from heated plate

i) Blank Drawing

Blank is a sheet metal piece of the required size which can be formed into a component. Components are produced from these blanks by a drawing process. The blank is placed over a die and held on it by a holder. A punch presses the centre portion of the blank into the die. The metal flows plastically by the action of the punch. Cup shaped articles are produced by this cold forming process. Sufficient clearance must be given between the punch and die.

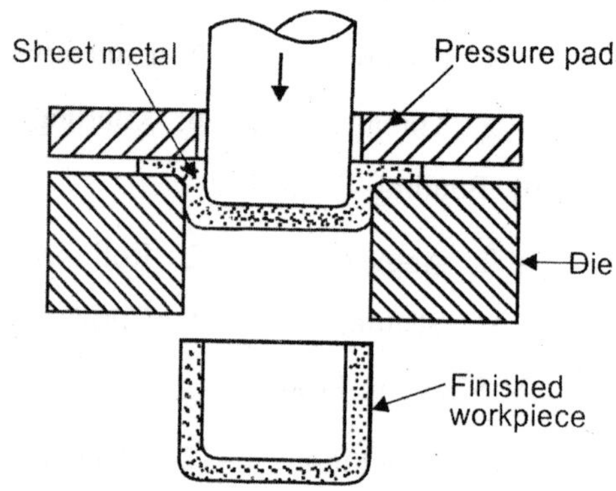

Figure 3.1 Drawing or Cupping

The size of the blank (diameter D) required to draw out a cup is calculated as below.

$$D \doteq \sqrt{d^2 + 4dh}$$

where, d is the diameter of the cup and h is the height of the cup.

ii) Cold drawing of tubes and wires

Cold drawing is the process of pulling metal at room temperature through the opening of a die. Bars or tubes that have been hot rolled or cold drawn in order to.

1. Reduce the size and increase the length 2. Improve the strength 3. Improve surface finish
4. Change the shape 5. Provide accuracy

In cold drawing, plastic deformation of the metal takes place due to the compressive force developed between the surfaces of the metal and the die. Before cold drawing, the bars or tubes will be first cleaned by acid picking. i.e.. immersing them in dilute sulphuric acid for 15 to 30 mts. This is done to remove scale from the metal surface. After the scale is removed, the metal is washed in lime to remove the acid sticking to it. Then the metal is cold drawn.

3.4.1. Wire Drawing:

This is the process of producing wires by pulling a rod through the opening of a die. In drawing, the cross section of the bar stock is reduced and its length is increased. The raw material for cold drawing of wires are hot rolled bars obtained from hot rolling mills. The rods are first cleaned in acid bath and then washed in lime solution. The end of the rod is shaped to a taper so that it can enter the die opening. The end of the rod is pushed through the die and gripped by tongs on the other side of the die. The rod is pulled to sufficient length. Then it is attached to a power driven reel. The reel is rotated to draw the wire at a given speed. (Refer figure 3.2)

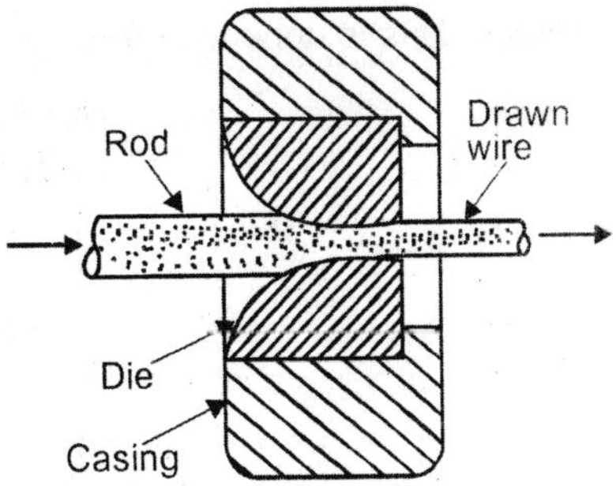

Figure 3.2 Wire Drawing

A lubricant (Soap) is applied on the rod surface and die. Some times the rod is pulled through successive dies with decreasing diameters to obtain thin wires.

The wire drawing dies are made of chilled cast iron, hardened alloy steel or tungsten carbide. Cold drawn wire will be in work hardened condition. So it is annealed before use.

3.4.2. Tube Drawing:

This is the process of finishing hot rolled by pulling them through a die opening with a stationary mandrel. Tube drawing is used for producing thin tubes with small diameters. They will have good surface finish and accurate size.

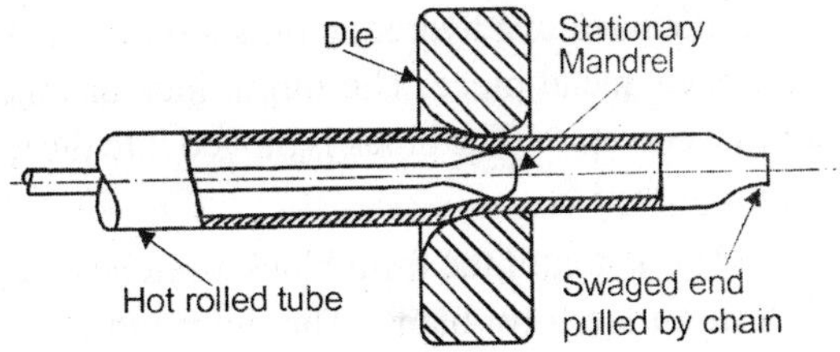

Figure 3.3 Tube Drawing

First the hot rolled tube is cleaned by acid and then by lime solution. The tube is coated with a lubricant. A cold drawing die is used. A stationary mandrel is held at the centre of the die opening.

The end of the tube is first reduced in diameter (by swaging) so that it can enter the die opening. The mandrel rod is introduced inside the tube from behind. The tube end is pushed through the die and gripped by jaws. (Refer figure 3.3)

The jaws are then pulled by a drawing mechanism. In drawing, the outside diameter is formed by the die and the inside diameter is formed by the mandrel. By one pass through a die, about 40% reduction in size can be obtained. For further thinning of tube, two or more passes with different mandrels are necessary. As cold drawing increases the hardness, the tube is softened by annealing before use.

3.4.3 Metal Spinning:

This is the process of producing circular shaped articles from very thin metal blanks by spinning action. The metal is worked at room temperature.

Cold metal spinning is used for thin metals, where as hot metal spinning is used for thick metals. Metals with high ductility can be spun in cold condition. Hot spinning is applied for metals with lesser ductility.

When cold spinning is to be done in a number of stages, the component is to be annealed between the stages. This is done to soften the metal which is work hardened. (Refer figure 3.4)

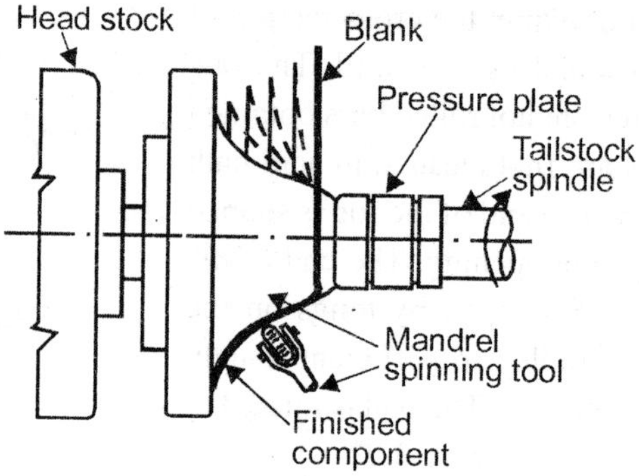

Figure 3.4 Metal Spinning

3.4.4. Stretch forming

In stretch forming, the sheet metal is stretched beyond its yield point over a form block to produce compound curves. The ends of the sheet metal are tightly gripped in jaws. The stationary jaws have serrations for effective gripping. A form block, mounded in a hydraulic ram is positioned at the middle of the sheet metal piece. The form block has the shape of the required curvature. The block is moved into the metal as shown in the figure 3.5.

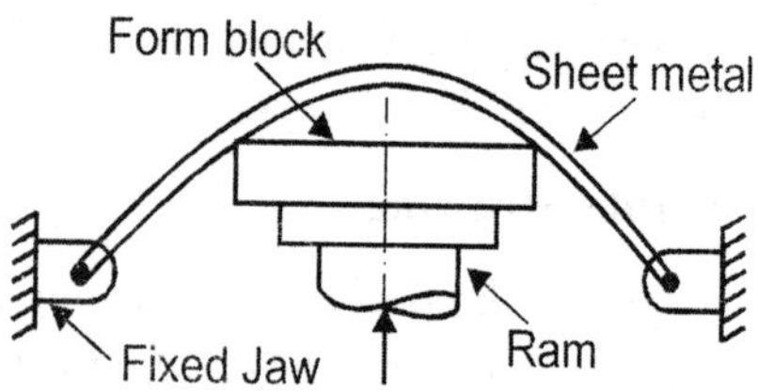

Figure 3.5 Stretch Forming

The action of the form block stretches the metal beyond its yield point. Hence the sheet metal takes the form of the form blank. The outer radius is stretched more than the inner radius. The formed sheet metal will have small spring back.

The form block can be made of any cheap material. Its contact surface must be smooth. Proper lubrication must be applied between the sheet metal and form block.

(a) Stretch wrap forming:

This is similar to stretch forming. However, the metal is first stretched beyond its yield point by the movement of the jaws. Then the form block mounted on a ram moves upwards into the metal. At the same time, the gripping jaws move downwards. Due to these movements, sheet metal is wrapped over a form block. (Refer figure 3.6).

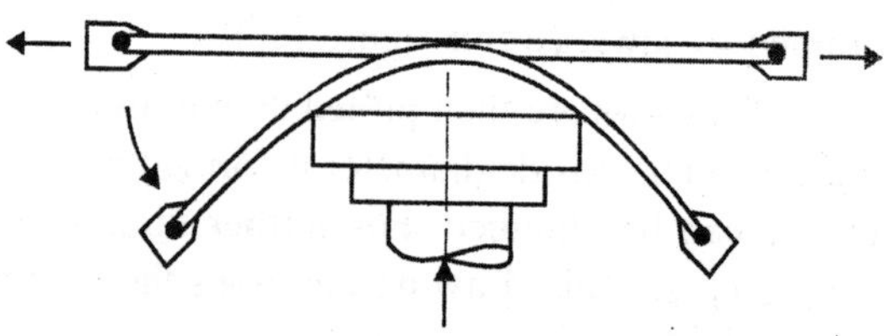

Figure 3.6 Stretch Wrap Forming

(b) Stretch draw forming:

In stretch draw forming, the sheet metal is stretched beyond its yield point to get the required shape by means of a drawing die and punch. The sheet metal piece is placed over the die. The ends of the sheet are gripped over the die by hydraulically operated serrated jaws. The punch is moved down into the metal. The sheet metal is stretched by the drawing action of the punch. (Refer figure 3.7).

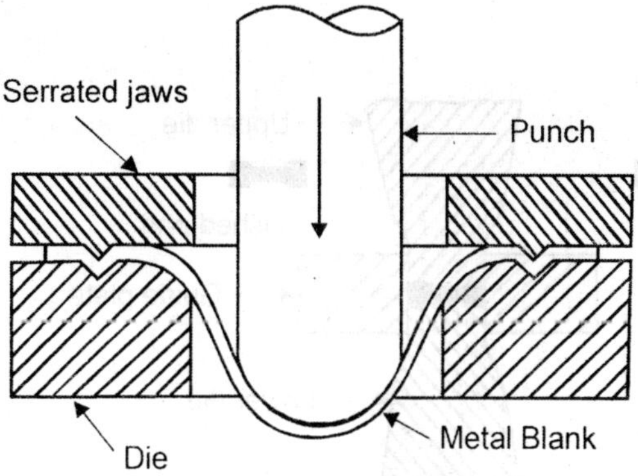

Figure 3.7 Stretch Draw Forming

Here the deformation of metal is restricted to the area within the die. This is because the ends of the sheet are held tightly by the jaws. Here all the fibres of the metal are stretched. Fibres on the inside radius are stretched less than the fibres on the outside radius.

The punch surface is blunt, highly, polished and lubricated.

3.4.5 Embossing

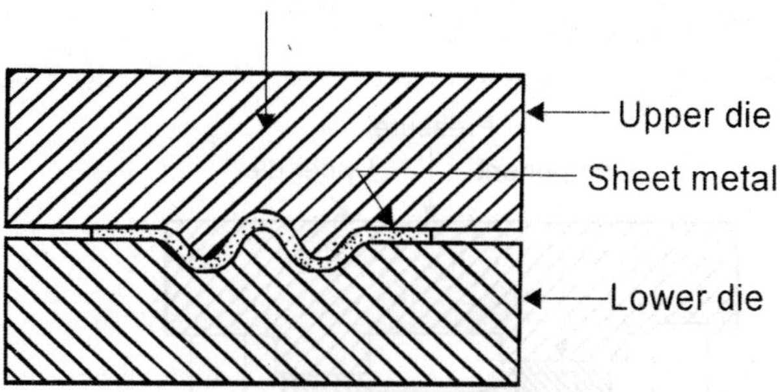

Figure 3.8 Embossing

Embossing is the operation of giving impressions of letters, figures or patterns of sheet metal parts. The blank is placed between the two dies. The dies will have the impressions required on the component. One die will have projection and other die will have the corresponding depressions. The upper die presses down the blank on the lower die. The blank metal flows in the cold state and gets the required impressions. The operation is used on tin containers, covers and name plates. (Refer figure 3.8).

3.5 SQUEEZING OPERATION

Squeezing is the operations of pressing the metal within the cavity of a die and punch to get the required shape. This operation is done when the metal is in cold state. For squeezing, very high pressure is required. So hydraulic press is used. Squeezing can be done only on ductile material having good plasticity.

3.5.1 Coining

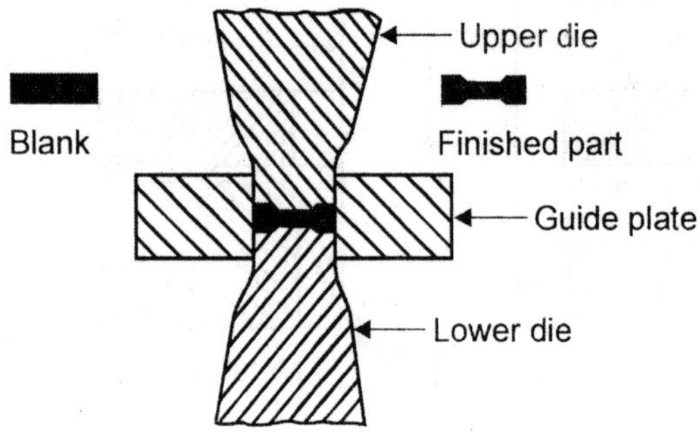

Figure 3.9 Coining

It is the operation of producing coins, medals and other similar parts by squeezing the metal between two dies. A blank of correct size is placed between the lower and upper die. Both dies are engraved to have the required pattern and shape. Both the dies press the blank with very high force. The movement of the dies is guided by a guide plate. The blank metal flows in the cold state and fills up the cavity between the dies. The pattern on the two dies are transferred to the component. (Refer figure 3.9).

3.5.2 Sizing

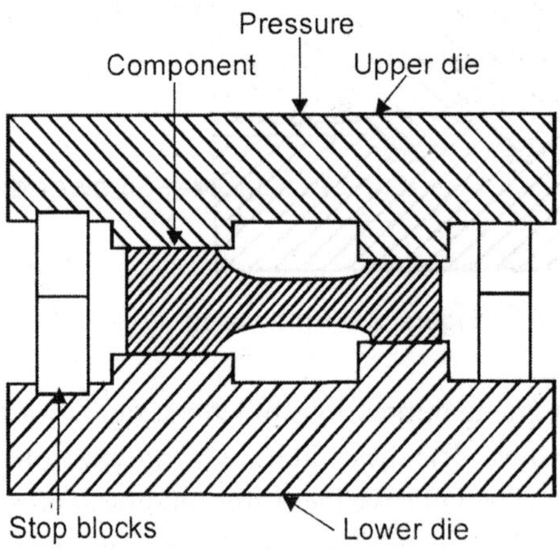

Figure 3.10 Sizing

This is a process used for bringing the thickness of a forged component to the required size by pressing in a heavy duty press. A special die is used for every job.

This process is used for sizing malleable iron castings, steel forgings and components made in Aluminium and other non-ferrous metals.

Sizing is a very fast process when compared to machining. So it is economical when applied in mass production. (Refer figure 3.10).

3.6 BENDING OPERATIONS

Bending is a non cutting operation, performed on a press. It is a process of plastically deforming a metal sheet along a line. The surface area remains the same even after bending. Bending can be done using die and punch or rolls. Bending should be done perpendicular to the direction of the grains. If bending is done parallel to the gains, cracks may develop. Bending operation can be done only on ductile materials. Bent parts will spring back slightly after bending. This factor should be considered while designing bending dies.

Different bending operations are:

1) Angle bending
2) Curling
3) Roll forming
4) Plate bending
5) Seaming.

3.6.1 Angle bending:

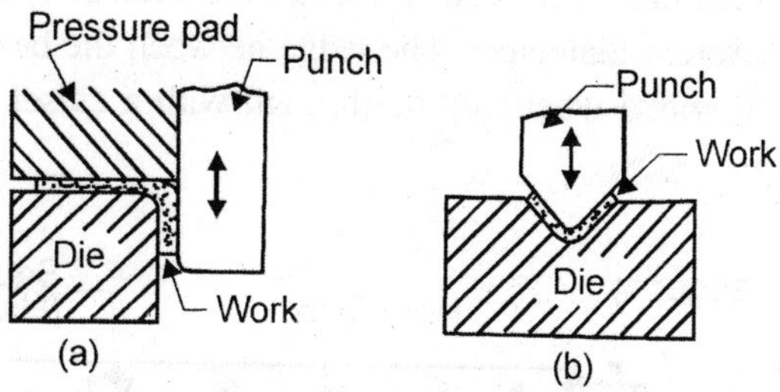

Figure 3.11 Angle Bending

Angle bending is the operation of bending a sheet metal to a sharp acute angle. The punch and die are shaped to the required angle. The angle of the die and punch is selected taking into account the spring back of the sheet metal. (Refer figure 3.11).

3.6.2 Curling:

Curling is the operation of rolling the edges of the sheet metal to a small circular form. Curling is done to strengthen the edges. By curling, the sharp edge of the sheet metal is avoided. The lower die is held stationary. The upper die or punch has got curved form at its end. When it moves down, the edge of the workpiece is curled into a circle. Curling is used in making sheet metal vessels and containers.

3.7 FORGING PROCESSES

The forging processes, according to the method used, are classified into the following two groups:

1. Flat die forging, and
2. Closed die forging.

The flat die forging is also known as smith forging. This may be done by hand or power. In hand forging, the work is heated to the required temperature, placed on an anvil and then shaped with suitable hand hammer. It is employed only for shaping small number of light forgings. When the work is shaped under power hammers, it is then known as power forging.

The closed die forging includes such processes as drop forging, press forging and machine or upset forging. In the closed die forging, the work is heated to the required temperature, placed in a two part or multipart hollow mould called die (depending upon the shape of work required) and then pressed.

3.7.1 Forging Tools

The hand tools and appliances used by the blacksmith in various forging operations are as follows:

1) Anvil:

The anvil is an important smith's tool. It is used for supporting the work while it is struck with hammer. It is made in several shapes but two most common forms are the English and French types. An English anvil has a single beak, as shown in Figure 3.12, while the French anvil has a beak at each end. The body of the anvil is made of mild steel to which a piece of high carbon steel about 2.5 mm thick is welded on the top to give a hard top face. The beak is soft like the anvil body and its shape makes it useful for bending round section of different diameters. The ledge between the beak and the anvil face is also soft and may be used for resting metal when cutting through with a chisel. The soft underneath metal does not damage the chisel edge.

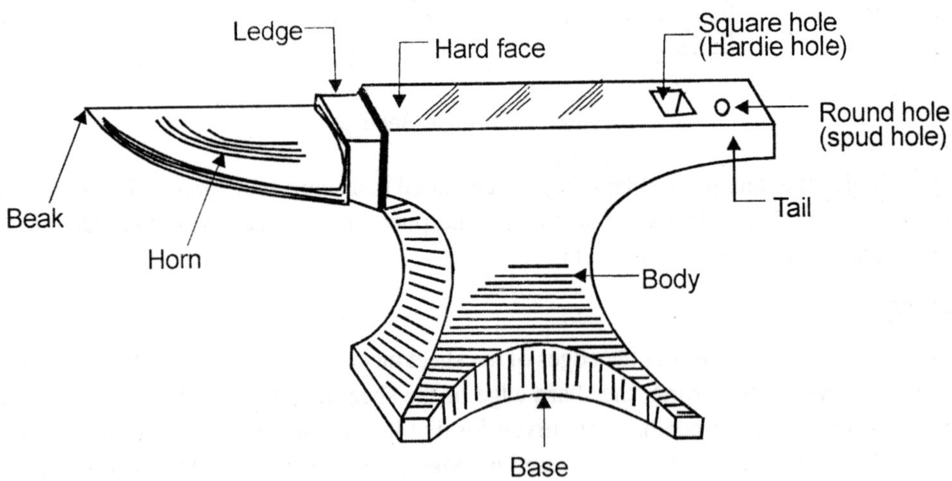

Figure 3.12 Anvil

In the top of the anvil, is a square and a round hole. The square hole (also known as Hardie hole) is used for holding square shank shaping tools like bottom fullers, swages and hardie (bottom chisel) etc. The round hole is used for admitting the ends of the punches or drifts. This hole is also used for bending round bars of different curvatures.

2. Swage block:

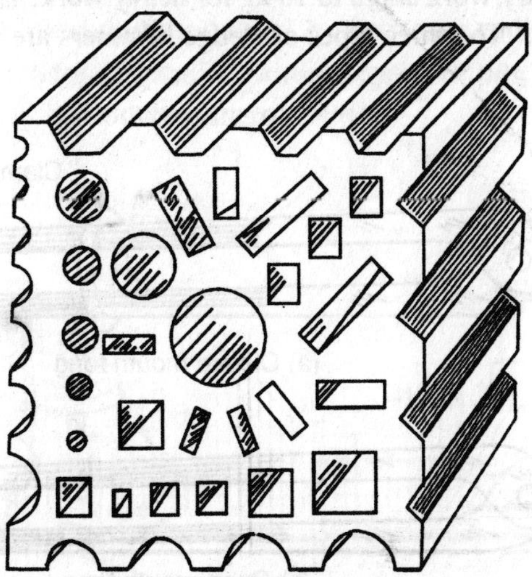

Figure 3.13 Swage Block

The swage block, as shown in figure 3.13 is generally made of cast iron and has round, square, rectangular and half round grooves. In addition to this, it is provided with holes which are useful for holding bars while bending and knocking up heads. The swage block is usually supported at a suitable height on a stand which is adaptable to hold it flatwise or on its edge.

3. Hammers:

The hammers commonly used in hand forging are of the following tow types:

(a) Hand hammer or Ball Pen hammer and

(b) Sledge hammer

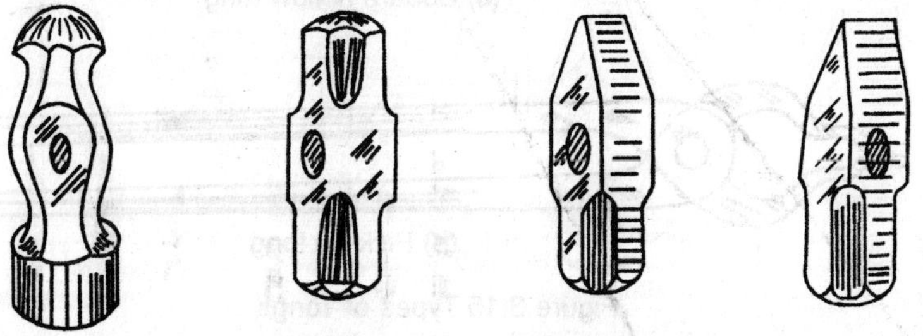

(i) Double faced (ii)Straight peer en

(a) Ball peen hammer (b) Sledge hammers

Figure 3.14 Types of Hammers

The hand hammers are generally ball peen hammers as shown in figure 3.14(a). They usually have a slightly convex striking face and should have mass from 1 to 1.5 kg. The hammer heads should be of cast steel, the ends are hardened and tempered with the portion round the eye soft.

The sledge hammers are used by the smith's helper or striker. The mass of the sledge hammer varies form 4 to 6 kg for ordinary work and 8 to 10 kg for heavy work. The shaft is about one meter long and it is mostly made of wood. The three types of sledge hammers are shown in figure 3.14(b)

4. Tongs:

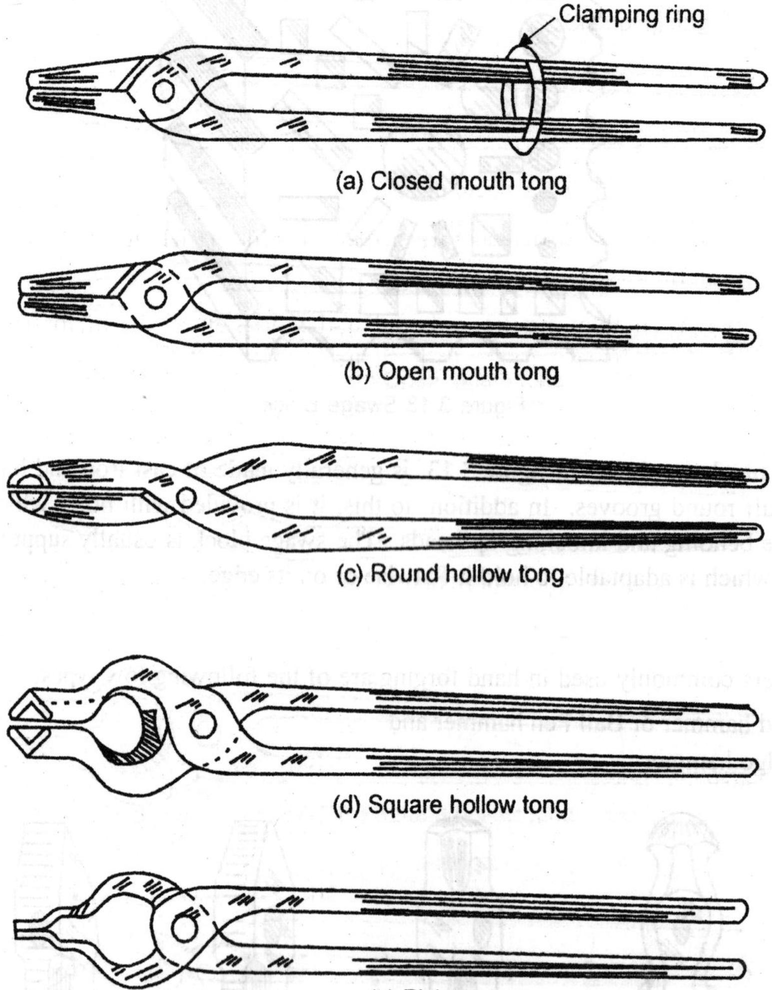

(a) Closed mouth tong

(b) Open mouth tong

(c) Round hollow tong

(d) Square hollow tong

(e) Pick-up tong

Figure 3.15 Types of Tongs

The blacksmith requires suitable tongs to handle the work while forging. These are made of mild steel. The various types of tongs commonly used for holding work are as follows:

(a) Closed mouth tong:

A closed mouth tong, as shown in Figure 3.15(a), is used for holding thin sections.

(b) Open mouth tong:

An open mouth tong, as shown in figure 3.15(b), is suitable for holding heavier stock.

(c) Round hollow tong:

A round hollow tong, as shown in figure 3.15(c), is used for holding round, hexagonal and octagonal section.

(d) Square hollow tong:

A square hollow tong, as shown figure 3.15(d), is used for holding square, hexagonal and ocatagonal work.

(e) Pick-up tong:

A pick-up tong, as shown in figure 3.15(e), is used for picking up round bars, but not for holding work during forging.

In order to relieve the smith of the strain of holding the tongs during a long forging operation, the tongs are provided with clamping rings as shown in figure 3.15(a)

5. Chisels:

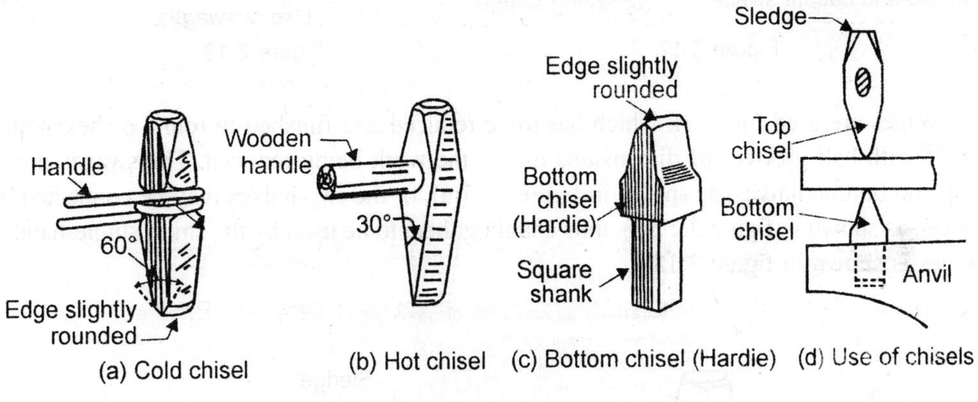

Figure 3.16 Chisels and their Uses

The chisels are used for cutting metals and nicking prior to breaking. They may be cold or hot depending upon whether the metal to be cut is cold or hot. The main difference between the two chisels in the edge.

The *cold chisel* is made of tool steel with a cutting angle of about 60 degrees and its edge is hardened and tempered as shown in figure 3.16(a). The handle of this tool is usually formed of an iron rod bent around a fullered recess made in the body of the chisel. The edge of a chisel should not be quite straight but slightly rounded.

The *hot chisel* is made of low carbon steel and has a cutting angle of 30 degrees, as shown in figure 3.16(b). The hardening of the edge is not necessary as the hot metal would soften it. This type of chisel is fitting with wooden shafts.

The chisels are generally used in pairs. The top chisel which is held by the smith may be hot or cold chisel. The bottom chisel (often called hardie), as shown in figure 3.16(c), has a square shank and fits in the square hardie hold in the anvil face. Figure 3.16(d) shows the use of top and bottom chisels.

6. Swages:

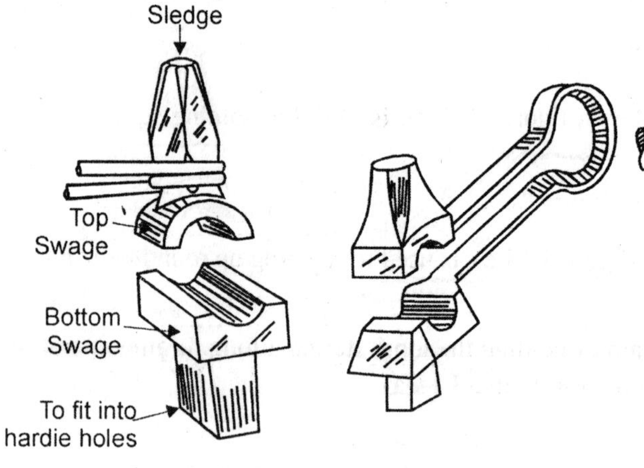

(a) Top and bottom swages (b) Spring swage

Figure 3.17

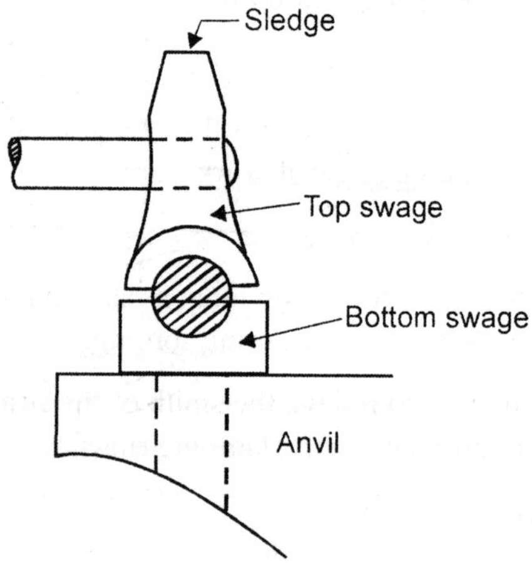

Use of swages

Figure 3.18

The swages are used for work which has to be reduced and finished to round or hexagonal form and are made with half grooves of dimensions to suit the work being reduced. The swages may be in separate top and bottom halves, as shown in figure 3.17(a), or the two halves may be connected by strip of spring steel, as shown in figure 3.17(b), thus enabling them to be used by the smith single-handed. The use of swages is shown in figure 3.18.

7. Fullers:

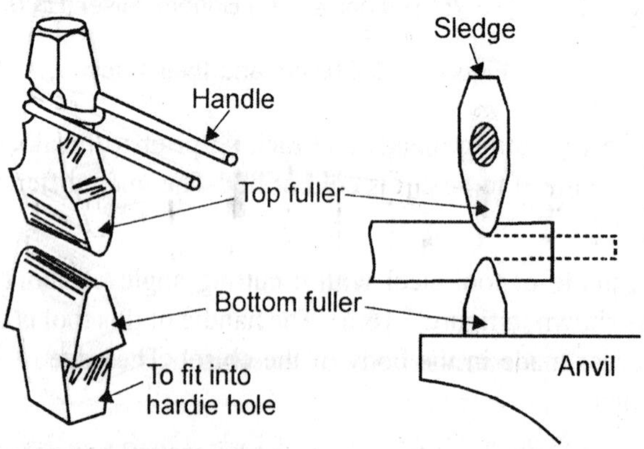

(a) Top and Bottom fullers (b) Use of fullers

Figure 3.19 Fullers and their Uses

The fullers, as shown in figure 3.19(a), are used for necking down a piece of work. It is mainly used to reduce the metal while making shoulders. They are made of tool steel. The fullers are made in pairs. The top fuller is held by the blacksmith and struck by the striker while the bottom fuller is fitted into the hardie hole of the anvil. The fullers are made in various sizes according to the requirement. The use of fullers to neck before setting down is shown in figure 3.19(b)

8. Flatters:

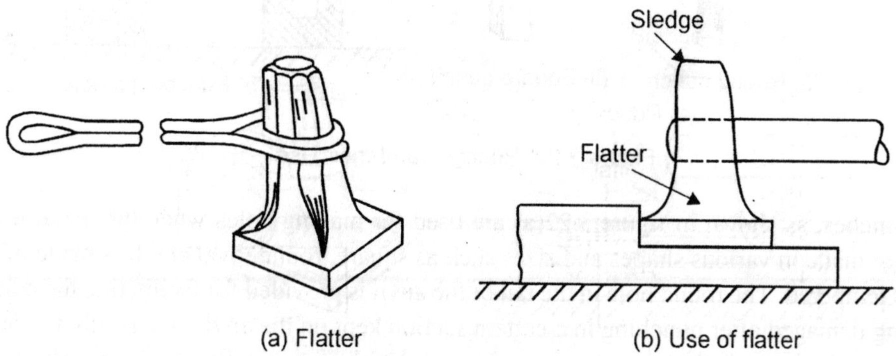

(a) Flatter (b) Use of flatter

Figure 3.20 Flatters and its use

The flatter, as shown in figure 3.20(a), is used for finished flat surfaces and are made with a perfectly flat face. It gives smoothness and accuracy to articles which have already been finished by swages and fullers. Its size is about 7.5 cm square or round. It is made of tool steel. It is struck by a hammer on the head. The use of flatter is shown in figure 3.20(b)

9. Set hammer:

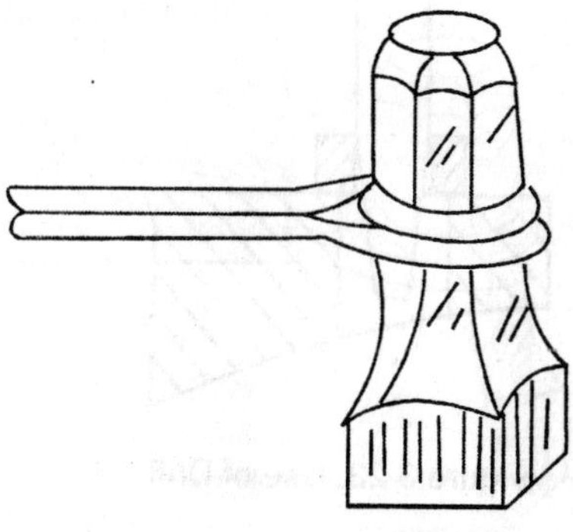

Figure 3.21. Set Hammer

The set hammer, as shown in figure 3.21 is similar to flatter but is a smaller tool. It is made up of forged steel and is hardened. It is generally available in 4 cm square size. It is used to set down shoulders or smooth out fuller marks near the shoulders before the blacksmith finishes with the flatter.

10. Punches:

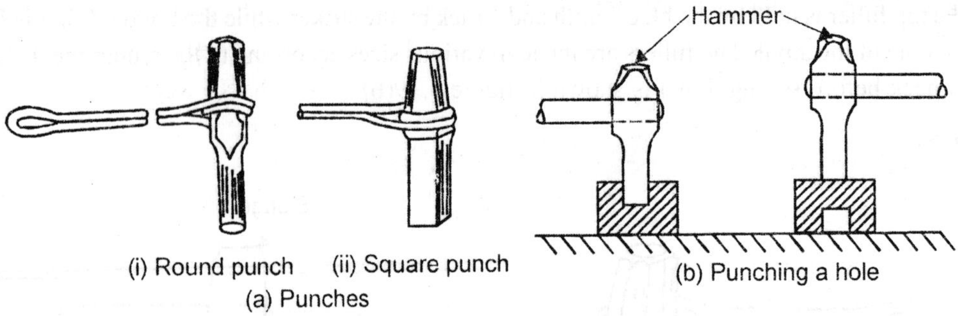

Figure 3.22. Punches and their Use

The punches, as shown in figure 3.22(a) are used for making holes when the metal is at forging heat. These are made in various shapes and sizes such as square, round, oval etc. It is made of tool steel, hardened and tempered. The round hole in the tail of the anvil is provided for facilitating the punch to pass without getting damaged after punching in a certain section kept on the anvil. Generally the punching of hole is carried out from both sides, as shown in figure 3.22(b), to avoid driving the punch onto the hard anvil face.

11. Drifts:

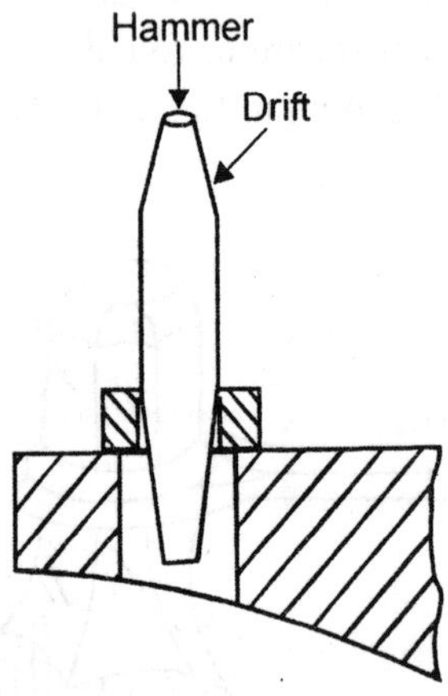

Figure 3.23. Use of Drift

The drift, as shown in figure 3.23, is used to finish the size of the punched holes. They are made taper and bulging in the centre. These are made in various shapes and sizes like punches. They are also made of tool steel, hardened and tempered.

3.8 HAND FORGING OPERATIONS (TYPICAL FORGING OPERATIONS)

The hand forging operations are usually carried out in smithy shop, where knowledge of hand operations is imparted to develop awareness about the plastic deformation of engineering materials. These operations in combination to other are essentially used to bring a work-piece into the final product after manipulating the work-piece by the use of different hand tools. The following are the different hand forging operations which have been out-lined here.

i) Upsetting

ii) Drawing -down

iii) Bending

iv) Punching and drifting

v) Swaging

vi) Fullering and flattening

vii) Forge-welding

viii) Cutting

3.8.1. Upsetting Operations:

This is carried out to increase the thickness or diameter of work-piece after reducing its length by hammering process. The work is held by tong at one end while the other end of work is supported on the anvil-face. Upsetting may be required on any end portion, central portion or complete length. This is accomplished by heating the work in a furnace. The portion of the work to be upset must remain hot and the rest may be cooled down by dipping it into the cold water. Figure 3.24 shows the upset forging operations.

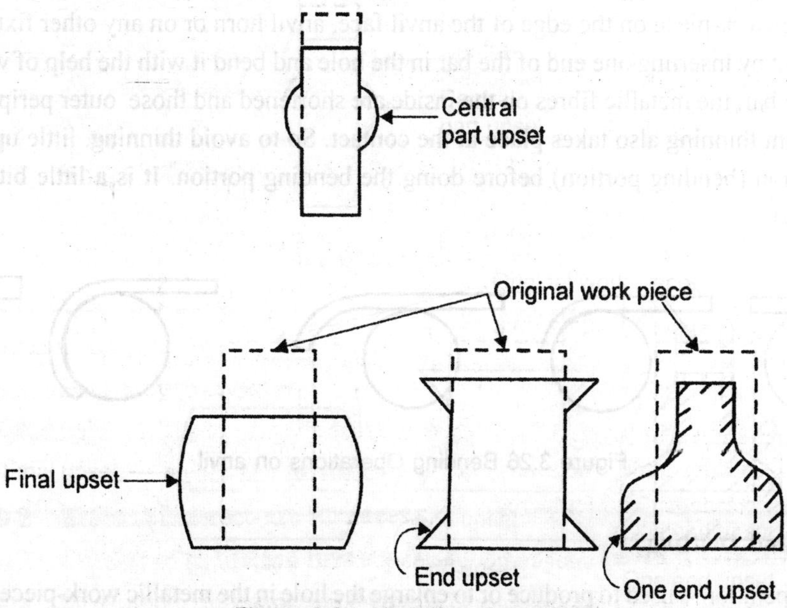

Figure 3.24. Upsetting Operations

3.8.2. Drawing-Down Operations:

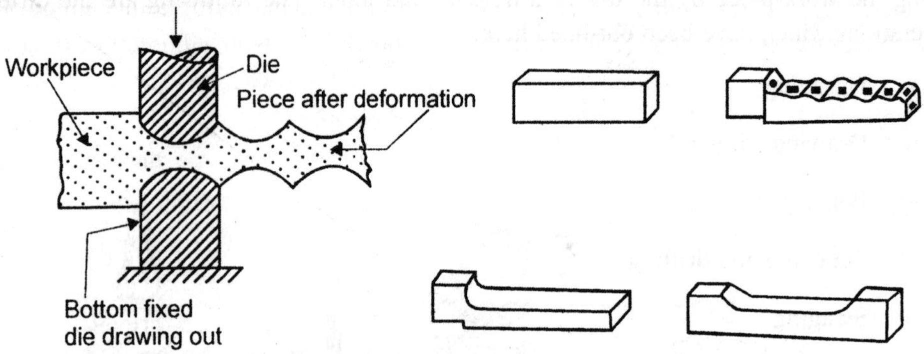

Figure 3.25 Drawing-Down Operations

This operation reduces the thickness of the work or diameter by increasing the length of the stock. This operation is carried out by hammering the hot work-piece while keeping it on the anvil (use straight peen hammer) and holding it by a suitable tong. Sometimes the fullers are also used to reduce thickness. It may be localized drawing-down also. The rectangular part of the fuller is mounted in the anvil square hole and keep the hot work piece held by tong on the fuller top. Counter part of the same set is kept on it and hammer the head of fuller. Usually it is followed by flatter to make the surface of work flat. So, this process is termed as fullering and flattering. (Refer figure 3.25).

3.8.3. Bending:

It is also a very common operation and frequently used in work-shop. Bending operation is carried out by keeping the work-piece on the edge of the anvil face, anvil horn or on any other fixture/support. A thin bar can be bent by inserting one end of the bar in the hole and bend it with the help of wrench or tong. While bending the bar, the metallic fibres on the inside are shortened and those outer periphery of the bar are extended. Slight thinning also takes place at the contact. So to avoid thinning, little upsetting can be done on that portion (bending portion) before doing the bending portion. It is a little bit slow process. (Refer figure 3.26).

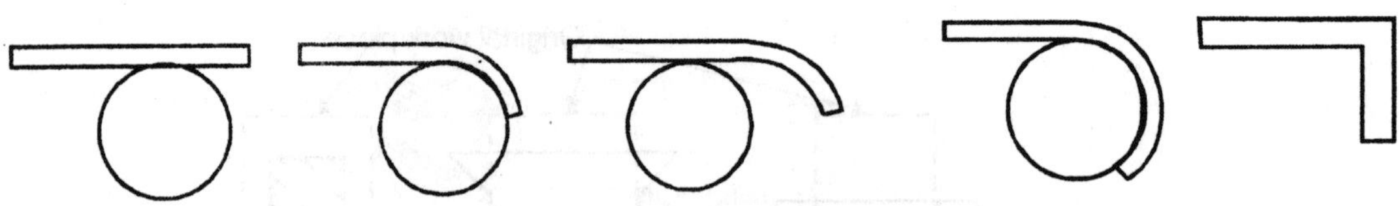

Figure 3.26 Bending Operations on anvil

3.8.4. Punching and Drifting:

Punching operation is used to produce or to enlarge the hole in the metallic work-piece by hammering a tool called punch. The hole may also be blind. Figure 3.27 shows the complete process of punching.

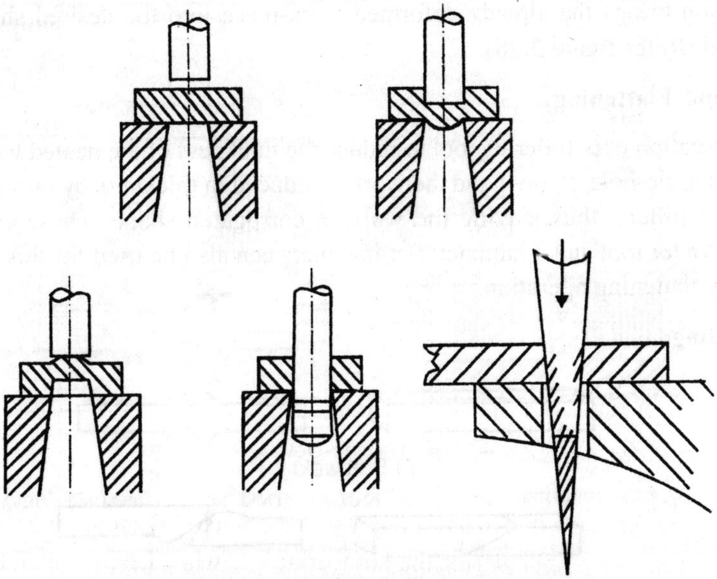

Figure 3.27 Punching and Drifting Operations

Drifting operation is to cut a slit in the work and the tool used is called 'Drift' which is having a very small cutting angle with sharp edge. Sometimes a circular drift is also used to enlarge the hole to the required dimension. Both the operations can be carried out at room temperature or at high temperature.

3.8.5. Swaging:

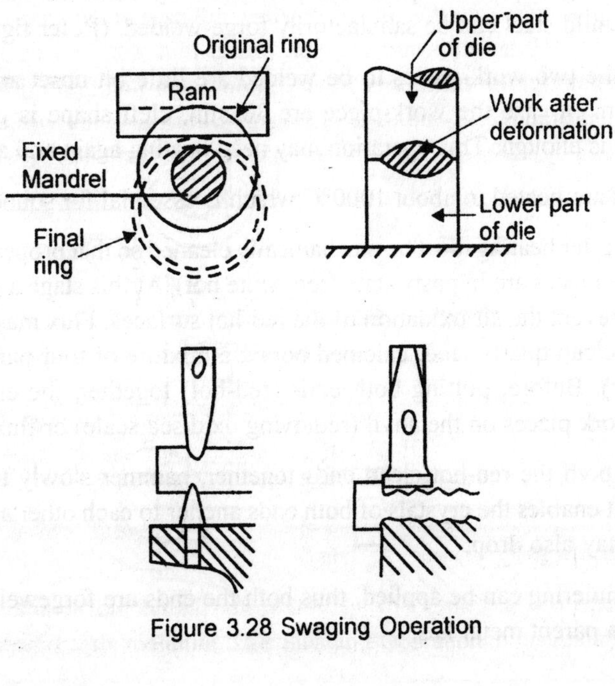

Figure 3.28 Swaging Operation

This operation brings the already deformed work-piece into the desired shape and size usually round or hexagonal (Refer figure 3.28).

3.8.6. Fullering and Flattening:

Fullering operation uses fullering tool to reduce the thickness of the heated work-piece. The fuller head is held in the hardie-hole of anvil and the work is reduced in thickness by moving the work between the top and bottom fullers, thus leaving the work in corrugated shape. These corrugations are later flattened by using flatter tool and a hammer. Set-hammers can also be used for this purpose. Fullering is always followed by flattening operation.

3.8.7. Forge-Welding:

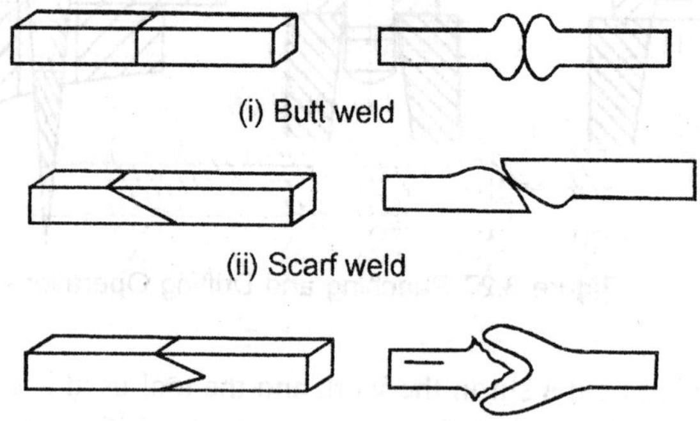

(i) Butt weld

(ii) Scarf weld

(iii) 'V' weld or splice weld

Figure 3.29. Various types of Forge-Welding Joints

It is one of the most important operations performed by black-smiths. Wrought iron and low carbon steels including mild steel can be satisfactorily forge-welded. (Refer figure 3.29).

a) The ends of the two work-pieces to be welded are little bit upset and cleft shape or tapered shape is given. In case the work-piece are 30 mm, cleft shape is given and in other cases tapered shape is enough. This operation may need heating again and agāin.

b) Both the ends are heated to about 1000°C which is essential for sound welding.

c) Both the ends after heating must be mechanically cleaned so that proper cohesion may takeplace while both the pieces are in pasty state (red white hot). At this stage a protection to the metal is provided to prevent the air oxidation of the red-hot surfaces. Flux may also some time be used (usual fluxes-clean quarts sand; calcined borax; a mixture of four parts of borax and one part sal-ammoniac). Before, putting both ends (red-hot) together, the ends must be cleaned by striking the work pieces on the anvil (removing oxidised scale) or flux can also be sprinkled.

d) After putting both the red-hot clean ends together, hammer slowly to avoid excessive metal spread. Thus it enables the crystals of both ends anchor to each other and during this period the temperature may also drop.

e) Now firm hammering can be applied, thus both the ends are forgewelded, which has as much strength as the parent metal has.

Four forms of welded-joint are employed(a) the lap or scarf; (b) the butt weld; (c) the T or jump weld and (d) the split, fork or V-weld or splice

3.8.8. Cutting:

It is sometimes termed as hot cutting or chiselling by the use of hot chisels (cutting edge 30° angle) and the work-piece is red-hot (light cherry red hot) by heating it into the furnace to the approximate temperature of 800°c to 900°c. The cutting process is accomplished by hammering the chisel into the notch of cherry-red hot work-piece.

3.9. TYPES OF FORGING

3.9.1 Impression-die Forging:

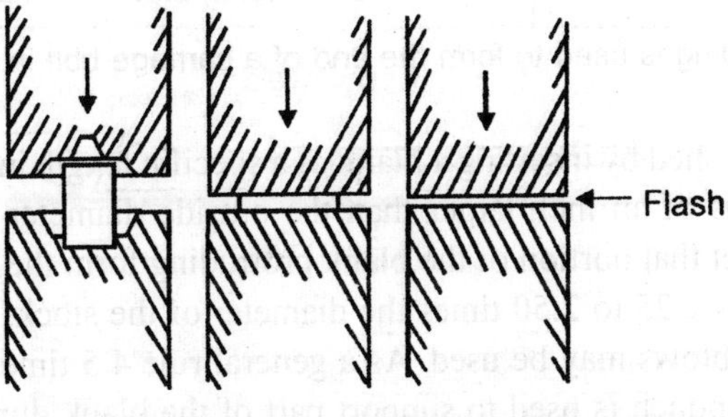

Figure 3.30 Closed-die Forging Sequence

In the simplest example of impression-die forging, two dies are brought together, and the workpiece undergoes plastic deformation until its enlarged sides touch the side walls of the die figure 3.30. A small amount of material is forced outside the die impression, forming flash that is gradually thinned. The flash cools rapidly and presents increased resistance to deformation, effectively becoming a part of the tool, and helps build up pressure inside the bulk of the workpiece that aids material flow into unfilled impressions.

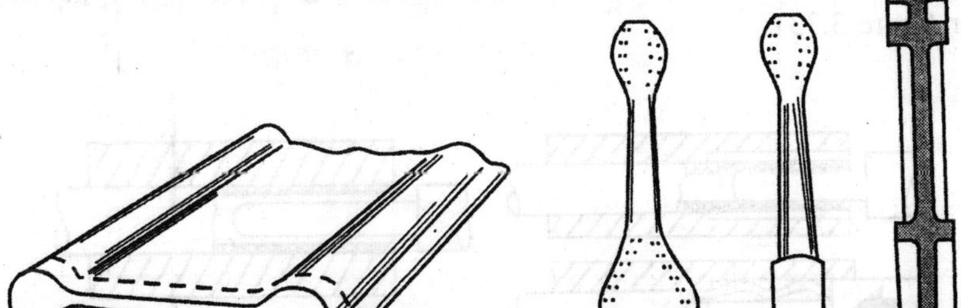

Figure 3.31 Extruded preform shape that can be sliced up and forged into connecting rods

Closed -die forgings, a special form of impression-die forging, does not depend on the formation of flash to achieve complete filling of the die. Thus closed-die forging is considerably more demanding on die design. Since pressing is often completed in one stroke, careful control of the workpiece volume is necessary to achieve complete filling without generating extreme pressures in the dies from over. One approach to getting the right amount of metal for the die cavity and reduce forging time is the use of filling roll-formed shapes or extruded preform shapes, as shown in figure 3.31.

3.9.2 Upset Forging

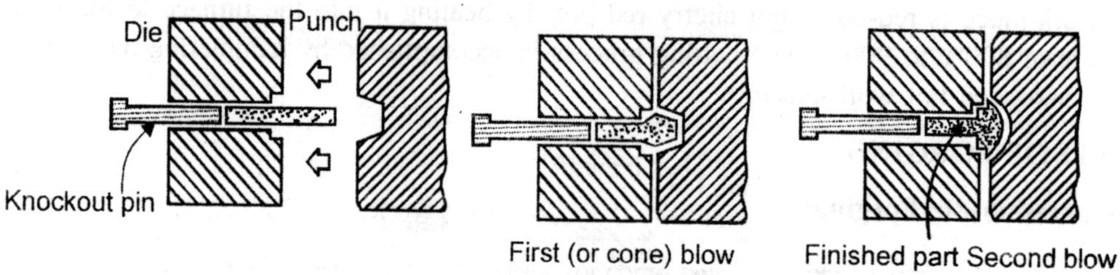

Figure 3.32 Cold heading is used to form the end of a carriage bolt in two blows in a single die

Upsetting is accomplished by inserting a blank of a specific length into a stationary die. The bore of the die is a few thousandths of an inch larger than the outside diameter of the blank. A punch moves toward the die blank to upset that portion of the blank protruding form the die. The maximum length that can be upset in one stroke is 2.25 to 2.50 times the diameter of the stock. If this length is not enough to form the part, two or more blows may be used. As a general rule 4.5 times the diameter can be upset in two blows. When a sliding punch is used to support part of the blank during upsetting, up to about 6.5 times the diameter can be formed. A rule of thumb used to estimate impact values required for upsetting is that the impact value must be seven times the yield strength of the material. (Refer figure 3.32).

The horizontal forging machine is also useful for producing hollow and pierced components as they can be made with very small or no taper, and piercing can be done with little or no materiel loss. In the simplest case, the ends of the tube are upset. If the material is gathered so as to increase the outer diameter, the same rules apply as for solid bars, except that the thickness of the tube is substituted for the bar diameter. Heavier deformation is permissible if the outer diameter is kept constant while the inner diameter is reduced, as the natural flow of the material tends to give sound flow patterns even at higher reductions.(Refer figure 3.33).

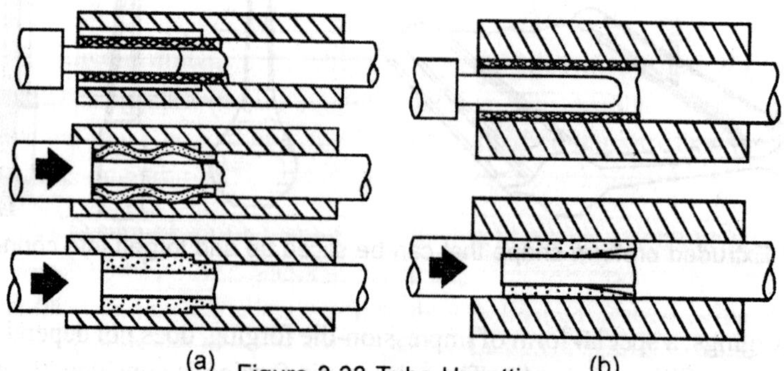

Figure 3.33 Tube Upsetting

3.9.3. Extrusion Forging:

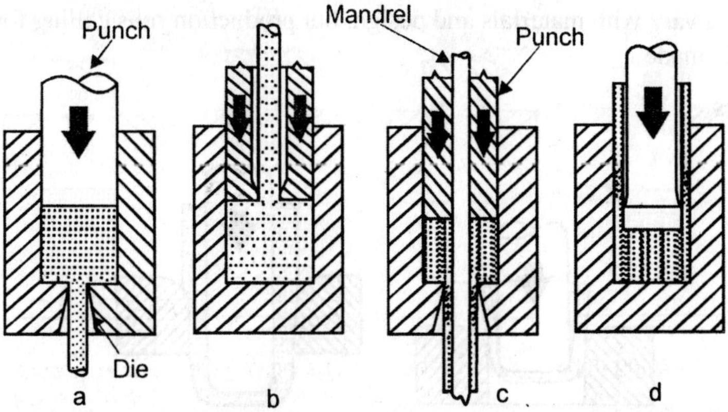

Figure 3.34 (a) Forward extrusion, (b) backward extrusion, (c) tube extrusion, and (d) container extrusion

As with upsetting, extrusion forging is often accomplished by cold working. Three principal types of metal displacement by plastic flow are involved. Backward and forward, tube, and impact extrusion are shown in figure 3.34. The metal is placed in a container and compressed by a ram movement until pressure inside the metal reaches flow-stress levels. The workpiece completely fills the container, and additional pressure causes it to leave through an orifice and form the extruded product.

Extruded products may be either solid or hollow shapes. Tube extrusion is used to produce hollow shapes such as containers and pipes. Reverse-impact extrusion is used for mass production of aluminum cans. The ram hits a slug of metal in the die at high impact, usually 15 times the yield strength of the metal, which causes it to flow instantaneously up the walls of the die. Other common hollow extrusion products are aerosol cans, lipstick cases, flashlight cases, and vacuum bottles. Secondary operations, such as beading, thread rolling, dimpling, and machining, are often needed to complete the items. Figure 3.35 shows a three-piece assembly that is currently produced as a one-piece combination upset and extrusion forging.

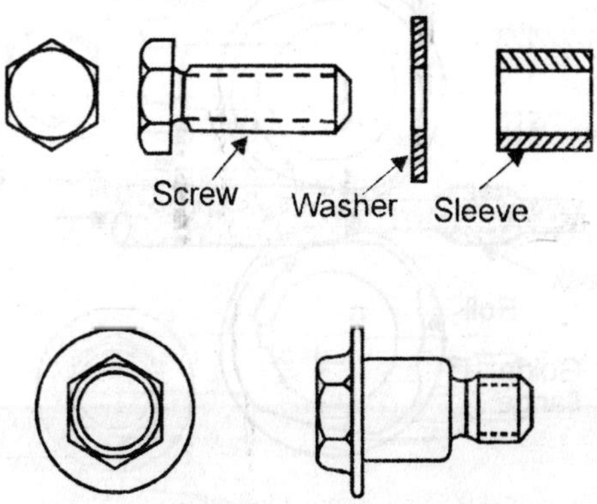

Figure 3.35 Schematic shows how three-piece assembly was replaced by single transfer headed part. Cost savings extend to raw materials, purchasing administration, inventory, and assembly

Generally, steel impacts are limited to 2.5 times the punch diameter. Hydraulic presses are used for loads of over 2000 tons because they have a greater variation in stroke length, speed, and other economic advantages. Tolerances vary with materials and design, but production runs calling for 0.002 to 0.005-in. tolerance are regularly made.

Auxiliary operations

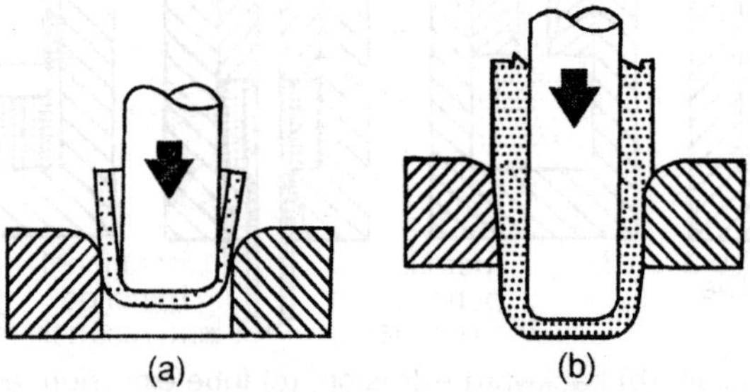

Figure 3.36. Auxiliary operations: (a) deep drawing, and (b) ironing

Closely related to the forging extrusion process is drawing and ironing, as shown in figure 3.36. Drawing through a die is a convenient way to eliminate forced draft. The mode of deformation is tangential compression. The diameter of the drawing ring can be slightly smaller than the outer diameter of the preforged shell to control or reduce wall thickness and increase the height of the shell in an operation known as ironing. (Refer figure 3.36).

3.9.4 Roll Forging

Roll forging in its simplest form consists of a heated billet passing between a pair of rolls that deform it along its length (Figure 3.37). Compared to conventional rolling processes, the rolls are relatively small in diameter and serve as an arbor into which the forging tools are secured. The active surface of the tool occupies only a portion (usually half) of the roll circumference to accommodate the full cross section of the stock.

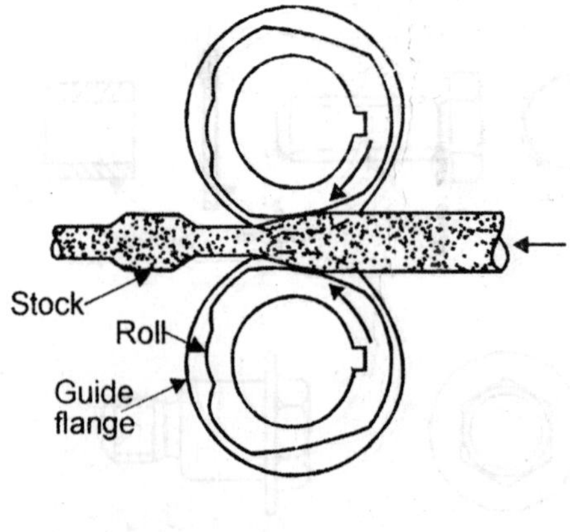

Figure 3.37 Roll Forging

The reduction of the cross section obtainable in one pass is limited by the tendency of the material to spread and form an undesirable flash that may be forged into the surface is a defect in the subsequent operations. The workpiece is introduced repeatedly with 90^0 rotation between passes.

3.9.5 Ring Rolling:

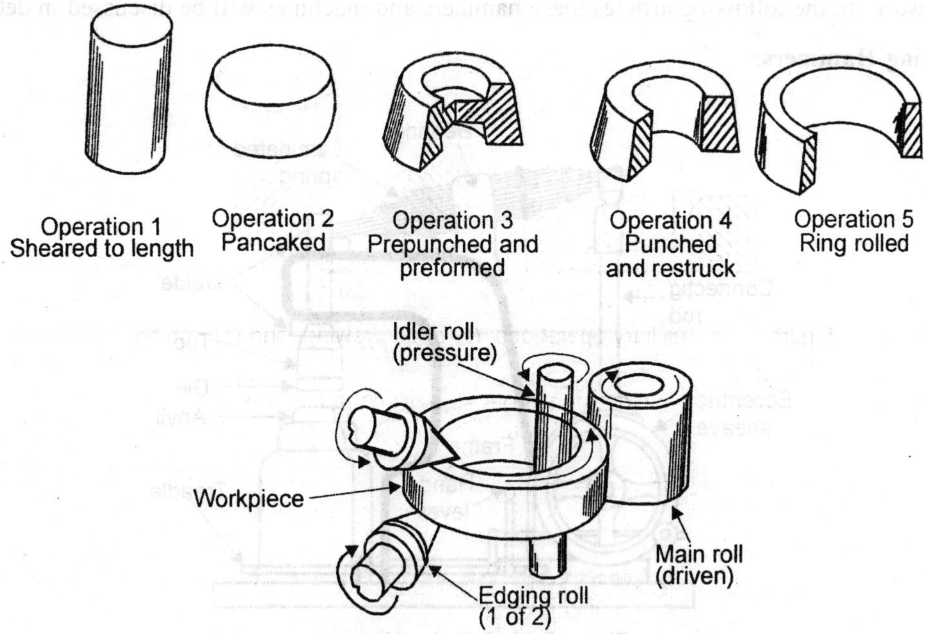

Figure 3.38 Sequence of steps in producing a small ring to be used in the horizontal ring rolling machine

Ring rolling offers a homogeneous circumferential grain flow, ease of fabrication and machining and versatility of material size. Manufacture of a rolled ring starts with a sheared blank, which is forged to a pancake, punched and pierced. (Refer figure 3.38).

There is no limit to the size of the rolled rings, ranging from roller-bearing sleeves to rings 25 ft. in diameter with face heights of 80 in. Various profiles may be rolled by suitably shaping the driven idling rolls.

3.10 FORGING MACHINE - HAMMERS

3.10.1. Power Hammers:

Until now we have confined ourselves to the study of hand tools used in smithy work. They certainly perform very well so far as the hand-forging is concerned, but their use for satisfactory production is limited to small forgings only. It would not be difficult to understand that the intensity of blows, however great one may try to achieve through hand hammering, will not be sufficient enough to effect the proper plastic flow in a medium sized or heavy forging. For this, a power hammer is usually employed. The capacity of these hammers is given by the total weight of their falling parts i.e., tup or ram and die. A 200 kg hammer will be one of which the falling parts weigh 200 kg. The heavier these parts and greater the height from which they fall, the higher will be intensity of blow the hammer will provide. Power hammers in common use are of different types, e.g. spring power hammers, pneumatic power hammers, Steam hammers and Drop or Forge hammers. These hammers are named partly after their construction, partly according to their way of operation and partly according to the type of fuel they use for getting the required power for operation. Apart from these, a large number of forging presses and machines are used in forging work. In the following articles these hammers and machines will be discussed in detail.

3.10.2 Spring Hammers:

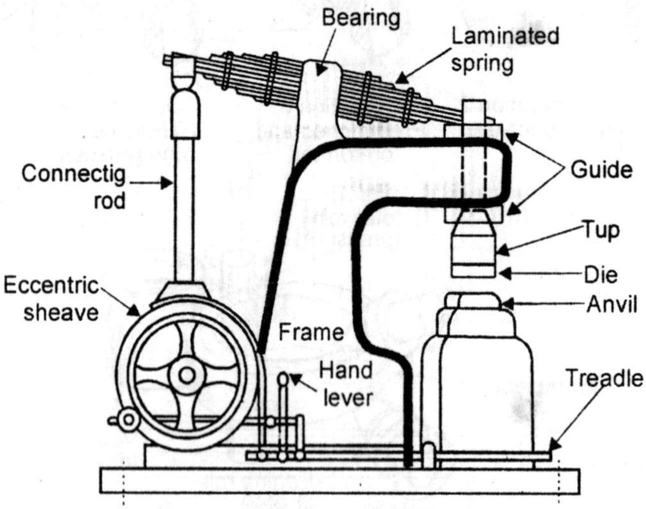

Figure 3.39. A Spring Hammer

It is a very light type of power hammer and is best suited for small forgings. These hammers are made in various designs having different systems of operation. A typical design of a spring hammer is shown in figure 3.39. It consists of a heavy rigid frame carrying a vertical projection at its top. This projection acts as a housing for bearing in which the laminated spring oscillates. One end(rear) of this spring carries a connecting rod and the other (front) end a vertical tup which carries weight and moves vertically up and down between fixed guides provided for this purpose. The connecting rod at its lower end is attached to an eccentric sheave, which is further connected to the crank wheel. For operating the hammer the treadle is pressed downwards, which makes the sheave to rotate through the crank wheel and, thus, the laminated spring starts oscillating in the bearing. This oscillation of the spring causes the tup to move up and down and thus, the required blows are provided on the job. The hand lever is operated to adjust the stroke of the connecting rod and, hence, the intensity of blows.

Another common design of spring hammers is the one in which a rotating eccentric disc is used for producing vibrations in the spring. It is operated by means of a foot ring, known as Treadle, provided at the bottom. This treadle is connected to the shaft at the top through a vertical bar having a clutch at its end. The shaft at the top carries a pulley and a solid disc at the end. The pulley is driven by means of a belt from the line shaft or an electric motor. The solid disc, at the end of the shaft, carries a crank connected eccentrically to it which has a laminated spring at is lower end. The tup carrying the weight is suspended on a toggle joint connecting the two ends of the laminated spring. When the foot treadle is pressed the clutch engages with the shaft and the disc carrying the crank starts rotating, which, in turn, produces fluctuations in the toggle joint. It makes the tup to fluctuate in vertical direction. The speed of blows entirely depends upon the speed of the driving pulley.

Spring hammers are made in various capacities having the tup weights from 30 to 250 kg. Those having tup weight from 50 to 100 kg. and speed of blows upto 300 per minute are in common use. The spring hammers, though relatively simpler in design and operation, have a common drawback in their springs getting broken very frequently due to severe vibrations during the operation.

3.10.3 Pneumatic power hammer

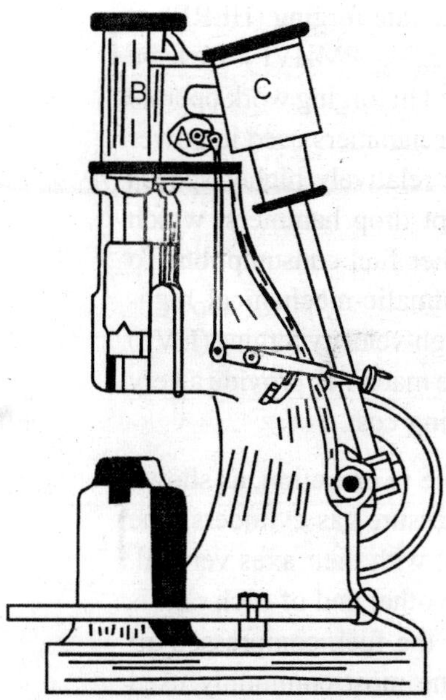

Figure 3.40. A Massey Pattern Single Acting Pneumatic Power Hammer

It is shown by means of a line diagram in figure 3.40. Although the design of these hammers varies with the different manufacturers, they all work on the same principle.

A typical form of pneumatic hammer, shown in figure 3.40, carries a cylinder C. A piston works inside this cylinder and is connected to the main motor shaft by means of a crank and connecting-rod mechanism, as shown in the diagram. A hand lever operates an air valve A which is proved on the air passage from cylinder C to cylinder B. Another piston carrying the tup at its bottom works inside cylinder B The tup is made to slide inside fixed guides.

Now suppose the tup is resting on the anvil. To start motion of the tup the piston in cylinder C is moved downwards creating a vacuum above the piston in cylinder B. This provides a suction effect on the piston and thus the tup starts lifting. A little before the end of upward stroke of the piston, in cylinder B, the air is forced from cylinder C to cylinder B, where it is compressed due to the said upward stroke of the piston. This, in turn, forces the piston and, hence, the tup, downwards at a considerably high velocity. This operation is repeated again and again and the required blows are obtained. By varying the opening of the air valve the air supply from cylinder C to B can be controlled and thus intensity of the blows regulated. For higher intensity of blows the air valve should be totally closed so that the pressure of the air inside the piston is very high. For lighter blows the valve opening can be varied according to the requirement. To stop the hammer, the air valve should be opened full so as to allow the inside air to escape into the atmosphere, thereby disconnecting the supply of the air to cylinder B. In latter developments the single acting cylinder is replaced by a double acting one, air being admitted on both sides of the piston and both the upward and downwards strokes of the piston are due to compressed air.

3.10.4. High velocity forging (HVF) Machine

It is also known as High-energy rate forging (HERF) or Pneumatic-mechanical high-velocity forging (PMHVF). We have seen earlier that most of the presses used in forging work operate at relatively slower speeds. The power hammers used in power hammer forging, although operate at relatively higher speeds but their strokes are repetitive, except drop hammers, which adds to the operating cost due to higher fuel consumption. To overcome these difficulties the Pneumatic-mechanical high-velocity forging machines, or simply High-velocity forging (HVF) machines, have been developed. These machines provide a very high impact rate at a very low operating cost.

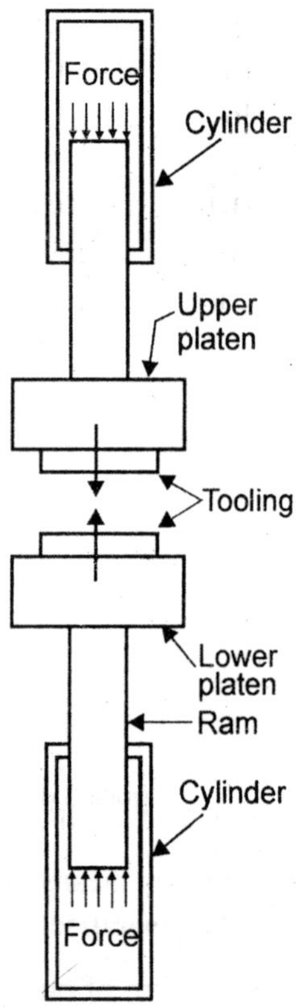

Figure 3.41. Principle of Operation of a High-Velocity Forging Machine

These machines are quite simple in operation, as shown in figure 3.41. There are two high pressure gas cylinders, one at the top and the other at the bottom, with their axes vertical. Each cylinder carries a ram and at the other end of each ram is mounted a platen. The cylinder carry the fuel, compressed air or nitrogen. However, nitrogen is the most commonly used fuel in most of the HVF machines.

Hydraulic power is employed for compressing the gas. The principle of operating the platens involves compressing of gas, sealing it and then releasing it to apply pressure on the ram to move it and, hence, the platen. A hydraulic kicker cylinder, suitably incorporated in the machine, enables this operation of application of force on both the rams simultaneously and, hence, the simultaneous movement of both platens in opposite directions at a very high velocity.

The entire gas system is fully sealed. The necessary energy to the platens to provide a high energy impact is provided by the adiabatic expansion of the highly compressed gas. The gas utilized in each blow is recompressed after the blow and reused for the next blow and so there is practically no loss for of gas during the operation. This results in very low operating costs. Apart from the high speeds and high energy impact the blows can be repeated, if needed desired number of times.

The tools and dies are held between the two platens. All the above factors, viz. high velocities of platen movements. high energy rate, etc. lead to the build up of a very high amount of force on the metal piece being worked and the metal undergoes plastic deformation (flow) under the action of this force at a fairly fast rate, enabling filling of all the sections of the die cavity quickly. The platen figure 3.41, velocities normally involved range from 10 metres/sec. to 20 metres/sec. The rapid flow of metal does not allow enough time to it to cool down due to its contact with the metal of the tooling (die and punch). This enables production of deeper parts without any appreciable amount of draft on its surfaces and, thus, minimising the need of further machining. This again contributes to cost reduction. The tooling used in high velocity forging should be made from a steel possessing high impact resistance. Low energy levels should be employed while starting a new job and the same gradually raised.

These machines can be effectively used forging, extruding, compacting and may other metal forming operations. The most efficiently performed operation, however, is the closed die forging of metals, although even ceramics can be worked on these machines. Almost all metals and metal alloys can be successfully forged on these machines. involing simple as well as complex shapes and with varying thicknesses of sections. Also, mechanical properties of the metals forged on these machines are found to have been substantially improved. Scaling and decarburisation of metal surface during the operation are almost eliminated. Extrusion of even such materials, which either can not be extrude on the conventional machines or will need conventional machines of giant sizes for this purpose, is also possible on these machines.

Some more advantages with these machines are that thus do not need any special foundation and can be installed in any area because problems of vibrations are not there. The products forged on these machines carry superior surface finish, need almost no draft in most cases, carry fairly close dimensional tolerances and are free of most of the conventional forging, defects. And, above all, the overall production cost is very low. However, there is a limitation of the size and weight of the product these machines can handle. Extremely large and very heavy forgings, say above 25 kg., can not be easily produced on these machines. A few examples of the type of products made on these machines include gears, engine housings, valve bodies, missile components, rocket components,etc.

3.11 ROLLING

3.11.1 Rolling mills and their Classification:

A rolling mill is a complex of machines for deforming metal in rotary rolls and performing auxiliary operations such as transportation of stock to rolls, disposal after, rolling, cutting, cooling, melting, pilling or coiling etc. A set of rolls of rolls in the massive housing is called a "Stand". Rolling mills are classified according to the number, of rolls in the working stand, the product rolled, arrangement of stands etc.

3.11.2 Classification based on number of rolls in the stand:

According to this criteria, rolling mills are of the following types:

i) Two high rolling mill

ii) Three high rolling mill

iii) Four high rolling mill

iv) Multi-roll rolling mill

v) Universal rolling mill

i) Two high rolling mill:

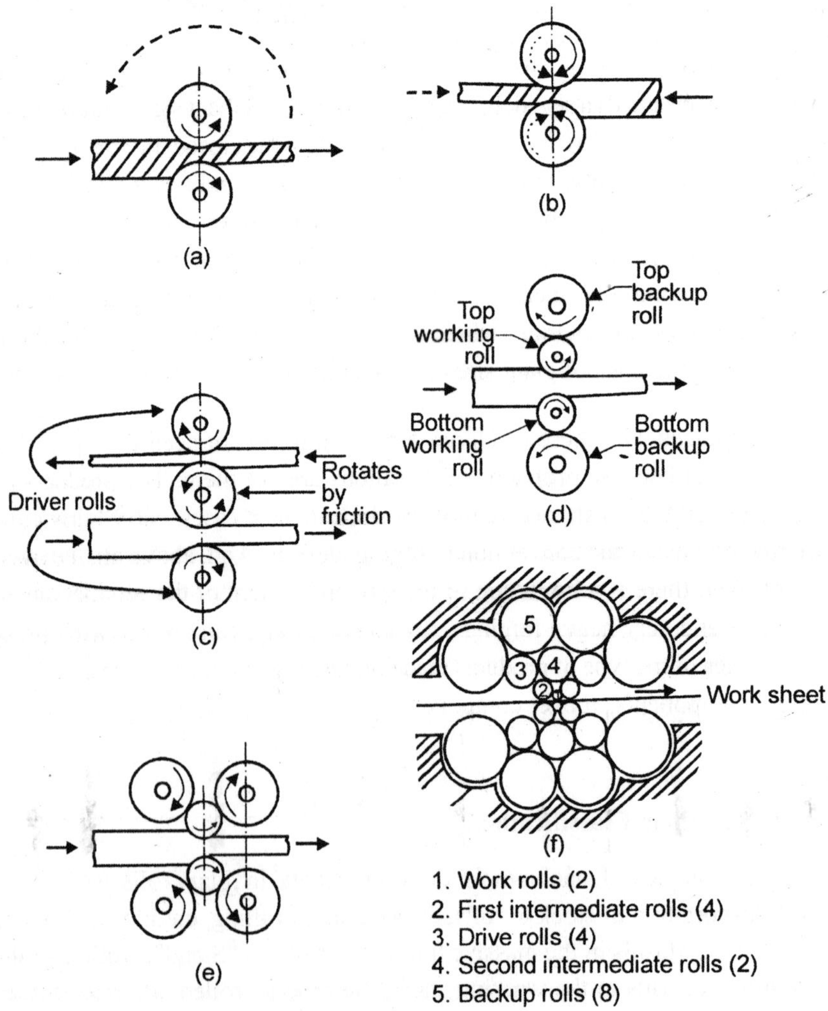

Figure 3.42 Rolling Mills

As the name implies. it has two rolls with a constant direction of rotating about the horizontal axes, (Figure 3.42(a)). For successive reductions, the stock is returned to the entrance of the rolls by hand carrying or by means of a platform which can be raised to pass the stock over the rolls. The upper roll may be raised or lowered to change the distance between the rolls. This method of successive reductions slows down the process. The faster method is to pass the stock through a series of rolls for successive reduction, but this method requires more investment in equipment. The alternative procedure is to use "two high reversing" rolling mill, (Figure 3.42(b)) where the direction of rotation of the rolls is reversed of the rolls is reversed after each pass. The rolls are brought closed together after each pass and the bar or plate reciprocates many times between the rolls before the final thickness is obtained. This design is limited by the length that can be rolled and by the intertial forces that must be over come each time a reversal is made. Also, much more power is consumed as compared to simple two high stand. Some time is also lost in reversing the mills.

A two high non-reversing mills are the most common for both hot and cold rolling since thus are leat expensive such stands are widely used in mills through which the passes only once (continuous mills)

ii) Three high rolling mill:

The disadvantages of two high reversing mill are over come by using a three-high rolling mill in which three rolls with a constant direction of rotation are arranged in a single vertical plane, (Figure. 3.42(c)). Lifting tables are provided on one or both sides of the stand to raise and lower the bar after each pass. In these mills, the top and bottom rolls are drive rolls and the middle roll rotates by friction. These mills are employed an blooming mills, for billet rolling and finish rolling. These stands also find extensive use in open-train section mills.

iii) Four-high rolling mill:

For a given width of plate/sheet and reduction per pass, the bending of the rolls will be less as their diameter is increased. However, increasing the roll diameter will increase the arc of contact between the roll and the plate/sheet. This will result in an increase in roll separating force (due to the reaction of the metal on to the rolls). which is given as,

P = mean specific pressure between metal and roll $\times$ contact area.

$\therefore$ In plate/sheet rolling (especially cold rolling), the diameter of work rolls should be as small as possible to reduce P and power requirements. To avoid bending of work rolls (due to their low strength and rigidity), large diameter back up rolls are installed, Figure 3.42(d) resulting in a four-high rolling mill. Four-high rolling mills are used in reversing mius for the hot rolling of armour and other plate as well as for the hot and cold rolling of sheet steel in continuous mills.

Note: The rolling load can be reduced by applying tensile force to the workpiece in the horizontal direction. This will lower the compressive yield strength of the material. Both "back tension" and front tension" can be applied.

iv) Multiple-roll mills:

In four-high rolling mill, the diameter of back up rolls can not be greater than 2 to 3 times that of the work rolls. As the diameter of work rolls is decreased more and more to accommodate processes with exceedingly high rolling loads, the size of the back up rolls must also decrease. A point is reached when the back up rolls themselves begin to bend and must be supported, hence the ultimate-design - cluster mill.

Stands with a cluster of six (Figure 3.42(e)), 12 and 20 rolls are intended to manufacture strips down to 0.001 mm thick and upto 2000 mm wide. As the diameter of work rolls is decreased more and more also decrease. A point is reached when the back up rolls themselves begin to bend and must be supported, hence the ultimate-design- cluster mill. In 12 or 20 roll stands, the very small dia. of working rolls (10 to 30 mm) allows no coupling of driving spindles, and the working rolls are idle running. They are supported by a row of driving rolls which, inturn are supported by a row of backup rolls. The work rolls are thus rotated by friction against the driving intermediate back up rolls.

Sendzimir mill:

This mill is a modification of the cluster mill with 20 rolls which is very well adapted to rolling thin sheet or foil stainless steel). Only the outer rolls are driven. The remaining rolls get the motion due to friction, from the back up rolls, (Figure 3.42(f)). High C-steel sheets of 5 to 50 mm can be made to an accuracy of 1 to 5 mm. This type of mill has been set up in Salem stainless steel complex in Tamil Nadu.

(v) Universal Rolling mill:

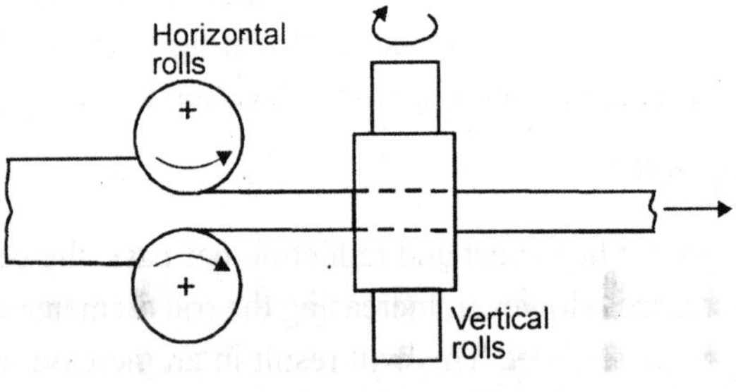

Figure 3.43. Universal Rolling Mill

In this rolling mill, the metal is reduced by both horizontal and vertical rolls, (Figure 3.43) The vertical rolls roll the edges of the bar even and smooth. Vertical rolls are mounted eiether on one side (front or back) or on both sides for the horizontal roll stand. The horizontal rolls may be either two, three, or four-high arrangement. These mills are used for rolling sheets, plates and slabs that requires rolled edges and also wide strip sheets, rolling of beams and H-sections.

Planetary Rolling mill:

This mill consists of a pair of heavy backing rolls surrounded by a large number of small planetary rolls, Figure 3.44. The main feature of this mill is that it hot reduces a slab to coiled strip in a single pass. Each pair of planetary rolls gives an almost constant reduction to the slab as it sweeps out a circular path between the slab and the backing roll. The total reduction is the sum of a series of such small reductions following each other in rapid succession. The feed rolls push the slab through a guide into planetary rolls. On its exit side, a 2 or 4 high planishing mill is installed to improve the surface finish and is followed by a coiler.

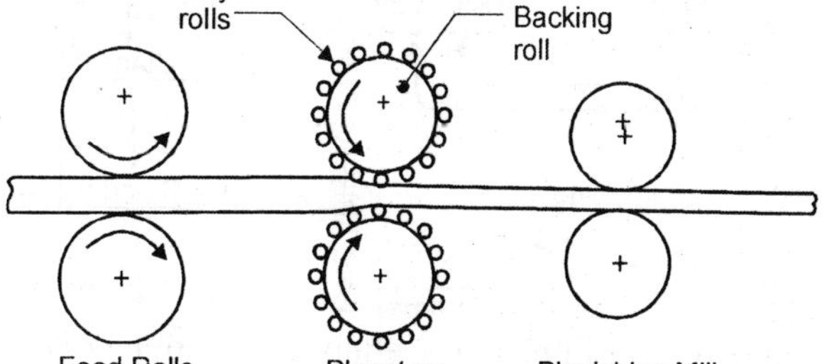

Figure 3.44 Planetary Rolling Mills

Steckel rolling:

In all the above methods, the rolls are powered and the job is pulled in due to the frictional force between the metal and the roll surface. In steckel rolling, the strip is drawn through idler rollers by front tension only, the tongue on the rolls being zero.

3.12. CLASSIFICATION BASED ON THE PRODUCTS ROLLED

According to the products rolled, the rolling mills may be classified as follows:

i) **Blooming and slabbing mills:**

These are heavy mills with rolls from 800 to 1400 mm in diameter. They are designed to roll ingots (2 to 25 tones in weight) into blooms and slabs. Since blooming mills and slabbing mills are quite similar, these are called as primary mills.

ii) **Billet mills:**

These mills have rolls from 450 to 850 mm in diameter and are designed to further reduce blooms into billets.

iii) **Rail and Structural mills:**

These mills have rolls from 750 to 800 mm in diameter and are used mainly to produce railroad rails, beams, channels and other heavy structural shapes.

iv) **Section mills:**

These mills have rolls from 250 to 750 mm in diameter, depending upon shape and section to be rolled.

v) **Rod mills:** have rolls about 250 mm in diameter and are use to produce wire rod,

vi) **Sheet and Plate mills**

These mills have barrel lengths ranging from 800 to 5000 mm for hot rolling and from 300 to 2800 mm for cold rolling.

vii) **Seamless Tube mills** produce seamless tubes as discussed later.

viii) **Tyre and Wheel mills** are used to manufacture rail-road wheels **and types.**

3.13 CLASSIFICATION BASED ON ARRANGEMENT OF ROLLING STANDS:

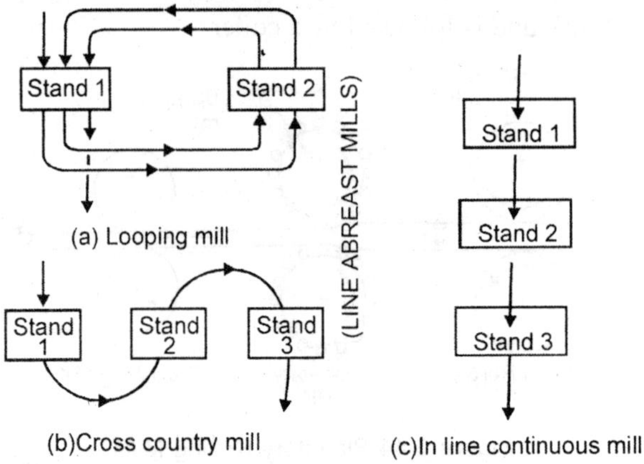

Figure 3.45 Arrangements of Rolling Stands

Single stand mills are used singly or in various floor-layout combinations for rolling metals.

i) Looping mill:

(Figure 3.45(a)). The products are looped from one mill stand to another by arranging the stands in a "train" in which the rolls of one mill stand are driven from the ends of the rolls in the adjacent stand.

ii) Cross-country mill:

(Figure 3.45(b)) It consists of individual roll stands placed some distance apart so that the workpiece must be leave one set of rolls before it enters the next set. to save spare and avoid complicating the mill drive mechanism. the roll stands are generally arranged in one or more parallel lines. Transfer and skid tables are used to reverse the direction of travel of the workpiece and to convey it from set of roll stands to another.

iii) Continuous mill:

(Figure 3.45(c)). It employs a number of individual roll stands arranged one behind the other in a straight line (in tandem). The work pieces continuously pass through various roll stands and emerge from the last roll stands as finished shape. the rolls of each successive stand must twin faster than those of the proceeding one by a precise amount so as to accommodate the increased length which accompanies each successive reduction in thickness.

All the above three mills require a large capital outlay and are only justified for mass production. Looping and cross country mills require the work piece to be bent or turned between the stands and are, therefore, used for rolling rods, rails or sections. Continuous mills are used for plates, strips or sheets. Looping mills and cross-country mills are also called as "Line abreast continues mills" and continuous mills are called as "In - line continuous mills".

3.13.1 Defects in rolled parts:

i) Rolling frequently involve non-uniform deformation (reduction) of the metal and this is most pronounced in the rolling of complex shapes such as rails, beams, channels, etc.,

ii) Fins may be formed on the rolled bars if the metal forces itself into the clearance between the rolls. Fins may lead to various defects such as laps, cracks etc.,

Laps may be caused by overfilling int he rolling mill passes, which will produce fins. In passes which follow, these fins are pressed back into the surface wihout welding, causing laps.

iii) When the metal is hot rolled, its surface is not smooth and it has scale (oxide) formed over the same.

iv) Cracks may form during cold rolling if the metal becomes too much work-hardened during the process.

3.14. PIERCING OR SEAMLESS TUBING

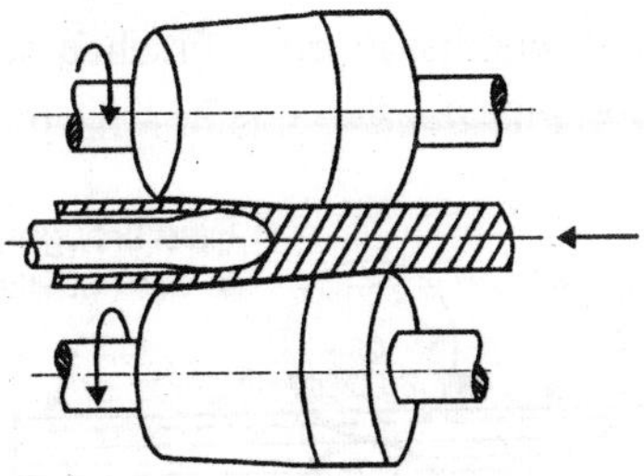

Figure 3.46 Tube Piercing

Piercing is employed to produce seamless tubing which is the natural form from which is made any thing-walled round subjects. Seamless tubing is a popular and economical raw stock for machining because it saves drilling and boring of parts. The process of making hot-pierced tubing consists of passing a hot-rolled billet between two conical-shaped rolls and over a mandrel which assists in the piercing and controls the size of the hole as the billet is forced over it as depicited is figure 3.46.

The solid billet is uniformly heated to about 1100 C, and the piercing action is actually started previous to placing in between the rolls by placing, punching, or piercing with oxygen a hole to about 25 mm deep. It is then pushed into the two piercing rolls which impart axial as well as rolling movement to the billet and force it over the mandrel. The alternate squeezing and bulging of the billet open up a seam in its centre. The first pass makes a rather thick walled tube, which is again passed over a tapered plug and through and through grooved rolls in a two-high mill where the thickness is decreased and the length in increased. While it is still up to a working temperature, it is passed onto a reeling machine which has two rolls similar to the piercing rolls, but with flat surfaces. The reeler further straightend and sizes the tubes. If more accuracy and better finish are desired, the tube may run through sizing dies or rolls. After cooling, the tubes are usually placed in a pickling bath of dilute sulphuric to remove the scale and oxide.

3.15 WELDED PIPE AND TUBE MANUFACTURING

Welded pipe is made form a flat strip of which one end is crimped. it is known as the skelp. Its edges are bevelled to butt properly when the skelp is formed into a round shape for welding the seam. The skelp is charged into a furnace and heated to its welding temperature. It is then drawn out of the furnace by gripping its crimped and with tongs which are attached to a drawn chain. The gripped skelp is pulled through a welding bell which forms it into the round shape and simultaneously enables pressing and welding together of the bevelled edges.

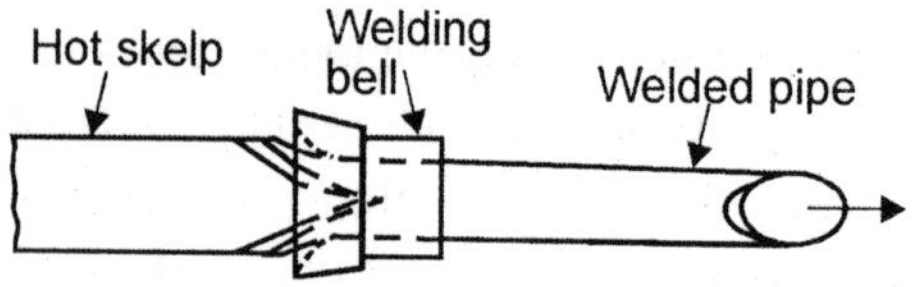

Figure 3.47 Production of a Welded Pipe

Thus, on one side of the bell enters the hot skelp and on the other side emerges the welded pipe, as shown in figure 3.47, to bring it to the correct outside diameter. Finally it is passed through the finishing and straightening rolls where it is correctly finished, its surface appearance improved by removing scale and it is correctly finished, its surface appearance improved by remove scale and it is straightened.(Refer figure 3.48). It is then cut into desired lengths, its ends finished and threaded. The above method is an intermittent method of forming welded pipes.

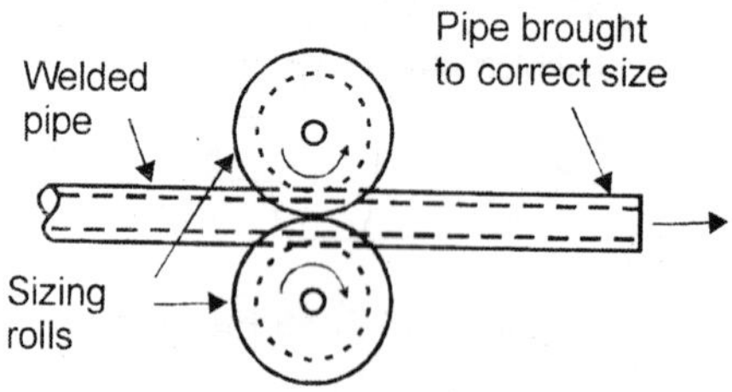

Figure 3.48 Sizing the Welding Pipe

A continuous butt welded pipe is produced by using the skelp in the form of a coil. The ends of the coiled strips are welded to form a contiuous strip. It is then passed through the furnace where the flames are directed towards the edges hot to bring them to welding temperature. As the hot skelp comes out of the furnace passed through a series of grooved rolls where it is rounded, welded and sized, followed by cutting to required lengths. The arrangement of rolls is shown in figure 3.49. As usual, the cut lengths of the pipe are finished in the same way as described in the above paragraph. This method is used for pipes are finished in the same way as described in the above paragraph. This method is used for pipes upto 75 mm dia.

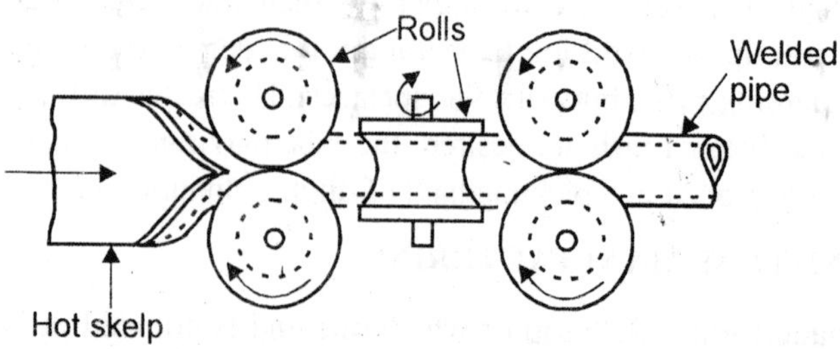

Figure 3.49 Production of a Continuous Butt-Welded Pipe

3.16 HOT EXTRUSION

The process of extrusion consists of compressing a metal inside a chamber to force it out through a small opening called die. Any plastic material can be extruded successfully. Most of the presses used for extruding metals are hydraulically operated horizontal presses. A large number of extruded shapes are in common use, such as tubes, robs, structural shapes and lead covered cables. The principles of operation are the for both hot and cold extrusion, and the choice of one of these is governed by factors like the metal to be extruded, thickness of the extrudes section, size of the raw material being used, capacity of the press and type of product, etc. Billets of 125 mm to 175 in diameter and 300 to 675 mm in length are in general use as raw material for extrusion. Extrusion of steel needs adequate lubrication around the billet. This is done by providing a coating of finely powdered glass over the surface of hot billet. The process of extrusion suits best to the nonferrous metals and alloys, although some steel alloys, like stainless steel, are also extruded.

3.16.1 Types of Extrusion:

The extrusion process can be classified as:

1. Direct of forward extrusion
2. Indirect of backward extrusion.

3.16.2 Direct or forward extrusion

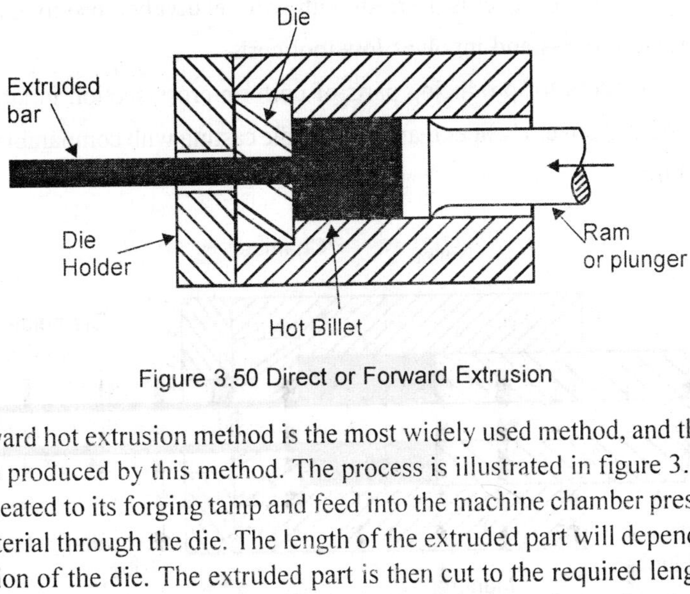

Figure 3.50 Direct or Forward Extrusion

Direct or forward hot extrusion method is the most widely used method, and the maximum number of extruded parts are produced by this method. The process is illustrated in figure 3.50. The raw material used is a billet, it is heated to its forging tamp and feed into the machine chamber pressure is applied to the billet forcing the material through the die. The length of the extruded part will depend upon the size of the billet and cross-section of the die. The extruded part is then cut to the required length. The overhanging extruded is fed into a long support called the run out table. It is a usual practice to leave the last nearly 10% length of billet as unextruded. This portion is known as discard and contains the surface impurities of the billet.

3.16.3 Backward or indirect extrusion:

For this type of extrusion the ram or plunger used is hollow, and as it presses the billet against the backwall of the closed chamber the metal is extruded back into the plunger. As the billet does not move inside the chamber, there is not friction between them. As such, less force is needed in this method in comparison on the direct extrusion. A more complicated type of equipment is required because the plunger becomes weak due to the reduction in its effective area of cross-section and difficulty is experienced in supporting there overhanging extruded part. The process is illustrated in figure 3.51.

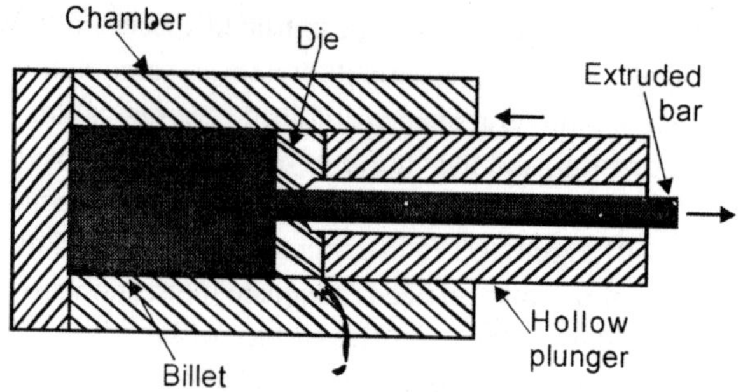

Figure 3.51 Backward or Indirect Extrusion

Advantages and limitations of hot extrusion

1. A very dense structure of metal is obtained because of high pressures used and compressive nature of the process.
2. Surface finish of the product is quire smooth with relatively close tolerances.
3. It is very rapid process and involves low tool costs.
4. It is an ideal process for producing parts of uniform cross-section in large quantities.
5. In many cases it prooves cheaper than pressure die casting with comparable finish and tolerances.

3.16.4 Tube extrusion:

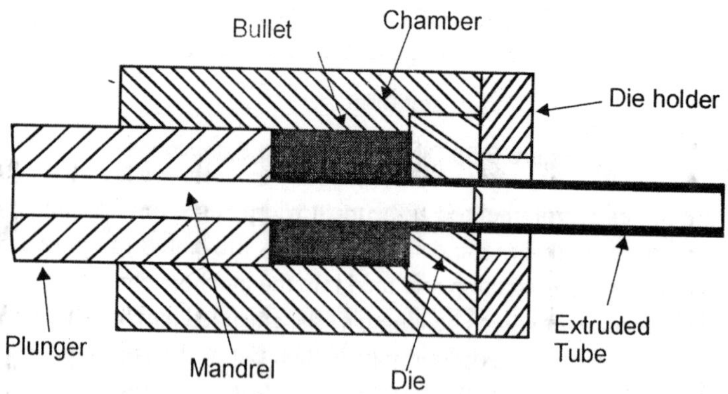

Figure 3.52 Tube extrusion

A common method of hot extrusion of tubes is shown in figure 3.52. It is actually a forward extrusion method using a mandrel to form the bore of the tube. First the mandrel is pushed through the centre of the billet and the die, followed by applying pressure on the billet by advancing the plunger. The metal is forced no flow through the opening between the die and the mandrel. The operation is performed quite rapidly. Most of the metals and alloys are hot extruded, although some of these are cold extruded also, for production of seamless tubes.

3.16.5 Hydrostatic extrusion:

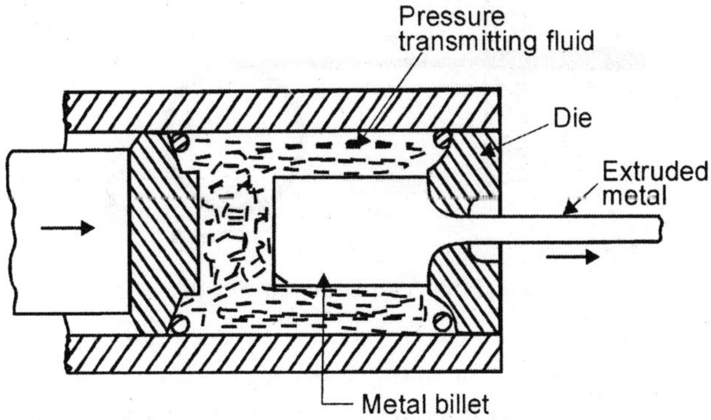

Figure 3.53 Hydrostatic Extrusion

Another extrusion process that is being used for special applications is the hydrostatic extrusion, in this, the metal billet is compressed from all sides by a liquid rather that the ram, The presence of liquid inside the container eliminates the need for any lubricant and also the material is more uniformly compressed from all sides throughout the deformation zone. Because of this, highly brittle materials such as grey cast iron can also be extruded. A typical hydrostatic extrusion operation is shown in figure 3.53. Some of the pressure transmitting fluids used are castor oil with 10% alcohol, SAE 30 mineral lubricating oil, glycerine, ethyl glycol and iso pentane. The hydrostatic pressure range is from 1110 to 3150 MPa. The commercial applications of the process are limited to the extrusion of reactor fuel rods, cladding of metals making wires of less ductile materials.

3.17 COLD EXTRUSION

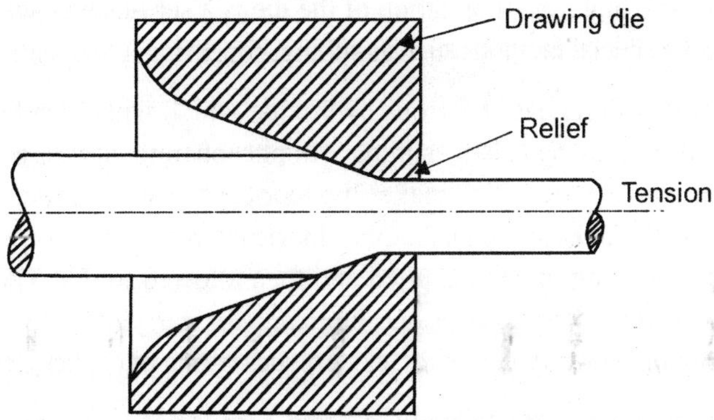

Figure 3.54 Wire Drawing

A wire by definition, is circular with small diameters so that it is flexible. The process of wire drawing is to obtain wires from rods of bigger diameter through a die. Wire drawing is always a cold working process. (Refer figure 3.54).

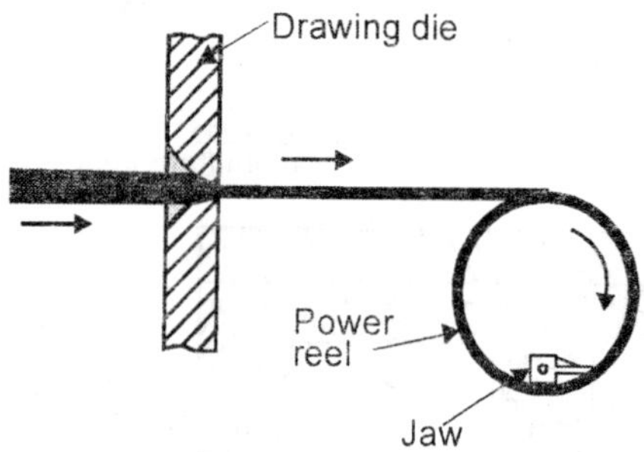

Figure 3.55 Wire Drawing Set up

A typical wire drawing die is shown in figure 3.55 and the wire drawing machine in figure 3.55. The wire drawing die is of conical shape of the rod or wire which is to be further reduced is made into a point shape and inserted through the die opening. This end is then gripped on the other side with a gripper which would then pull the wire through the die. The wire thus drawn is then coiled round a power reel as shown in figure 3.55.

Before the wire is drawn, the stock needs to be prepared for wire drawing. The material should be sufficiently ductile since have to it is pulled by the tensile forces. Hence will may be annealed properly to provide the necessary ductility. Further, the wire is to go through the conical portion and then pulled out through the exit by the gripper. In this process, there is no force applied for pushing the will into the die from the entrance side. To make for an easier entrance of wire into the die, the end of the stock is made pointed to facilitate the entry. Thus pointing is done by means of rotary or swaging by simple hammering.

The other aspect of preparation needed is the cleaning of the wire and lubricating it as it flows through the die. Cleaning is essentially done to remove any scale and rest present on the surface which may severely affect the die. It is normally done by acid pickling. The pressures, acting at the interface of the die and the metal being very high, the lubrication of the die is a serious problem. Therefore, to carry the lubricant throughout the die special methods such as sulling, coppering, phosphating and liming are used.

The wire is coated with a thin coat of ferrous hydroxide which when combined with lime acts as a filler for the lubricant. This process is called "sulling". In phosphating, at thin film of manganese, iron or zinc phosphate is applied on the wire which makes the lubricant to stick to the wire, thereby reducing the friction and consequently, the drawing load. Another lubricant vehicle that is used in wire drawing is a coating of lime. After acid pickling, lime is applied and then allowed to dry. The lime neutralises any amount of acid left on the surface and absorbs the lubricant for carrying it to the die. The lubricant normally used is the soap solution. For very thin wires, electrolytic coating of copper is used to reduce friction.

The dies used for wire drawing are severely affected because of high stresses and abrasion. The various die materials that are used are chilled cast iron, tool steels, tungsten carbide and diamond. The cast iron dies are used for small runs. For very large sizes, alloy steels are used in making the dies. The tungsten carbide dies are used most commonly for medium size wires and large productions. The tungsten carbide dies are preferred because of their long life which is 2 to 3 times that of alloy steel dies. For very fine wires, diamond dies are used. Some of the commercially available tungsten carbide die sizes are presented in figure 3.56.

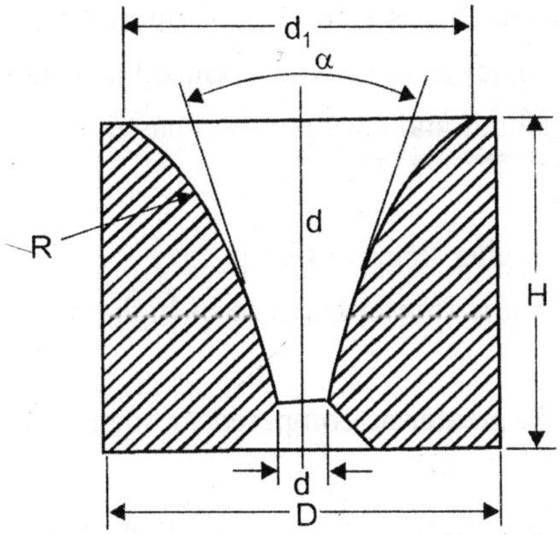

Figure 3.56 Wire Drawing Pellet

Wire drawing improves the mechanical properties because of the cold working. The material loses its ductility during the wire drawing process and when it is to be repeatedly drawn to bring it to the final size, intermediate annealing is required to restore the ductility.

The drawing machines can be arranged in tandem so that the wire coming from one die is coiled upto a sufficient length before it is reentered into the subsequent dies and so on. This coiling of sufficient wire helps for any discrepancy in the speed of wire drawing in any die. Since there is no change in volume successive drawing have to be done at higher speeds.

3.18 ROD AND TUBE DRAWING

Rod drawing is similar to wire drawing except for the fact that the dies are bigger because of the rod size being larger that the wire. But the rod drawn in coiled form is to be straightened and then cut into proper lengths.

For larger size stock called bars, the heavy equipment which generally keeps the drawn product straight, is used since bar cannot be coiled. The straight drawing equipment consists of a table which contains rollers on which the bar stock is fed into the die head after pointing. The point of the bar is then pulled out through the die and put on a carriage. The carriage will be somewhat like an end less chain with grips which grasp the metal protruding from the die and pull along as the chain moves. The bars coming out of the die are generally of short lengths, so that when they are completely drawn, they can be transported to other place from the draw bench.

Tube drawing is also similar to the other drawing processes. The main difference is that it requires a mandrel of the requisite diameter to form the internal hole as shown in figure 3.56. The tubes are also first pointed and then entered through the die where the point is griped in similar way as the bar drawing and pulled through in the form desired along a straight line. There may be more than one pass required to get the final size and when the final size is obtained, the tube may be annealed and straightened.

Advantages of Extrusion:

★ The range of extruded items is very wide.

★ No time is lost when changing shapes since the dies may be readily removed and replaced.

★ Dimensional accuracy of extruded parts is generally superior to that of rolled ones.

★ Automation in extrusion is simpler as items are produced is a single passing.

★ Extrusions are lighter, more sound and stronger than castings.

Drawbacks:

★ Process waste is extrusion is higher than in rolling.

★ Service life of extrusion tooling is shorter because of high contact stresses and slip rates.

★ Relatively high tooling cost.

★ Cost of extrusion are generally greater as compared to other techniques.

Applications:

★ Working of poorly plastic and non ferrous metals and alloys.

★ Manufacture of sections and pipes of complex configuration

★ Medium and small batch production.

★ Manufacture of parts of high dimensional accuracy.

SUMMARY:

❖ ***Cold working:*** Cold working may be defined as the process of plastically deforming a metal piece below its recrystallisation temperature.

❖ ***Cold drawing of tubes and wires:*** Cold drawing is the process of pulling metal at room temperature through the opening of a die.

❖ ***Wire Drawing:*** This is the process of producing wires by pulling a rod through the opening of a die.

❖ ***Tube Drawing:*** This is the process of finishing hot rolled by pulling them through a die opening with a stationary mandrel.

❖ ***Metal spinning:*** This is the process of producing circular shaped articles from very thin metal blanks by spinning action.

❖ ***Embossing:*** Embossing is the operation of giving impressions of letters, figures or patterns of sheet metal parts.

❖ ***Squeezing operations:*** Squeezing is the operation of pressing the metal within the cavity of a die and punch to get the required shape.

❖ ***Coining:*** It is the operation of producing coins, medals and other similar parts by squeezing the metal between two dies.

❖ ***Sizing:*** This is a process used for bringing the thickness of a forged component to the required size by pressing in a heavy duty press.

- ***Angle bending:*** Angle bending is the operation of bending a sheet metal to a sharp acute angle.

- ***Curling:*** Curling is the operation of rolling the edges of the sheet metal to a small circular form.

- ***Upsetting Operations:*** This is carried out to increase the thickness or diameter of work-piece after reducing its length by hammering process.

- ***Bending:*** Bending operation is carried out by keeping the work-piece on the edge of the anvil face, anvil horn or on any other fixture/support.

- ***Punching:*** Punching operation is used to produce or to enlarge the hole in the metallic work-piece by hammering a tool called punch.

- ***Drifting:*** Drifting operation is to cut a slit in the work and the tool used is called 'Drift' which is having a very small cutting angle with sharp edge.

- ***Fullering and Flattening:*** Fullering operation uses fullering tool to reduce the thickness of the heated work-piece.

- ***Roll Forging:*** Roll forging in its simplest form consists of a heated billet passing between a pair of rolls that deform it along its length.

- ***Ring Rolling:*** Ring rolling offers a homogeneous circumferential grain flow, ease of fabrication and machining and versatility of material size.

- ***Wire drawing:*** A wire by definition, is circular with small diameters so that it is flexible.

- ***Rod and tube drawing:*** Rod drawing is similar to wire drawing except for the fact that the dies are bigger because of the rod size being larger that the wire.

- ***Common high velocity forming processes:*** The common high velocity forming process in use having deformation velocities several times greater than conventional forming processes are: i) Explosive forming, ii) Electrical discharge forming, iii) Electro-magnetic forming, iv) Pnemo-mechanical forming (Dynapak).

- ***Explosive forming:*** This is a process of forming a sheet metal to the required shape in a die by the impact force produced by an explosive. Both low explosives and high explosives can be used in this process.

- ***Classification of Explosive forming:*** The explosive forming of metal is classified mainly into the following categories: i) Stand off operation or open type operation, ii) Contact operation or closed type operation, iii) Multiple type operation.

REVIEWS

1. Define cold working.
2. What are the effects of cold working?
3. Name the coldworking operations.
4. Briefly explain drawing processes.
5. Briefly explain squeezing and bending processes.
6. Briefly explain shearing processes.
7. Name some forging tools.
8. What are the type of hammers?
9. Name the types of tongs.
10. What is the use of drift?
11. Name the hand forging operation and explain briefy about them.
12. What is fullering and flattening?
13. Name the types of forging process.
14. Name the types of extrusion forging process.
15. Explain with a neat sketch roll forging process.
16. What are rolling mills and classify them?
17. Explain briefly about the rolling mills.
18. Explain hot extrusion processes.
19. Explain wire drawing process with neat diagram.
20. Draw neatly wire drawing pillet.
21. Briefly explain rod and tube drawing process.
22. How will you classify the rolling process based on the product rolled?
23. How will you classify the rolling process based on arrangement of rolling stands?
24. Explain seamless tube producing process with neat diagram.
25. Draw and explain the production of continuous butt-welded tube.

CHAPTER - 4

SHEET METAL FORMING PROCESS

4.1 INTRODUCTION TO METAL FORMING PROCESS

Metal forming can be defined as a process in which the desired size and shape are obtained through the deformation of metals plastically under the action of externally applied forces. The stresses induced are greater than the yield strength, but less than the fracture strength of the material. The type of loading may be tensile, compressive, bending or shearing, or a combination of these. This is a very economical process as the desired product is obtained without any loss of material.

4.2 SHEET METAL CHARACTERISTICS

After a blank is cut from a larger sheet, it is formed into various shapes. Basically, all sheet forming processes employ various dies and tooling to stretch and bend the sheet. Before we consider these processes, however, certain characteristics of sheet metals must be reviewed, because of their important effects on the overall operation. (Table 4.1)

i) Elongation:

Although sheet-forming operations rarely involve simple uniaxial stretching like that in a tension test, observations concerning tensile testing can be useful for understanding the behaviour of sheet metals. When a specimen subjected to tension first undergoes uniform elongation; only when the load exceeds the ultimate tensile strength does the specimen begin to neck.

Because the material is usually being stretched in sheet forming, high uniform elongation is desirable for good formability. The true strain at which necking begins is numerically equal to the strain-hardening exponent (n) thus, a high n value indicates large uniform elongation.

Table 4.1 Characteristics of Metal during the Sheet Metal Forming Process:

Characteristics	***Importance***
Elongation	Determine the capability of the sheet metal to stretch without necking and failure; high strain-hardening exponent (n) and strain-rate sensitivity exponent (m) desirable.
Yield-point elongation	Observed with mild-steel sheets; also called Lueder's bands and stretcher strains; causes flamelike depression on the sheet surfaces; can be eliminated by temper rolling, but sheet must be formed within a certain time after rolling.
Anistropy (planar)	Exhibits different behaviour in different planar directions; present in cold-rolied sheets bacause of preferred orientation or mechanical fibering; causes earing in drawing; can be reduced or eliminated by annealing but at lowered strength.
Anistropy (normal)	Determines thinning behaviour of sheet metals during stretching; important in deep-drawing operations.
Grain size	Determines surface roughness on stretched sheet metal; the coarser the grain, the rougher the appearance (orange peel); also affects material strength.
Residual stresses	Caused by nonuniform deformation during forming; causes part distortion when sectioned and can lead to stress-corrosion cracking; reduced or eliminated by stress relieving.
Springback	Caused by elastic recovery of the plastically deformed sheet after unloading; causes distortion of part and loss of dimensional accuracy; can be controlled by techniques such as overbending and bottoming of the punch.
Wrinkling	Caused by compressive stresses in the plane of the sheet; can be objectionable or can be useful in imparting stiffness to parts; can be controlled by proper tool and die design.
Quality of sheared edges	Depends on process used; edges can be rough, not square, and contain cracks, residual stresses and work-hardened layer, which are all detrimental to the formability of the sheet; quality can be improved by control of clearance, tool and die design, fine blanking, shaving and lubrication.
Surface condition of sheet	Depends on rolling practice; important in sheet forming as it can cause tearing and poor surface quality.

Necking may be localized or it may be diffuse, depending on the strain-rate sensitivity of the material. The higher the value of m the more diffuse the neck becomes; diffuseness is desirable in sheet-forming operations.

In addition to uniform elongation and necking, the total elongation of the specimen (in terms of that for a 50 mm (2 in.) gage length) is also a significant factor in the formability of sheet metals. Obviously, the total elongation of the material increases with increasing values of both n and m.

ii) Yield-Point Elongation:

Low-carbon steels exhibit a behaviour called yield-point elongation, one having upper and lower yield points [Figure 4.1(a)]. This behaviour indicates that, after the material yields, the sheet stretches farther in certain regions without any increase in the lower yield point, while other regions in the sheet have not yet yielded. Aluminium-magnesium alloys also exhibit this behaviour.

This behaviour produces Lueder's bands (stretcher strain marks or worms) on the sheet [Figure 4.1(b)]; they are elongated depressions on the surface of the sheet, such as can be found on the bottom of cans used for common household products [Figure 4.1(c)]. They may be objectionable in the final product, because coareseness in the surface degrades appearance and causes difficulties in subsequent coating and painting operations.

The usual method of avoiding these marks is to eliminate or to reduce yield-point elongation, by reducing the thickness of the sheet 0.5% to 1.5% by cold rolling (temper or skin rolling). Because of strain aging, however, the yield-point elongation reappears after a few days at room temperature - or after a few hours at higher temperatures. To prevent this undesirable occurrence, one should from the material within a certain time limit (which depends on the type of the steel).

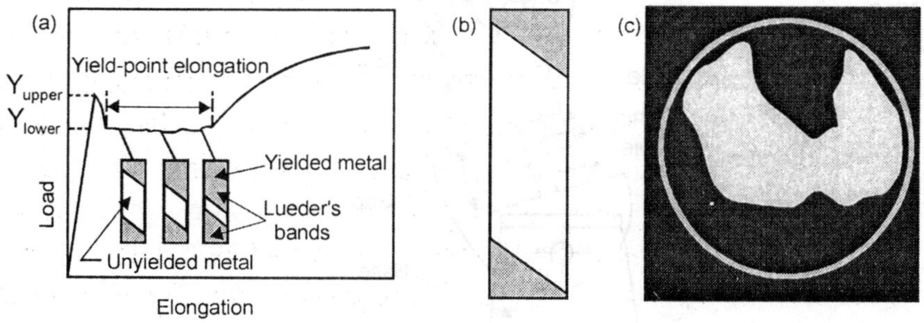

Figure. 4.1 (a) Yield-popint elongation in a sheet-metal specimen. (b) Lueder's bands in a low-carbon steel sheet. (c) Stretcher strains at the bottom of a steel can for household products.

iii) Anisotropy:

An important factor that influences sheet-metal forming is anisotropy (directionality) of the sheet. Recall that anisotropy is acquired during the thermo-mechanical processing of the sheet, and that there are two typcs of anisotropy. Crysallographic anisotropy (preferred orientation of the grains) and mechanical fibering (alignment of impurities, inclusions and voids throughout the thickness of the steel).

iv) Grain Size:

The grain size of the sheet metal is important for two reasons. a) The grain size affects mechanical properties, and it influences the surface appearance of the formed part (orange peel). b) The coarser the grain, the rougher is the surface appearance. An ASTM grain size of 7 or finer is preferred for general sheet-metal forming operations.

4.3 TYPES OF PRESSES

Presses are classified as follows:

1. According to the source of power
   a) Hand press-ball press and fly press b) Power press-mechanical, hydraulic and pneumatic.
2. According to the type of frame.
   a) OBI b) Straight side c) Adjustable d) Pillar
3. According to the driving mechanism
   a) Crank and connecting rod b) Eccentric c) Knuckle joint d) Toggle lever
   e) Screw f) Rack and pinion g) Hydraulic
4. According to the number of slides
   a) Single acting b) Double acting c) Triple acting

4.3.1 OBI press

OBI press means Open Back inclinable press. The press has a frame pivoted to the base. The frame can be adjusted to any inclinded position and clamped. There is a gap in the frame. This gap is meant for the removal of the parts. The parts will slide down through the gap due to gravity.

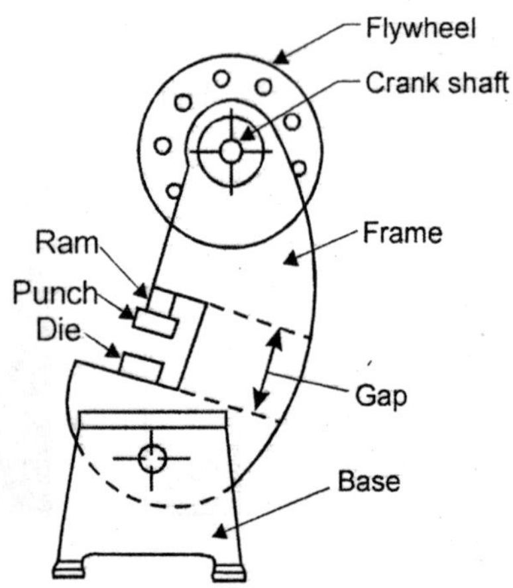

Figure 4.2(A) OBI Press

The die is fitted to the bottom of the frame. The punch is connected to the bottom of the ram. Ram slides along the guide ways fitted to the frame. A crank shaft is fitted to the top of the frame. The ram is connected to the crank shaft by means of a connecting rod. The crank shaft is driven by a motor. The crank shaft carries a fly wheel at the end. The rotation of the crank shaft makes the slide to move up and down. The fly wheel stores energy and makes the ram to move at constant speed. (Refer figure 4.2(A)).

4.3.2 Adjustable bed press

In this type of press the bed can be adjusted vertically. This is done by an elevating screw. The bed is lowered or raised to suit the different heights of the work and the dies. The press is not very rigid. The adjustable bed is known as knee, so this press is also called knee press. (Refer figure 4.2(B)).

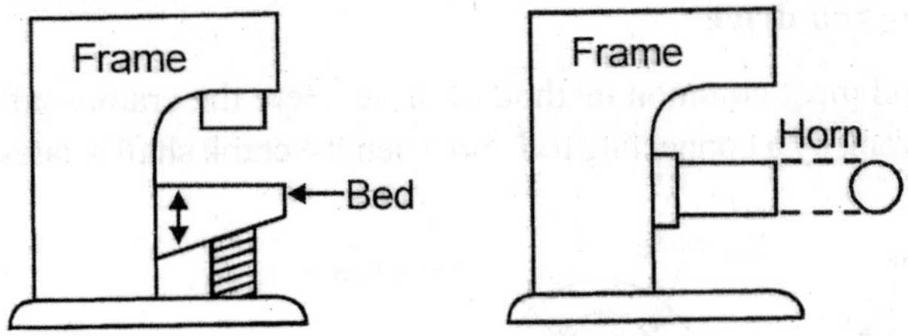

(a) Adjustabled Bed (b) Horn Press

Figure 4.2 (B)

4.3.3 Horn Press

Refer figure 4.2(B). This press has a cylindrical projection from the machanic frame. This projection is called horn. The horn supports round hollow workpieces. Horns are of removable type. Depending upon the size of the workpiece horns can be changed.

4.3.4 Straight side press

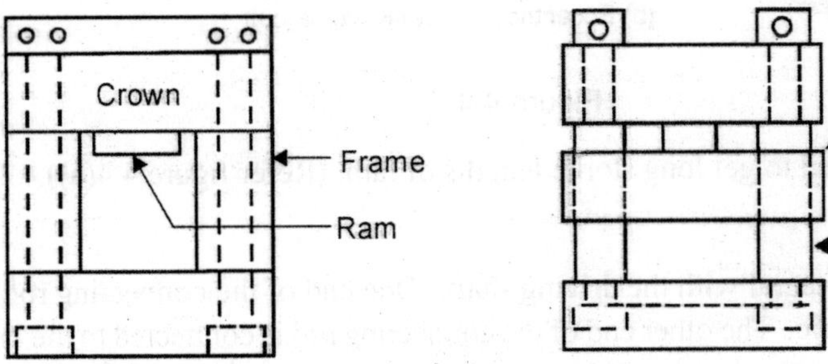

(a) Straight side (b) Pillar side

Figure 4.3

The straight side press has two vertical frames. The frames are mounted on to the two sides of the base. The two vertical frames are connected by means of a crown at the top. The ram is fitted to the crown. The frames and the crown are rigid and absorb the heavy load. So this press is suitable for heavy work. The sheet metal can not be fed from the sides. It can be fed only from the front.(Refer figure 4.3(a)).

4.3.5 Pillar type press

Refer figure 4.3(b). This press has four pillars. The pillars are mounted vertically on four corners of the square base. The four pillars are connected by crown at the top. The pillars guide the movement of the ram. This type of press is always of hydraulic type. The sheet metal can be fed from all the four sides. So this press is also known as open frame press.

4.4 TYPES OF DRIVES

In a press, the ram reciprocates vertically. For this movement of the ram, power drive is given. The type of drive depends upon the length of stroke needed and the load on the ram. The various types of drives are explained below.

4.4.1. Crank and connecting rod drive

This is the simplest and most common method of drive. Here the crank shaft is driven by a motor. The ram is connected to the crank by a connecting rod. So when the crank shaft rotates, the ram reciprocates.

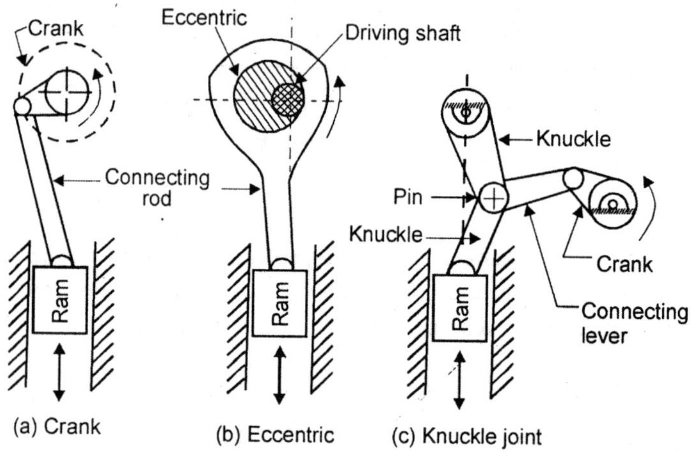

Figure 4.4

This type of drive is used to get long storke lengths of ram. (Refer figure 4.4(a)).

4.4.2 Eccentric drive

An accentric is made integral with the driving shaft. One end of the connecting rod has a housing. This housing fits on the eccentric. The other end of the connecting rod is connected to the ram. The driving shaft rotates with the eccentric. So the ram reciproctes.

This type of drive is used shorter storke lengths of ram. (Refer figure 4.4(b)).

4.4.3 Knuckle joint drive

The knuckle joint drive has two knuckle levers and one connecting lever. One end of the upper knuckle lever is pivoted to the crown of the machine. One end of the lower lever is connected to the ram. The other ends of the knuckle levers and one end of the connecting lervers are joined by a pin. The other end of the connecting lever is connected to the crank of the crank shaft. When the crank shaft rotates, the ram reciprocates.

At the bottom of the stroke, the knuckle levers become vertical. So the complete load acts on the crown through the knuckle levers.The crank shaft is not stressed much. So heavy loads can be applied in this press.

This drive is used for coining and squeezing operations. (Refer figure 4.4(c)).

4.4.4 Cam drive

In this press, the driving shaft has a cam. The upper end of the connecting rod has a housing. The housing fits over the cam. The other end of the connecting rod is connected to the ram. When the driving shaft rotates, the cam actuates the ram to reciprocate.

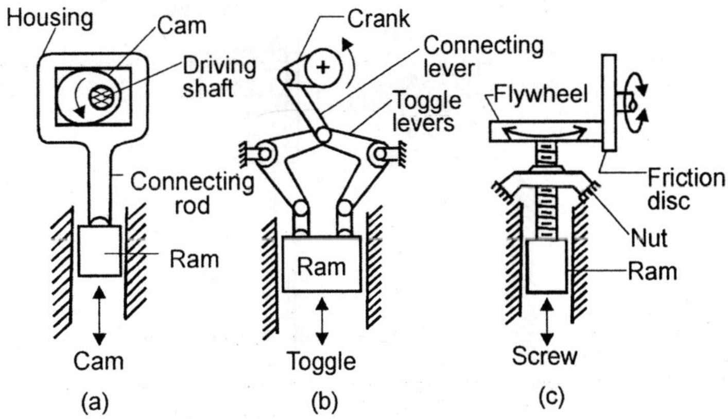

Figure 4.5

The cam profile is so designed that the ram remains idle for some time at the bottom of the stroke.

The cam drive gives short stroke length for the ram. This drive is used for actuating the outer ram of a double acting press. (Refer figure 4.5(a)).

4.4.5 Toggle lever drive

In this drive, there are two toggle levers. These lever are pivoted at their centres to the machine frame. The bottom of the toggle levers are connected to the ram through small links. The upper part of the toggle lever and one end of the connecting lever are joined together by a pin. The other end of the connecting lever is fitted to the crank. The crank is rotated by the driving shaft. When the crank rotates, the ram is actuated to move up and town. At the end of the stroke, the ram will remain stationary for some time.

This toggle drive is used for actuating the outer slide of the double acting press. This drive gives a shorter stroke length for the ram. (Refer figure 4.5(b)).

4.4.6 Screw drive

Here the ram is loosely connected to a screw. The screw passes through a stationary nut. The top end of the screw is connected to a fly wheel. The periphery of the fly wheel is in contact with a friction disc. When the friction disc rotates the screw also rotates. The screw moves down. So the ram moves down. It does not rotate. When the ram reaches the bottom of the storke, the friction dics reverse and the ram moves up. This screw drive is used for longer stroke length. The movement of ram is uniform in this drive.(Refer figure 4.5(c)).

4.4.7 Rack and pinion drive

In this drive, the ram is connected to a long rack. The rack is in mesh with a position. The pinion rotates about a fixed axis. When the pinion is rotated, the rack moves down the ram. At the end of the stroke, the pinion reverses. So, the ram moves up Rack and pinion drive is used when a long stroke length is required. (Refer figure 4.6(a)).

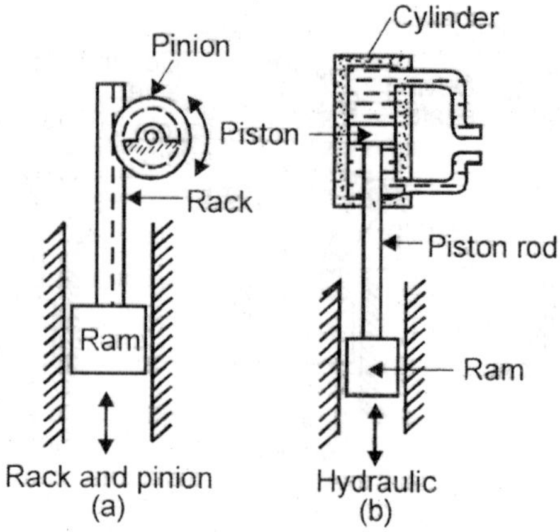

Figure 4.6

4.4.8 Hydraulic drive

In the hydraulic drive, the ram is connected to a piston through a piston rod. The piston moves up and down in a hydraulic cylinder. The oil at high pressure is pumped through the top side of the piston. the ram moves downwards. At the end of the stroke, the oil is pumped through the bottom side of the piston. The ram moves up. (Refer figure 4.6(b).

The hydraulic drive is used when a very heavy pressure is required on the ram. This is used for drawing and forming operations. Hydraulic drive gives a noiseless and smooth operation. The pressure on the ram can be easily changed. The pressure appied is uniform. The ram movement takes place at uniform speed. This drive is used when more force is required on ram.

4.5 PRESS ACCESSORIES

Press accessories (also known as die accessories) are essential for production work to facilitate easy and quick operation. These accessories are essential for easy and removal of finished material and for the correct location of the blank in between the punch and die. The different press accessories with their specific function are listed below.

Sl.No	Accessories	Function
1.	Step	to stop the work material at the correct length
2.	Pilot	for correct location of workpiece
3.	Stripper	for removal of workpiece from the punch or die after blanking or punching operation.
4.	Knock-out	for forcing out the cut out blank from the die opening.
5.	Pressure pad	to hold the sheet in its position while drawing. This is to prevent wrinkling of work piece surface.

i) Stops

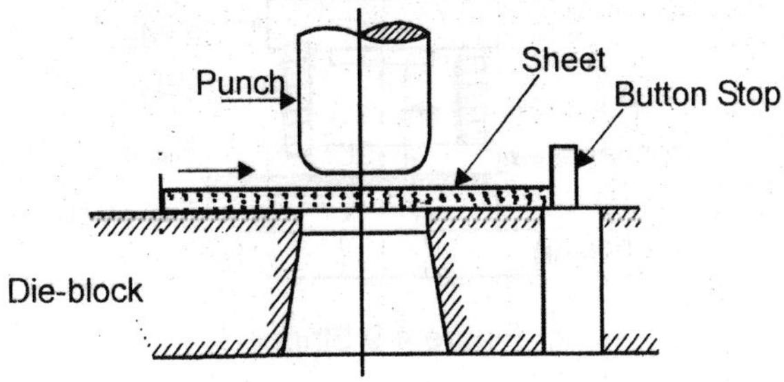

Figure 4.7 Stops

Stops are used to stop the sheet metal at the correct length. This prevents the wastage of sheet and reduces scrap. The stops may be made in the form of button or lever or pin. These stops also help to speed up the operation. (Refer figure 4.7).

ii) Pilot

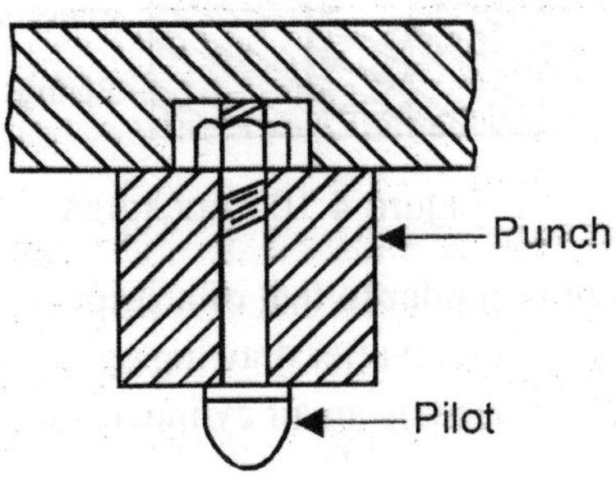

Figure 4.8 Pilot

Pilot is fitted to the lower end of the punch concentrically.

The function of the pilot is to locate the blank accurately with respect to the previously pierced hole for succeeding blanking and other operations. (Refer figure 4.8).

The correct spacing and position of the sheet is obtained without using stops. The pilot will have a sliding fit with the already pireed hole in the sheet.

iii) Strippers

The main function of the stripper is to remove the workpiece from the punch or die after a blanking or piercing operation. Strippers may be of fixed type or spring operated type. Fixed strippers are rigidly attached to the die block. The spring operated strippers travel up and down with the shank of the punch.

When the punch moves down, the springs compress the stripping plate after it comes into contact with the sheet metal. So the stripper plate holds the sheet. After the end of the operation, the punch goes up. The workpiece is stripped off from the punch cutting edge by the stripper plate. This prevents the workpiece from being lifted along with the punch. Thus the strippers knock out the work or scrap and speed up the operation. (Refer figure 4.9).

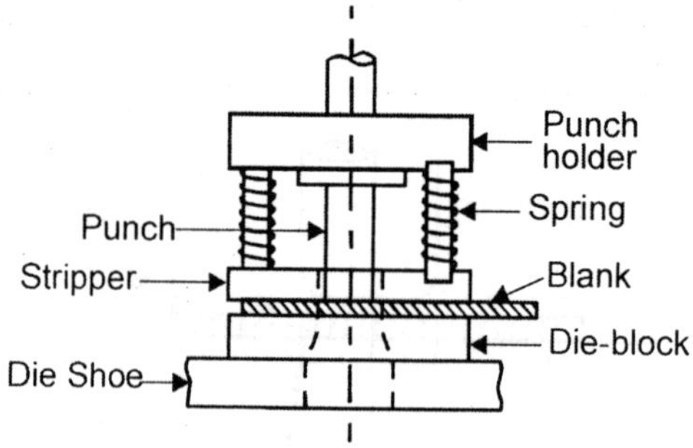

Figure 4.9 Strippers

iv) Knockouts

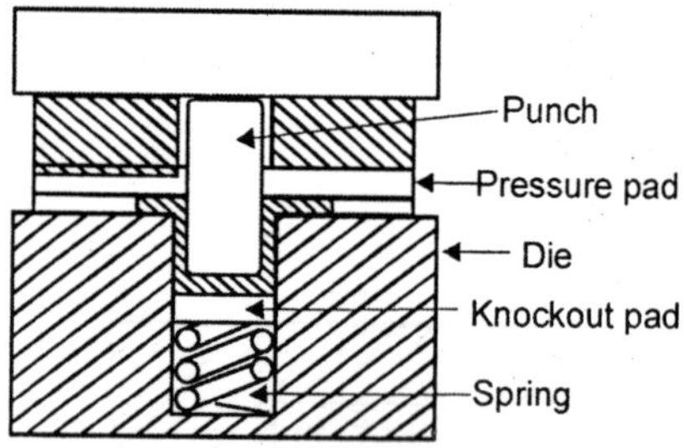

Figure 4.10 Knockouts

The function of the knockout is similar to that of a stripper. It forces the cut blank out of the die. Knockout is used to kick out the flanged shell after drawing operation which can not pass through the die. Knockout may be spring loaded or actuated by an air cylinder. (Refer figure 4.10).

v) Pressure pad

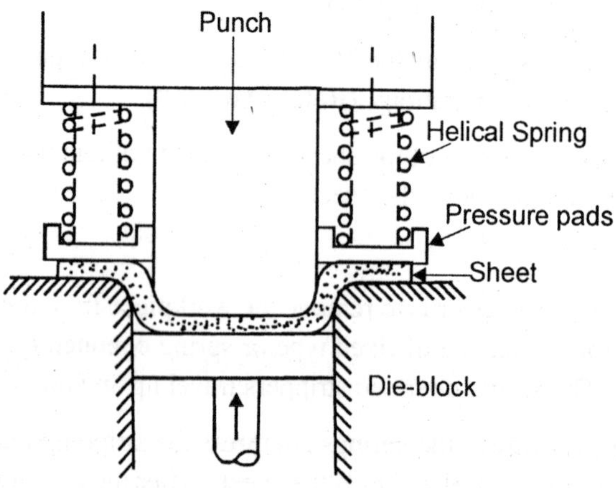

Figure 4.11 Pressure pad

Pressure pad is used for applying sufficient pressure on the metallic sheet and to hold it in its position. It prevents wrinkling during drawing process. The pressure may be applied by springs, pneumatic or hydraulic means.

When the punch moves downward, the springs are compressed and hence holds the sheet with sufficient pressure. This results in ironing of the metal as it flows plastically between the punch and die and eliminates wrinkling. To maintain the flat surface at the bottom, similar spring loaded or hydraulic loaded pad is used at the bottom of the cup. (Refer figure 4.11)

4.6 TYPES OF DIES

The sheet metal dies can be broadly categorized into three types;

i) Progressive dies,

ii) Compound dies and

iii) Combination dies.

4.6.1 Progressive dies

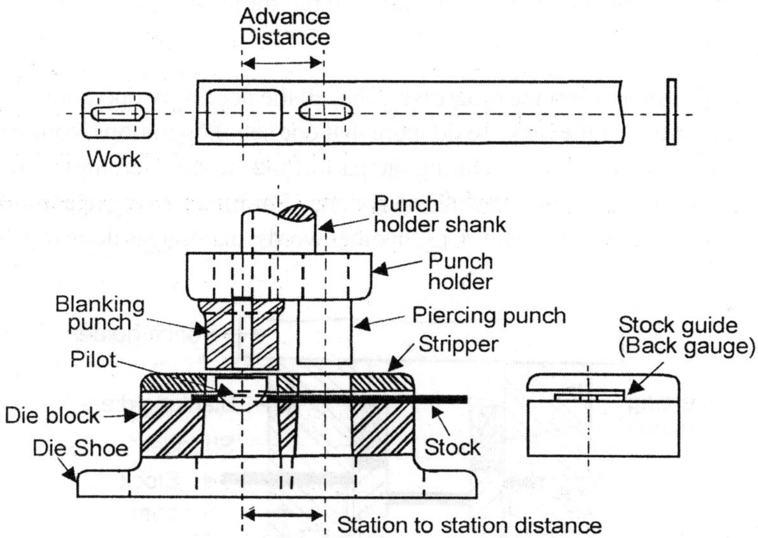

Figure 4.12 Pressure dies

The progressive dies perform two or more operations simultaneously in a single stroke of a punch press, so that a complete component is obtained for each stroke. The place where each of the operations are carried out are called stations. The stock strip moves from station to station undergoing the particular operation. When the strip finally leaves the last station, a finished component is ready. An example is presented in figure 4.12.

At the start of the operation, the sheet is fed into the first station. After undergoing the operations at this station the ram of the press moves to the top and the stock is advanced from the first station move to the second station, while a fresh portion of the stock comes under the first station. The distance moved by the strip from station one to two so that it is properly registered under the stations is called advance distance.

This advance distance should be the same between any two stations in sequence. Another variable called the feed distance is the amount of stock fed under the punch when ram comes for the next stroke. The feed distance may or may not be the same as the advance distance. The reason for the variance between the feed distance and advance distance, that is sometimes observed is because, often the sheet is overfed against a stop. The strip is, there-fore positioned correctly under the punch by pulling it backwards with the use of pilots. This helps in preventing the stop from interfering with the part geometry.

Progressive dies contain large numberof stations. It is generally preferred to have piercing operation first in the sequence and a blanking or cutoff operation in the end to get the final component. Any of the pierced hole may be advantageously used as pilot holes if none of the pierced holes are satisfactory as pilot holes, then special pilot holes may have to be made in the scrap part of the stock. It is also feasible to have blanking done in the first station, and the blank returned on to the die by means of spring plates and then carried for subsequent operations.

The choice of progressive dies is made only when, the production is of large numbers so that the handling costs are saved; stock material is not very thin, so that movement of the strip by pilots is convenient; stock material is not too thick so as to avoid the problems of stock straightening; and the overall size of the die or the press capacity are large and no available press in the shop can be utilized.

4.6.2 Compound dies

In a compound die, as distinct from the progressive die, all the necessary operations are carried out at a single station, in a single stroke of the ram. To do more that one set of operations, compounds die consists of the necessary sets of punches and dies. During the part of the stroke, piercing of holes is done in the stock and upon further travel, the blanking operation is done. For the blanking operation, the punch used for piercing becomes a die as shown in figure 4.13. In other words, blanking is done in a direction opposite to that of piercing.

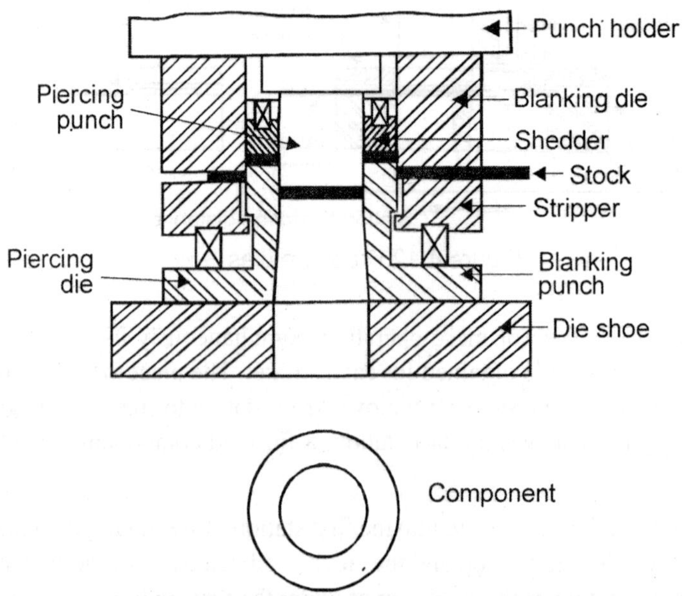

Figure 4.13 Compound die for making a washer

Compound dies are somewhat slower than the corresponding progressive dies in operation. But higher tolerances can be achieved in them than in the progressive dies. This is mainly because the part located in one position undergoes all the operations. Also in compound dies, small strips can be advantageously used, whereas in progressive dies very long strips are required to cover all the stations.

4.6.3 Combination Dies

A combination die is same as that of a compound die with the main difference that here non cutting operations such as bending and forming are also included as part of the operation. Often the nomenclature compound and combination are interchangeably used for the same type of die.

4.7 SHEARING OPERATIONS

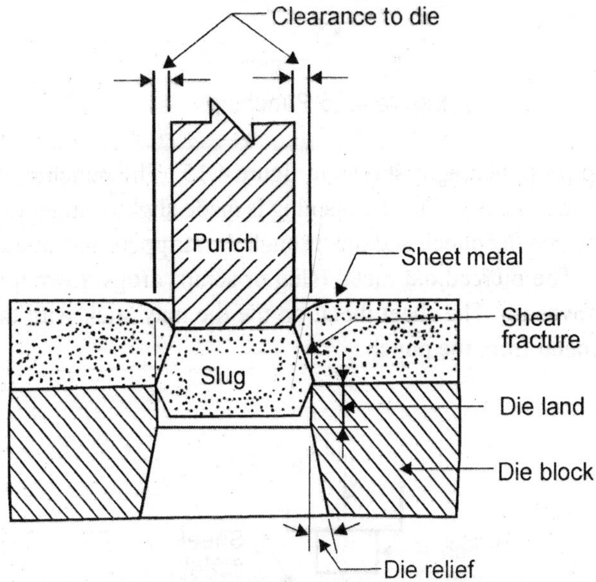

Figure 4.14 Shearing Operations

In shearing operation, the punch moves down and presses the workpiece in the opening of the die. There is a clearance between the die opening and punch. The space between the die opening and the punch is the clearance. The amount of clearance depends upon the type and thickness of material. By this pressure, the workpiece metal is deformed plastically. This plastic deformation takes place in a small area between the cutting edges of punch and die. So the metal in this area is highly stressed. When the stress exceeds the ultimate strength of the material, fracture takes place. The cutting edge of the die starts the fracture in the metal from the bottom. The cutting edge of the punch starts the fracture of the sheet metal from the top. These fractures meet at the centre of the plate. As the punch continues to move down, the metal under the punch is completely cut off from the sheet metal. The cut out portion to the die opening of the metal drops down through the die opening. To make the metal to drop down freely, a die relief is given. (Refer figure 4.14)

4.7.1 Punching

Punching is a piercing operation. It is the operation of producing a circular hole in a sheet metal using a punch and die. The material pushed out through the die is the scrap.

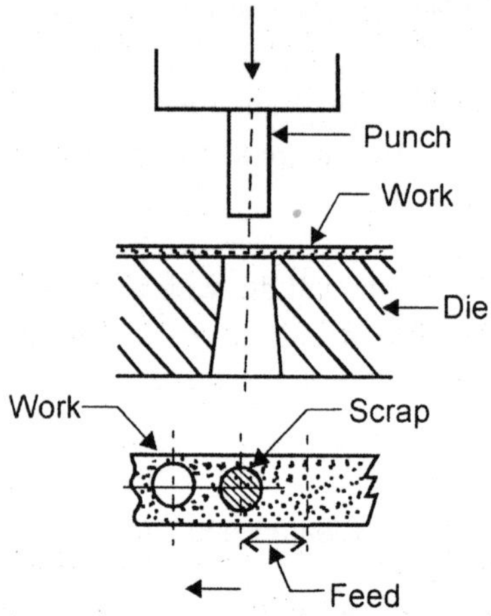

Figure 4.15 Punching

The punch and die set-up for punching is shown in figure 4.15. The punching size will be exactly the same as the size of the hole to be pierced. The die opening is made slightly larger to get the clearance. The sheet metal is introduced between the punch and die through the stripper. The punch pieces the hole in the metal when it moves down. The pierced out metal (slug or scrap) drops down through the die opening. After piercing, the punch moves up. The sheet metal on the die may stick with the punch surface. The stripper strips off the sheet metal from the punch.

4.7.2 Blanking

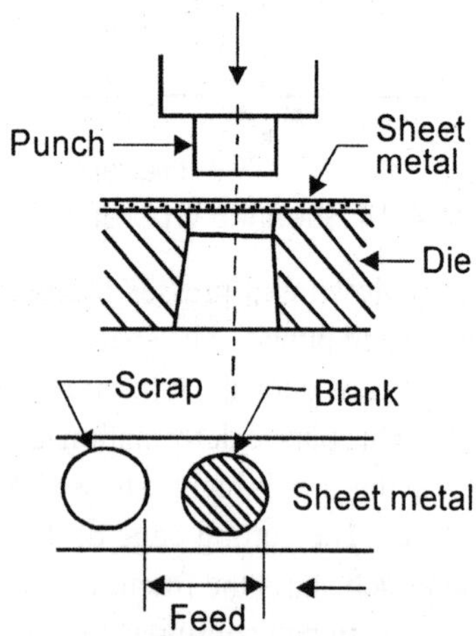

Figure 4.16 Blanking

Blanking is the operation of cutting off a flat sheet of required shape. The metal blanked out through the die is the required product. The sheet metal left on the die is the scrap. The blank is further processed-bending or drawing is done on the blank. The size of the blank depends on the size of the die. So the size of the die opening is equal to the blank size. Clearance is given to the punch. (Refer figure 4.16)

4.7.3 Cutting Off

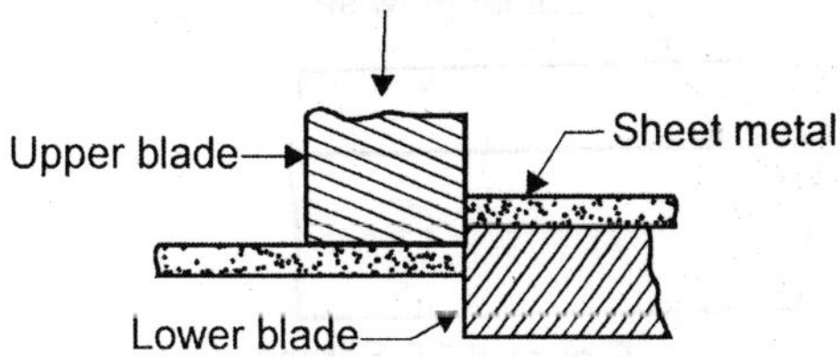

Figure 4.17 Cutting off

Cutting off is the operation of cutting a piece from a sheet metal. The cut is made along a straight line or a curve. The lower blade is fixed to the machine frame. The upper blade is connected to the ram. It slides vertically. The workspiece is placed between the two blades. When the upper blade moves down, it cuts off the sheet metal. A slight clearance is given between the cutting edges of the blades. The clearance depends upon the thickness of the workpiece. In cutting off; there is no scrap. (Refer figure 4.17)

4.7.4 Parting

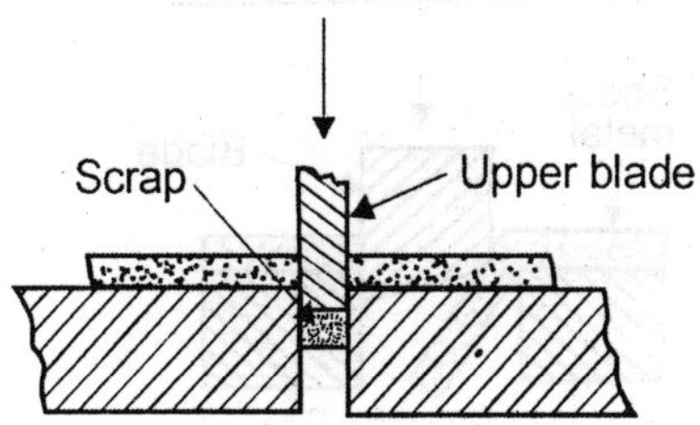

Figure 4.18 Parting

Parting is the operation of cutting the sheet metal into two pieces. There are two lower blades fixed to the machine frame. A gap is provided between the cutting edges of the lower blades. This gap depends upon the thickness of sheet metal. The gap can be adjusted. The upper blade is connected to the ram. The workpiece is placed between the upper and lower blades. The upper blade moves down and cuts the sheet metal. Some scrap is removed in this operation. (Refer figure 4.18)

4.7.5 Notching

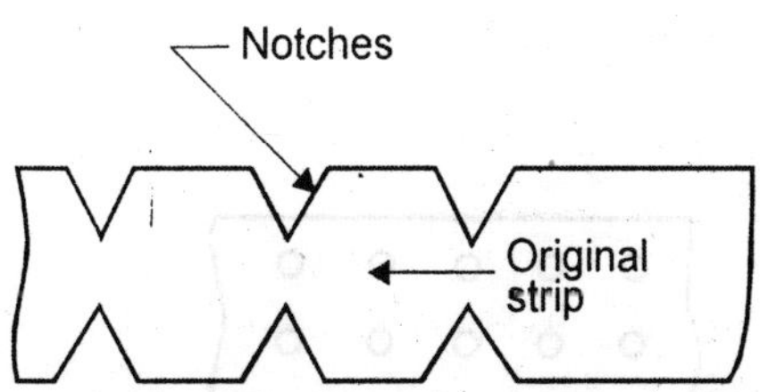

Figure 4.19 Notching

It is the operation of cutting small notches at the edge of the sheet metal. A notching die and punch are used. (Refer figure 4.19)

4.7.6 Slitting

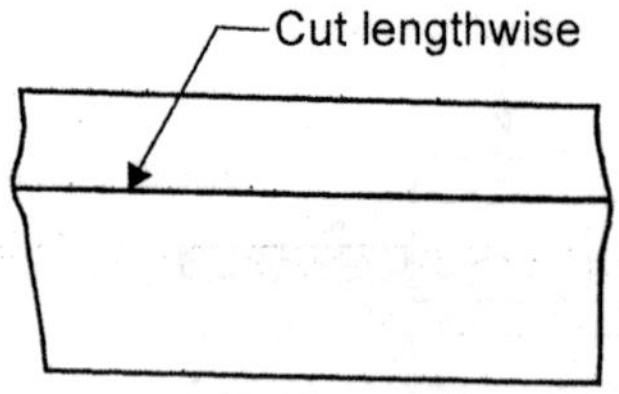

Figure 4.20 Slitting

It is the operation of cutting a sheet metal in straight line. The cut takes place along the length of the sheet metal. (Refer figure 4.20)

4.7.7 Lancing

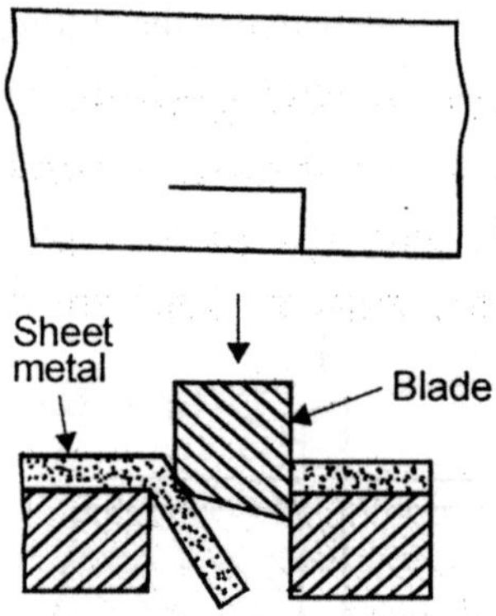

Figure 4.21 Lancing

Lancing is the operation of cutting a sheet metal thorough a small length and then bending the cut portion. (Refer figure 4.21)

4.7.8 Trimming

During drawn operations, the blank is held over the die by a holder. Impressions are formed on the area gripped by the pressure pad or holder. This unwanted metal surface should be cut off. Cutting off of the excess metal edge is called trimming. Dies and punches similar to blanking dies are used for trimming operation.

4.7.9 Perforating

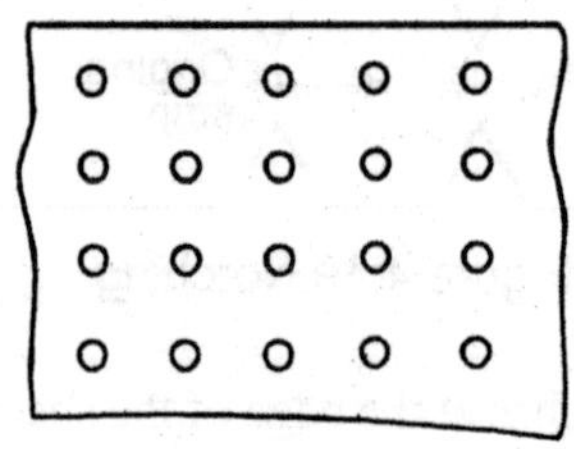

Figure 4.22 Perforating

Performating is a process of producing a number of small holes evently spaced in a regular pattern on a sheet metal. The dies and punches used here are similar to that used in punching operation. (Refer figure 4.22)

4.7.10 Shaving

The edges of components produced in sheet metal operations may not be smooth. They may have burrs and irregularities. These edges are finished by a shearing operation. This operation is called shaving.

The component to be finished is located over a die. A shearing punch is used to shear off the burrs.

4.8 BENDING OPERATIONS

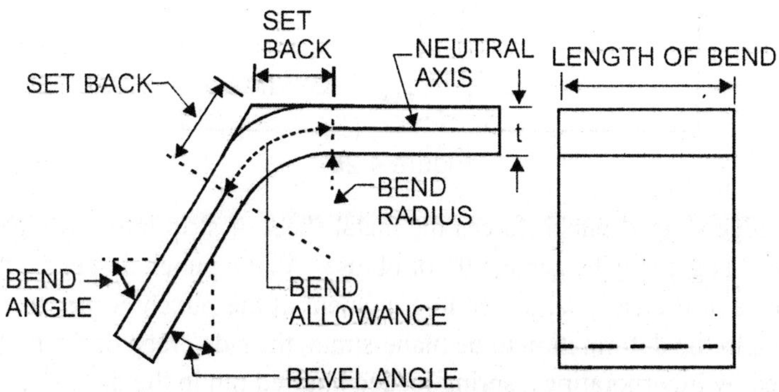

Figure 4.23 Bending Terminology

Bending is the metal working process by which a straight length is transformed into a curved length. It is a very common forming process for changing sheet and plate into channels, drums, tanks etc. During the bending operation, the outer surface of the material is in tension and the inside surface is in compression. The strain in the bent material increases with decreasing radius of curvature. The stretching of the bend causes the neutral axis of the section to move towards the inner surface. In most cases, the distance of the neutral axis from the inside of the bend is 0.3 t to 0.5 t, where 't' is the thickness of the part. Bending terminology is illustrated in figure 4.23.

Bending Methods: The three bending methods commonly used are : V-bending, Edge bending, and U-bending (channel bending).

4.8.1 V-Bending

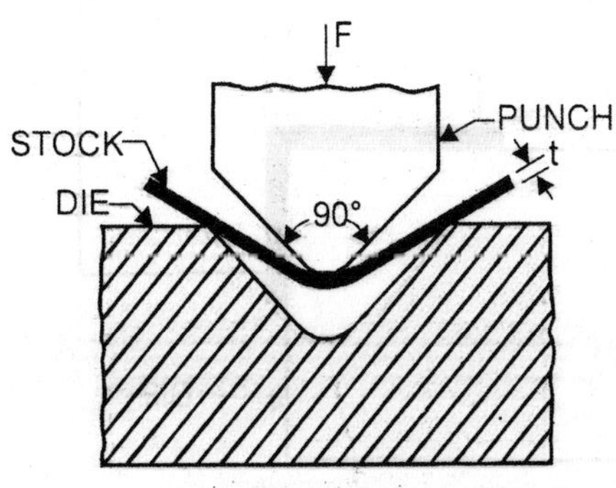

Figure 4.24 V-Bending

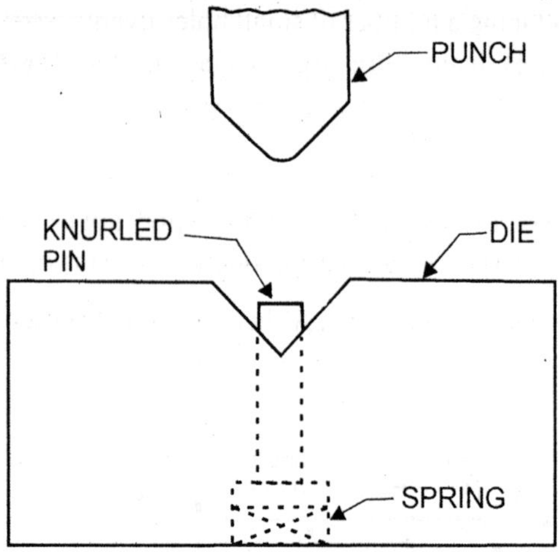

Figure 4.25

In V-bending, a wedge-shaped punch forces the metal sheet or strip into a wedge shaped die cavity, (Figure 4.24). The bend angle may be acute, 90°, or obtuse. As the punch descends, the contact forces at the die corner produce a sufficiently large bending moment at the punch corner to cause the necessary deformation. To maintain the deformation to be plane-strain, the side creep of the part during its bending is prevented or reduced by incorporating a spring loaded knurled pin in the die. The friction between the pin and the part will help to achieve this, (Figure 4.25). Plane strain conditions will also be established in the centre of the sheet if its width is more than 10 time its thickness.

4.8.2 Edge Bending

In edge bending, Figure 4.26 a flat punch forces the stock against the vertical face of the die. The bend axis is parallel to the edge of the die and the stock is subjected to cantilever loading. To prevent, the movement of the stock during bending, it is held down by a pressure pad before the punch contacts it.

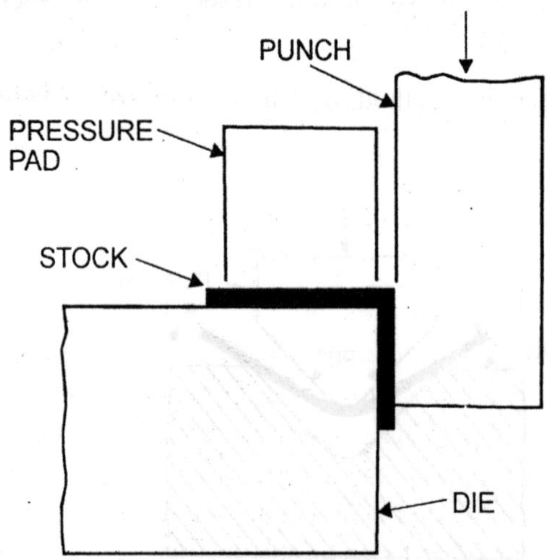

Figure 4.26 Edge bending

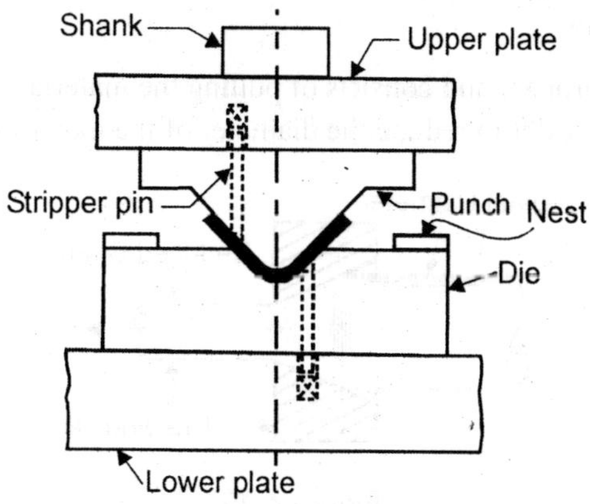

Figure 4.27 V-bending die

A V-bending die and a U-bending die are shown in figure 4.27 and figure 4.28 respectively. Stripper pins are provided to shed the part from the punch.

4.8.3 Spring back

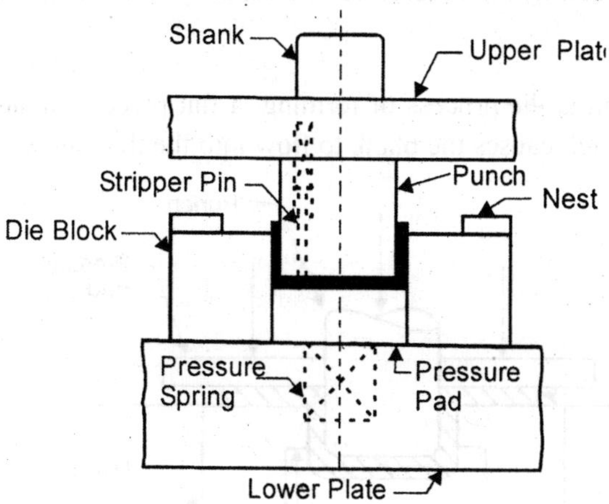

Figure 4.28 U-bending die

At the end of the bending operation, when the pressure on the metal is released, there is an elastic recovery by the material. This causes, a decrease in the bend angle and this phenomenon is termed as spring back. For low carbon steel, it can be 1° to 2° and for medium carbon steel, it can be 3° to 4°, for phosphor bronze and spring steel the spring back can be from 10 to 15°. To compensate for spring back, the wedge-shaped punches and the mating dies are made with included angles somewhat less than required in the formed component. Due to this, the component will be bent to a greater angle than desired, but it will spring back to the desired angle. For other types of bending, the part is overbent by an angle equal to spring-back angle by having the face of the punch undercut or relieved. The values of spring back given above are for 90° bends and are usally greater for greater angles.

4.9 DRAWING OPERATIONS

Drawing is a cold working process and consists of pulling the material to be drawn through a hole in a hard steel block, called a die in order to reduce the diameter of the materials as shown in figure 4.29.

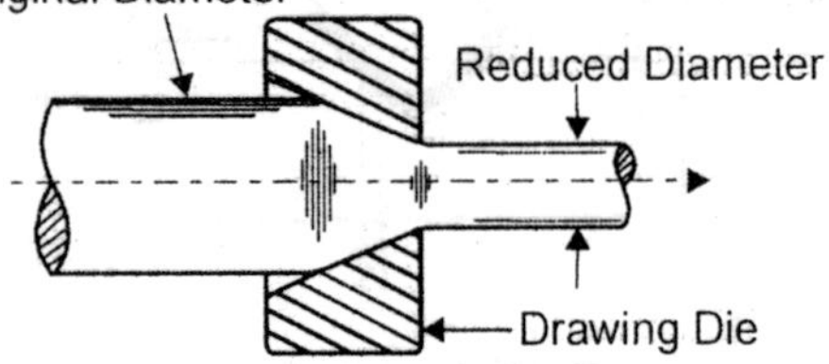

Figure 4.29

Drawing is used to produce bright drawn bars and solid drawn tubes, and is the main process of manufacture for the production of wires.

For cold drawing purposes, the material should be highly ductile. In cold drawing since the material being drawn becomes hard, it is given intermediate annealing.operations between successive drawing stages in order to restore the ductility and, therefore, the capacity of the material for further cold work.

4.9.1 Deep Drawing

Deep Drawing operation is the process of forming a flat piece of material (blank) into a hollow shape by means of a punch which causes the blank to flow into the die-cavity, Figure 4.30.

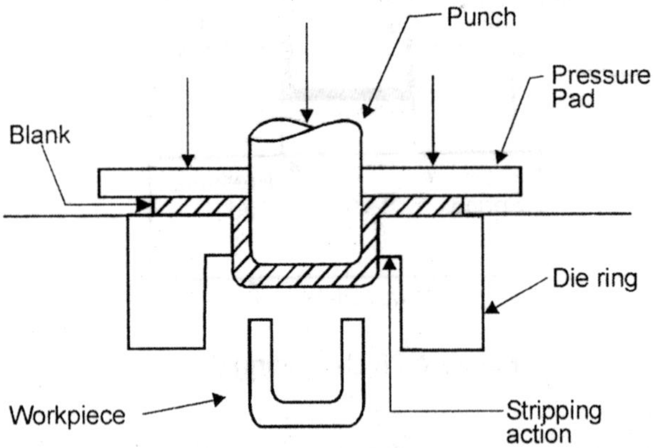

Figure 4.30 Deep drawing operation

The depth of draw may be shallow, moderate or deep. If the depth of the formed cup is upto half its diameter, the process is called "Shallow drawing". If the depth of the formed cup exceeds the diameter, it is termed as "Deep Drawing". Parts of various geometries and sizes are made by drawing operation, two extreme example being bottle caps and automobile panels.

As the drawing progresses, i.e. as the punch forces the blank into the die cavity, the blank diameter decreases and causes the blank to become thicker at its outer portions. This is due to the circumferential compressive stresses to which the material element in the outer portions is subjected. If this stress becomes excessive, the outer portions of the blank (flange) will have the tendency to buckle or wrinkle. To avoid this, a pressure pad or blank holder is provided. The holding down pressure is obtained by means of springs, rubber pad, compressed air cylinder or the auxiliary ram on a double action press.

The portion of the blank between the die wall and the punch surface is subjected to nearly pure tension and tends to stretch and become thinner. The portion of the formed cup, which wraps around the puch radius is under tension in the presence of bending. This part becomes the thinnest portion of the cup. This action is termed as 'necking' and in the presence of unsatisfactory drawing operation, is usually the first place to fracture.

As noted above, the outer portions of the blank under the blankholder become thicker during the operation. When these portions are drawn into the die cavity, 'ironing' of this section will occur if the clearance between the punch die is not enough to accommodate this increased thickness of the workpiece. This ironing is useful if uniform thickness of the product is desired after the drawing operation.

4.9.2. Redrawing

In deep drawing, the percentage reduction in one draw is defined as:

$$\% \text{ reduction} = \frac{D-d}{D} \times 100$$

now $$\frac{D}{d} = 1.6 \text{ to } 2.3$$

$$\therefore \frac{d}{D} = 0.435 \text{ to } 0.625 = 0.50 \text{(average)}$$

∴ The average reduction in deep drawing

$$= \left(1 - \frac{d}{D}\right) \times 100 = 50\%$$

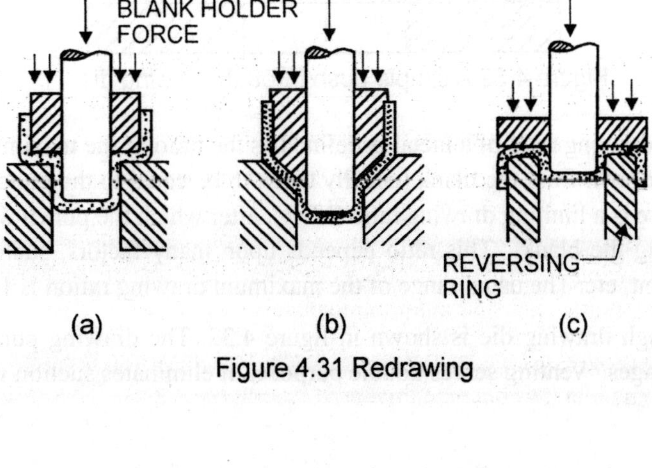

Figure 4.31 Redrawing

To make tall cups of smaller diameter (Such as cartridge cases and closed-end tubes), it is necessary to use successive drawing operations. Reducing a drawn cup to a smaller diameter and increased height is known as "Redrawing". The two methods used for redrawing are : direct redrawing and indirect or reverse redrawing.

Figure 4.31(a and b) are direct redrawing methods. In figure 4.31(a), the metal bends twice inducing high strain hardening in it. In figure 4.31(b), the metal again bends twice, but the angle through which it bends is less than 90°. This reduces strain hardening as compared to figure 4.31(a) and punch load is reduced. But the drawback is that the initial drawn cup must be made with a taper corner.

In reverse drawing (Figure 4.31(c)), the basic phenomenon that a cold-worked material exhibits greater ductility when the deformation direction is reversed in successive operations (strain softening) is made use of.

Redrawing is extensively used for food containers, fountain-pen caps, oil filter housings and shock absorber pistons etc.

Practical reductions in diameter for successive redraws are: 30,25,16 and 13%. Metal becomes work-hardened as it is drawn and redrawn and must be annealed to prevent failure before it reaches its limit.

4.9.3 Deep Drawability

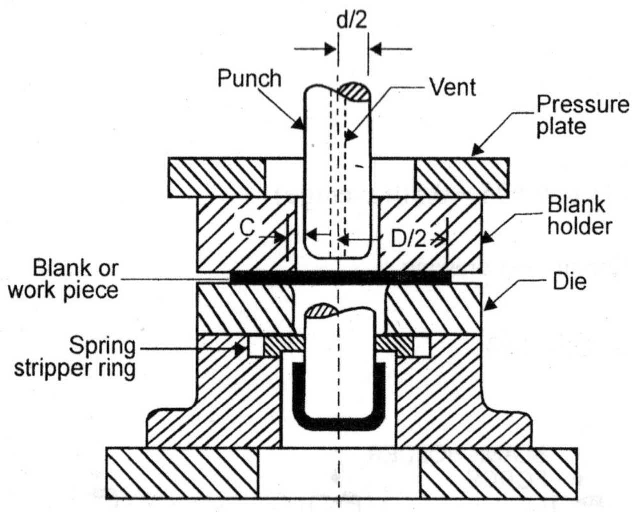

Figure 4.32 A simple push through drawing die

Deep drawability or drawing ratio of a metal is defined as the ratio of the maximum blank diameter to the diameter of the cup drawn from the blank (usually taken to be equal to the punch diameter), i.e. *D/d*. For a given material there is a limiting drawing ratio (LDR), after which the punch will piece a hole in the blank instead of drawing the blank. This ratio depends upon many factors, such as type of material, amount of friction present, etc. The usual range of the maximum drawing ration is 1.6 to 2.3.

A simple push through drawing die is shown in figure 4.32. The drawing punch must be properly vented with drilled passages. Venting serves double purpose. It eliminates suction which would hold the cup on the punch and damage the cup when it is stripped from the punch. Secondly, venting provides passages for lubricants.

As already discussed, the presses used for deep drawing are : double action cam press or double action toggle press. Many presses used for deep drawing operations are hydraulically operated and the these presses have an additional hydraulic cylinder and piston for the additional slide.

Product applications of deep drawing process are : Cups, shells, short tubes, automotive bodies, gas tanks, appliance covers and household hardware (spoons, utensils etc.,) beverage cans, seamless can bodies.

4.9.4 Ironing

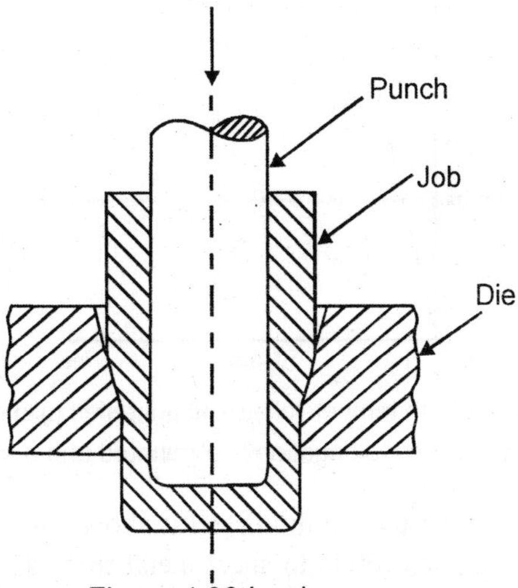

Figure 4.33 Ironing

It is the process of smoothening or thinning the wall of a shell or cup (cold or hot) by forcing the shell through a die with a punch. If the thickness of the sheet is more than the clearance between the punch and die, the thickness will get reduced. This is known as ironing and it produces cups with thinner walls than the base, Figure 4.33 as in the case of beverage cans and household saucepans.

4.10 STRETCH FORMING OPERATIONS

This process is used for forming of smoothly contoured parts, or those having double curvatures on the same curved surface, out of large and thin metal sheets. The operation can be performed in two ways. One method is to grip the two edges of the metal sheet firmly and stretch the same over a fined from block. This sheet is gripped on both sides by two jaws. These jaws are attached to the rods of pistons which work inside respective hydraulic cylinders. By providing pressure on these pistons the jaws can be pulled away and tensile forces provided on the sheet. At the same time the cylinders can be moved in guides along a predetermined path to stretch the sheet along the contour of the form. In another method, the jaws are mounted on two sides which are moved apart horizontally to provide tension on the seat. The form is attached to the ram of a hydraulic press which moves upwards. The combination of these two movements enables the desired forming of the sheet.

4.11 FORMABILITY OF SHEET METALS

In recent years there has been an unprecedented interest in the formability of metals. This is due in part to the continuing demands on industry to product lighter weight, yet stronger, more rigid components. Ever higher strength-to-weight ratios are being used in forming parts for missiles, aircraft, and more recently for automobiles. As new metal are tried, such as aluminum alloys and high-strength, low-alloy steels, they are found to be less forgiving than mild steels. Higher forming forces are required, which often result in ore fractures.

4.11.1 Uniaxial Tensile Test

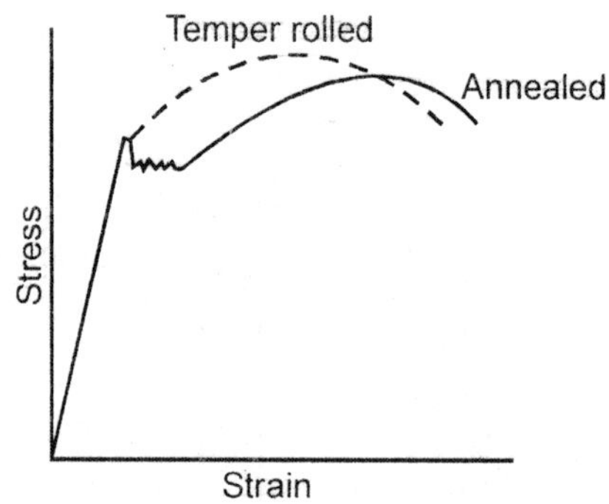

Figure 4.34 Stress-strain curve for both temper-rolled and annealed mild steel sheet. When a material has been temper passed sufficiently, the yield-point elongation is eliminated and a smooth stress-strain curve will develop.

A stress-strain curve of a temper-passed mild steel is shown in comparison to an annealed mild steel sheet in figure 4.34. Temper-passed refers to sheet metal that has been lightly cold-rolled to improve flatness, minimize stretcher strains, and obtain the desired texture and mechanical properties. Note that the annealed sheet shows a yield-point elongation. It is this yield-point elongation that produces the stretcher strains in sheet metal forming. The stretcher strains, often referred to as Luder's lines, are irregular surface patterns or ridges and valleys that develop during the forming of annealed or aged steel (Figure 4.35).

Aging is a term applied to changes that take place in the mechanical properties of temper-rolled sheets with the passing of time. Sheets become less ductile, partly because of the dissolved nitrogen in the metal. Aluminum-killed steels (AK) have very little dissolved nitrogen, the nitrogen being combined with the aluminum in the form of stable nitrides, and thus have practically no susceptibility to strainaging.

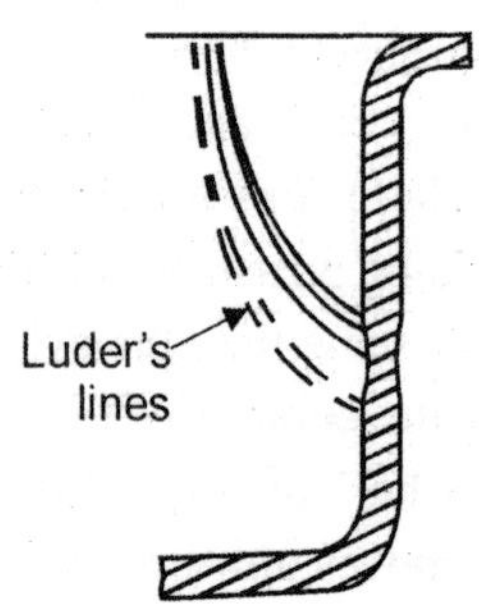

Figure 4.35. Luder's lines develop during sheet metal forming operations because of yield-point elongation.

When a material has been temper-passed sufficiently, the yield-point elongation is eliminated and a smooth stress-strain curve will develop, as shown in figure 4.35.

Several properties associated with the stress-strain curve, such as yield stress, ultimate tensile strength, and total elongation, have been used to evaluate formability. Yield stress has the disadvantage of being sensitive to speed of deformation. It increases with increasing speed. Tensile tests are run at speeds lower than production stamping operations, and therefore the measured yield stress will be lower than that encountered in stamping. The method of specimen preparation also affects the yield stress. Mechanical properties such as ultimate tensile strength and total elongation are also very sensitive to test conditions, surface scratches, and specimen geometry.

4.11.2 Strain-hardening Coefficient

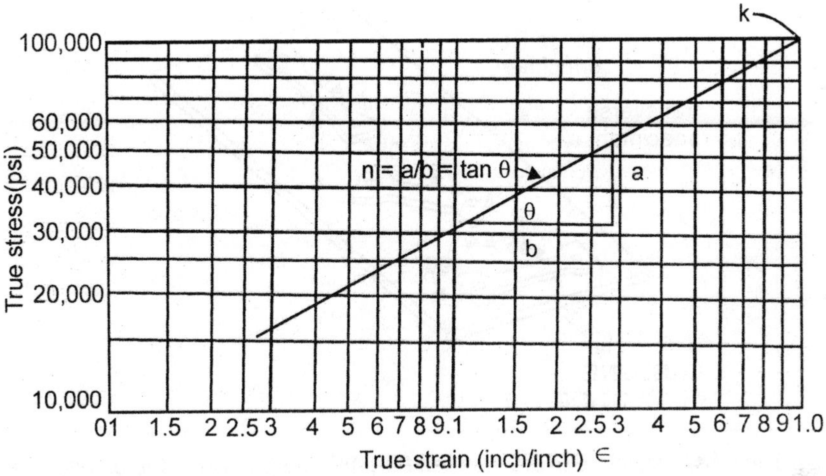

Figure 4.36 The true stress-strain curve is plotted on log-log paper. The n value is the slope of the line a/b and represents the strain-hardening coefficient.

Forming limits relate primarily to the exhaustion of ductility. To achieve better formability, a material must be capable of straining uniformly.

Research has shown that two properties strongly influence formability: the strain-hardening coefficient, or n value, and the anisotropy coefficient, or r value. The n value determines the ability of the material to be uniformly stretched. The K value represents the thickness-to-width strain.

Figure 4.36 illustrates a true stress-strain curve plotted on log-log paper. The n value represents the slope of the line a/b. The n value also shows the rate of work-hardening that took plays in the metal as it was being strained.

Strains describe the actual dimensional changes in a sheet. Strain hardening refers to the fact that, as a metal deforms in some area, dislocations occur in the microstructure. As these dislocations pile up, they tend to strengthen the metal against further deformation in the area. Thus the strain is spread throughout the sheet. However, at some point in the deformations, the strain suddenly localizes and necking, or localized thinning, begins. When this occurs, little-further overall deformation of the sheet can be obtained without its fracturing in the necked region.

Thus the strain-hardening coefficient reflects how well the metal distributes the strain throughout the sheet, avoiding or delaying localized necking. The higher the strain-hardening coefficient, the more the material will harden as it is being stretched and the greater will be the resistance to localized necking. Necks in the metal may affect structural integrity or harm surface appearance.

For most practical stamping operations, stretching of the metal is the critical factor and is dependent on the strain-hardening coefficient. Thus stampings that entail much drawing should be made with metal having high average strain hardening coefficients.

4.11.3 Normal anisotropy coefficient

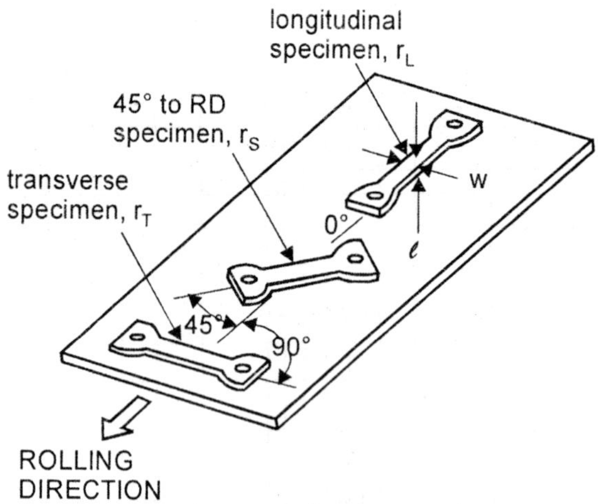

Figure 4.37 The tensile specimens are oriented 90°, 45° and parallel to the direction of rolling for determining the average coefficient.

A single-phase metal who microstructure and mechanical properties are the same in all directions is termed isotropic. If a metal is continuously rolled, the grain structure is oriented in the direction of rolling and is termed anisotropic, as in the case of sheet metal. Tensile specimens cut from the rolled sheet, as shown in figure 4.37 will show different mechanical properties because of their different orientation to the metal rolling direction. These differences can have a significant effect in deep-drawing operations.

The ratio of the plastic width strain (T) to the longitudinal strain (L) is the r value and can be measured in a tensile test. A material that has a high plastic anisotropy also has a greater thinning resistance. In general, the higher the anisotropy coefficient, the better the metal deforms in drawing operations. Although the r values may vary in relating to the rolling direction, they may still average out at unity. This conditions is termed planar anisotropy and is one that would lead to earing. Earing refers to a localized elongation that takes place when deep drawing, as in deep drawing a cup.

When tensile speciments oriented at 90°, 45° and parallel to the direction of rolling are pulled, an average $\overline{r}$ value is obtained.

$$\overline{r} = \frac{r_L + 2r_{45} + r_T}{4}$$

If $\overline{r}$ is greater than unity, the material is charaterized as having resistance to thinning and has a good thickness strength throughout.

The following Table shows r values for several common metals used in forming.

Normalized steel	1.0
Rimmed steel	1.0-1.35
Aluminum-killed steel	1.35-2.00
Copper, brass	0.8-1.0
Lead	0.2
Hexagonal, close-packed metals	3-6+

4.11.4 Formability evaluation

Methods developed to evaluate the formability of sheet metal include Olsen cupping values, the Fukui cup test, the Ericksen deep-drawing test, the forming limit diagram, and the drawability tester.

(i) Olsen cupping values and fukui tests:

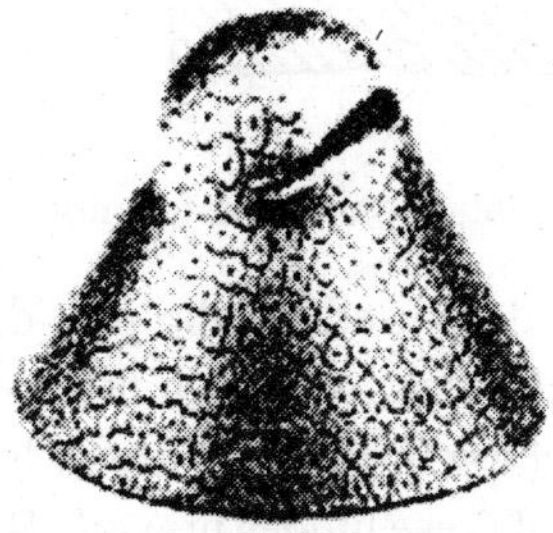

Figure 4.38 A Fukui cup test used to predict the formability of sheet metal. No blank holder is needed and wrinkles can be avoided by proper attention to the dimension ratios in the test disk.

The Olsen cupping values and Fukui tests are similar. In the Olsen test, a steel ball is pressed into the metal. Values are assessed as to the depth the ball can be pressed without fracture. In the Fukui test, a rounded-end punch is used along with a die (Figure 4.38). Values are assessed by measuring the distortion of the circles inscribed on the surface. In either case, no hold-down pressure is exerted on the surface of the metal surrounding the punch.

ii) Ericksen deep-drawing test:

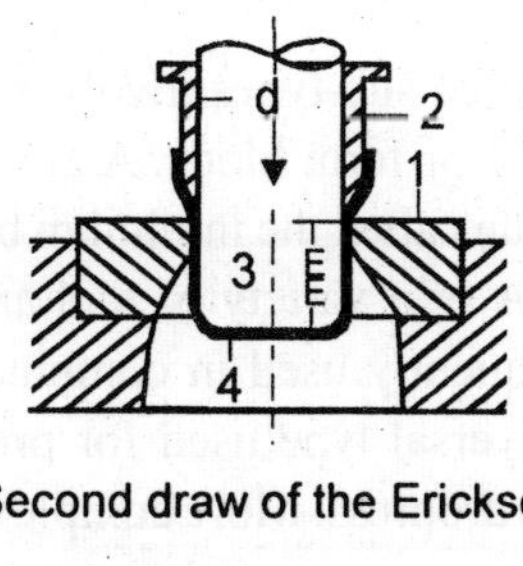

Figure 4.39 Second draw of the Ericksen deep-drawing cup test

The Ericksen test is based on how deep the metal can be drawn over a standard-type punch. It is also called the dome height test. Figure 4.39 shows an Ericksen test using a second or redrawing die. Formability is determined by the height of the final dome.

iii) Forming limit diagram (FLD):

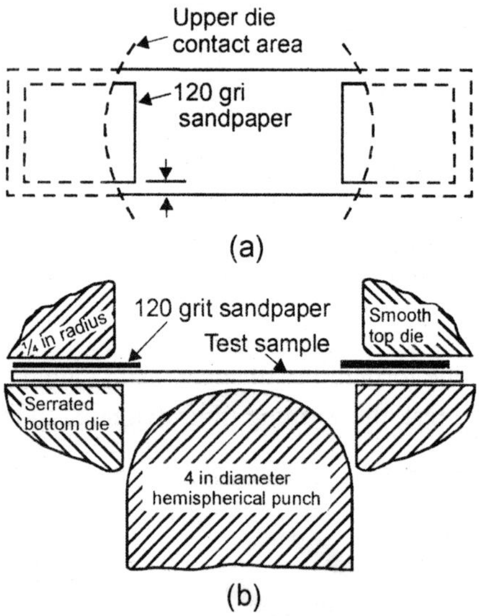

Figure 4.40 Test setup for establishing the forming limit diagram (FLD)

The forming limit diagram is method widely used for the past decade in the automotive industry to assess the formability of aluminum and steel sheet. Although the dome height test already discussed are useful in correlating formability to the stretching mode, they are of little value in assessing formability in bending and drawing or when combinations of these modes prevail. By contrast, the forming limit diagram characterizes formability over a wide range of forming conditions, including drawing, bending, and stretching. It also facilitates the solution of formability problems that might arise in production. The analysis involves primarily the construction and use of a forming-limit diagram, which indicates the material's local ductility for forming.

The tooling required to obtain the information for the diagram is a hemispherical punch, upper and lower die set and a means of securely clamping the sample of material being evaluated so that it can be deformed to the point of failure, as shown in figure 4.40(a) and (b).

4.12 FORMABILITY TESTS FOR FLANGED PARTS AND FOR STRAIGHT AND CURVED SECTIONS

The forming of singly curved parts of all three sub-types involves only the bending of a flat piece of metal of rectangular section over a radiused die or form block. A given blank in a given material can be bent without failure to radii above a certain value called the minimum bend radius. This can be determined with practically sufficient agreement by means of a variety of tool arrangements (Figure 4.41). For the bend test a 90 degree, V-shaped punch is frequently used in combination with one of various types of female die. It may be a square die of the universal type used for press-brake work, a more accurately forming, mating V-die, or the rubber pad of a hydropress. More complex tools, performing several bends in a single operation and also a bend of continuously varying bend radius, have been suggested. Either the bend radius, in terms of multiples of the metal thickness, or this multiple, or ratio between minimum bend radius and metal thickness is the forming limit for straight bending and flanging.

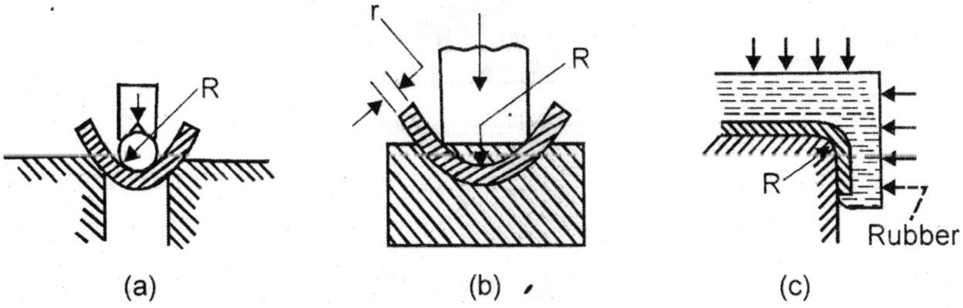

Figure 4.41. Bend Tests.

Minimum bend radii reported in design manuals generally relate to sheet or strip with a bend length that is at least 10 times the metal thickness. However, the narrower the section, the more sharply it can be bent, other conditions being identical. The minimum bend radii for square sections cut from plate, therefore, are considerably smaller than those for an extended length of plate of the same material and thickness. Reliable values are obtained only if the surface of the metal is smooth (particularly, if it is free from marks parallel to the bend line) and if the edges of the bent length are carefully deburred. Otherwise, the metal will crack at considerably greater bond radii. The minimum bend radius is generally expressed in multiples of the metal thickness. For many metals the minimum bend radius is proportional to the thickness of the sheet and strip. The minimum bend radius, however, may be different in the different directions of the sheet or strip surface, and this applies therefore, the minimum bend radii are generally given for the longitudinal direction or with the fiber, the bend line being perpendicular to the rolling direction; and for the transverse direction, or, across the fiber, with the bend line parallel to the rolling direction.

Contoured flanges are of two distinctly different basic types, namely, stretch flanges and shrink flanges. These are subjected to very different stresses and strains.

The stretch-flanging test consists of forming a 90-degree flange on a circular hole in a sufficiently large, usually square blank (Figure 4.42). The principal variation in the test is obtained by varying the hole diameter, D_0. One or more dies with circular holes having a diameter D_1 may be used. If these dies are used in combination with a press or with a cupping testing device of the Olsen or Erichsen type (which is a small, hand-operated press), several punches must be available for various sheet thicknesses to limit the clearance between the die and punch to a small value. The punch can also be replaced by the rubber in the platen of a rubber hydropress. The variations obtainable with this test can be expressed in terms of a theoretical stretch, *e*, which is defined as follows:

$$e = \frac{D_1}{D_0} - 1 = 100\left(\frac{D_1}{D_0} - 1\right) \text{ per cent}$$

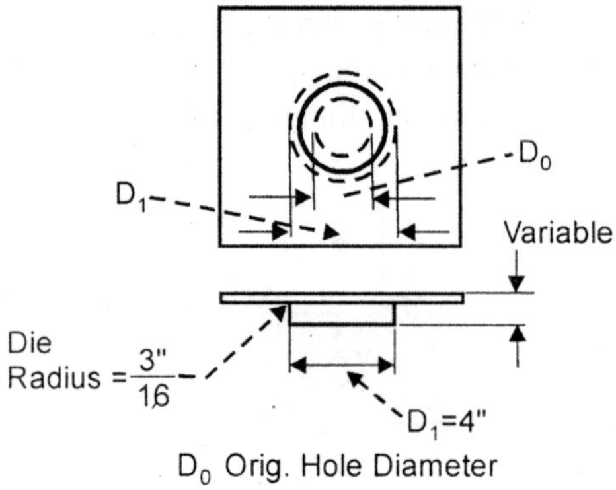

Figure 4.42. Test for stretch flanging.

This quantity corresponds approximately, but not exactly, to the circumferential stretch along the flange edge. The forming limit in stretch flanging is the maximum value of stretch that can be obtained without cracking or necking at the flange edge. To avoid cracking at the die radius this radius must exceed the minimum bend radius for straight bends; this also applies to stretch flanging.

The maximum stretch obtainable on stretch flanging a given metal depends primarily on three factors: (1) the ratio of flange diameter to metal thickness; (2) the length or angle of the circular segment subjected to flanging; and (3) the condition of the edge, whether deburred and polished, routed (milled), or sheared. To determine the effect of flange length the blank can be subdivided into several segments of different lengths by means of radial slots.

The shrink-flanging test consists of forming a flange on a circular disc or on the circular periphery of a segment of such a disc. The principal variation in this test is obtained by varying the diameter of the circular die, D_1, or radius of curvature at the die edge, $R_1 = D_1/2$, rather than by varying (only) the disc diameter D_0, or radius of curvature of the blank edge, $R_0 = D_0/2$. Because the flange length is another important variable in shrink flanging, the disc can also be subdivided, with advantage, into segments of different length by means of cutouts extending beyond the bend line (Figure 18.30). Test equipment for shrink flanging has been used to far only in conjunction with the rubber hydropress where the rubber acts as a punch. The strain in this test can be expressed as a theoretical shrink, *e*, (negative strain) as follows:

$$e = 1 - \frac{D_1}{D_0} = 100\left(1 - \frac{D_1}{D_0}\right) \text{ per cent}$$

The forming limit in stretch flanging is the maximum theoretical shrink that can be obtained without buckling; a slight wrinkling can either be tolerated or removed readily in a straightening operation. The maximum shrink that can actually be obtained in shrink flanges without wrinkling is very small. The test is therefore devoted primarily to the problem of determining the maximum flange length or spacing of cutouts that will permit a flange of given dimensions to be formed.

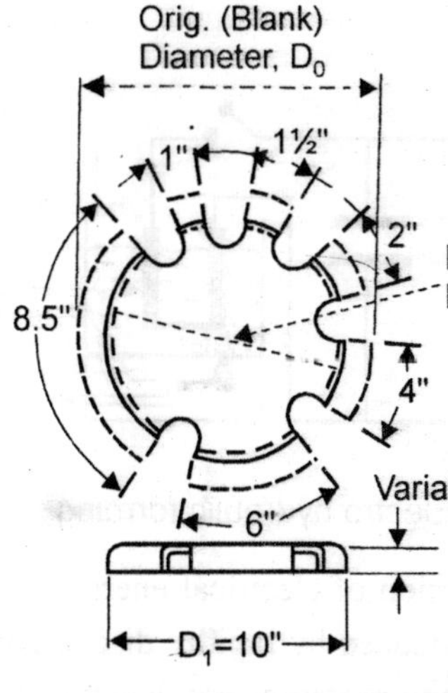

Figure 4.43. Test for shrink flanging.

The ease of contour forming a section cannot be predicted at presen on the basis of any simple formability test. Some research work has shown that the radii to which such parts can be curved vary within very wide limits, depending primarily on the cross-sectional contour and dimensions of such a part. In addition, the forming limits in contour forming also depend on the method of forming. Consequently, no simplified procedure has yet been devised which might reduce the work involved in the development of such parts. It may be stated, however, that fundamentally two types of failures must be considered in contour forming, namely, necking and fracturing at the outer or convex surface and buckling or wrinkling at the inner or concave surface.

4.13 WORKING PRINCIPLE AND APPLICATION OF SPECIAL FORMING PROCESSES

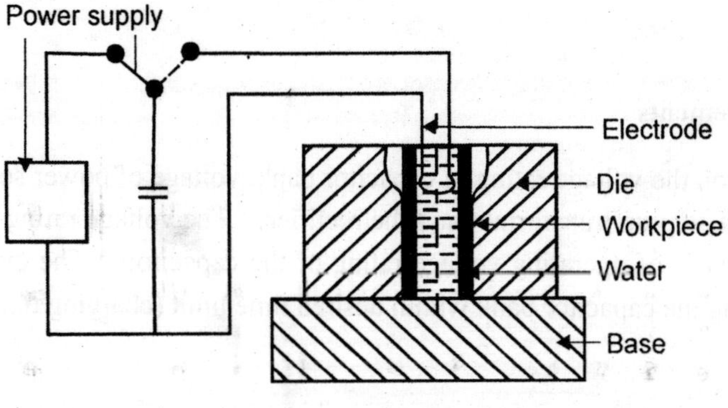

Figure 4.44 Working principle and application of special forming procecces

Electro-hydraulic forming is an unconventional forming process. In this process a high intensity shock wave is generated by discharging stored electrical energy across electrodes submerged in an electrolyte. A potential difference of 30 kV is discharged though a wire 1 mm diameter which is immersed in water. The vapouring action of the wire due to intense heat generation in turn creates high pressure shock wave. The process can be used for bulging, forging, drawing, blanking and piercing. Figure 4.44 shows the set-up.

4.13.1 Electro hydraulic forming (EHF)

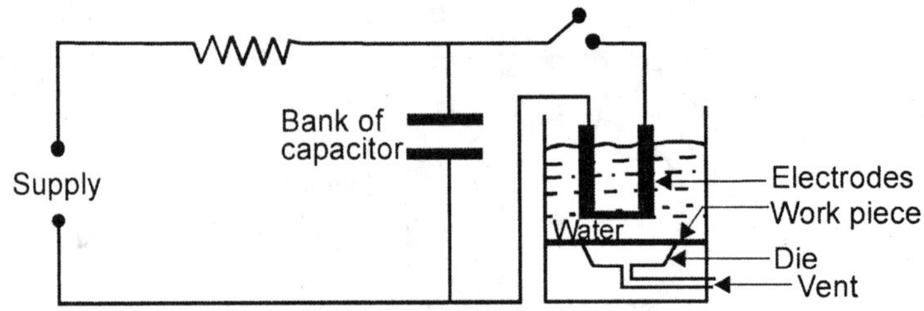

Figure 4.45 Electro hydraulic forming

Basically, this process involves the conversion of electrical energy to mechanical energy in a liquid medium. When a high energy discharge is produced by rapidly discharging a capacitor bank across an under water spark gap or wire, shock waves and pressures are developed and the same can be used for metal forming.

When capacitor discharges through a gap, a potential difference of 50000 volts jump an air gap of approximately 12 mm under water / liquid. The length of gap to affect a discharge depends on the voltage, conductivity of liquid and shape of electrodes. Voltage of the order of 10000 to 30000 is generally used when the spark discharge method is used. Refer figure 4.45.

When capacitors discharge through a wire, a better control and efficient energy conversion can be obtained, since the path of electrical discharge can be predetermined and shaped. Further lower voltage (4000-7000) can do the job. However, after each shot, loading of a new wire is required. Use of magnesium and aluminium of 0.75 mm to 1.5mm diameter is very effective.

Electrical requirements

(i) Power supply requirements

These are dependent on the voltage rating of capacitor bank, voltage of power source and the desired charging time. It needs a step-up transformer and the rectifiers. The voltage rating of the power supply must be chosen on the basis of operating voltage rating of the capacitors. The current rating must be enough to permit charging the capacitor bank within desired time limit (charging time 6 to 30 seconds).

(ii) Storage system

It needs a suitable bank of capacitors connected together with heavy conductors. The storage system size would depend on the amount of energy desired. One side of the capacitor must be connected through a triggering device to the gap.

(i) Application

The effects and deformation - rates are very similar to explosive forming, however, the cost of storing the energy is relatively large. A 50 KJ set would give equivalent deformation results to 14g high explosive. Application of the process is therefore limited to small, thin components and is particularly suitable for small, medium batch work.

4.13.2 Electro-magnetic forming (EMF)

The process of electromagnetic forming is increasing in popularity in industry because of its relative inherent simplicity and safety. Briefly this process works in the following manner. (Refer figure 4.46).

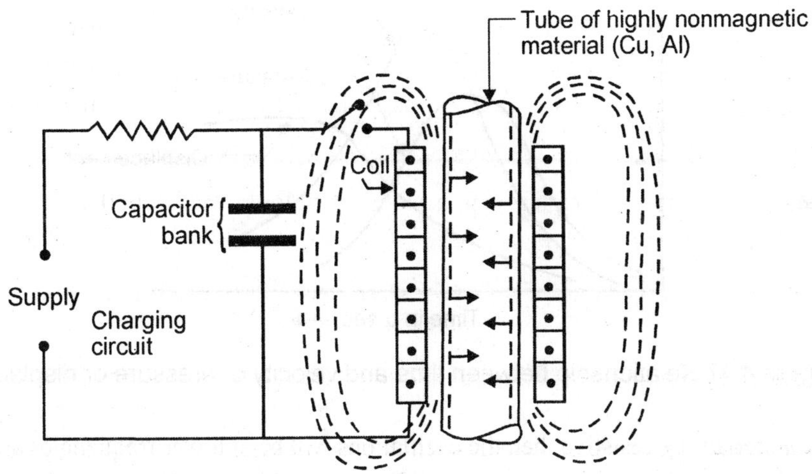

Figure 4.46 Electro Magnetic Forming

On discharging a capacitor bank a time varying primary current in the work coil sets up a transient axial magnetic field in the air gap between the coil and the work. Simultaneously, a secondary current is induced in a conducting workpiece, and owing to the high frequency involved, these currents tend to be confined to the skin of the workpiece. The interaction of the transient currents in the workpiece and the eddy current field induces radial inward forces in the work which can cause it to deform at a high rate. Using this method, it is possible to apply to a metallic workpiece a powerful, uniform impulse through the field of magnetic field. A pressure of 340 MPa for a period of microseconds is a typical of magnitude of impulse that is used. Since the magnetic field does interact with insulators, the magnetic impulse can be transmitted to the workpiece through the walls of non-conducting sheath, which may maintain the workpiece in a sterile, and an inert condition. These magnetic impulses have been used to swage and expand tubular forms, as well as to coin, shear, and form flat sheets. The deformation process can be completed in some thing of the order of 50 micro-seconds, and peak pressures of 200 MPa can be obtained. The figure 4.46 indicates the schematic arrangement of the process.

It is reported that the deformation may take place in one half to two cycles of the induced current depending on the damping of the system, but, generally in less than 90 micro-seconds. Figure 4.46 shows the relationship of time and other parameters, to indicate how much high production-rate is possible.

One of the major problems is the coil design, obviously very small coils are impractical on a strength basis. In practice, coils tend to be of a very small number of turns or even one turn, i.e. machined from solid tube with a slit along the axis. The success of magnetic forming depends on the magnetic pressure being greater on the 'work-coil' side of the workpiece than on the other side. This is only true when the system exhibits very high frequency transients. (Fefer figure 4.47).

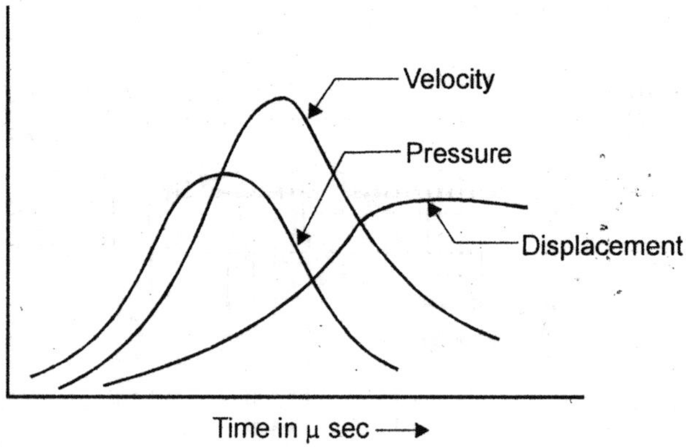

Figure 4.47 Relationship between time and velocity or pressure or displacement

If the work-materials are resistive then the oscillations will be at lower frequencies and skin effect will not occur. Consequently the magnetic pressure difference will be insufficient for deformation. To form stainless steel, it is necessary to either provide a separate high frequency supply to the work or discharge the capacitor through both the work-coil and workpiece in series or provide high frequency transients i.e. 250-400 KHz compared with 50 KHz for copper. These are generally inconvenient or expensive. Recent work at United Kingdom Atomic Energy Authority (1970) has shown that relatively standard component can produce the high frequency necessary for forming resistive materials with suitable circuit design.

There have been many exciting ideas suggested for the application of electromagnet force in metal forming, some of these now of industrial significance.

i) Application:

Magnetic-Hydraulic Forming:

Deformation is caused by a compressive shock front set up by a piston (flat or radial) making contact with water under the action of electromagnetic impulsive loading. The process efficiency is less than electro hydraulic, but potential cycling time is above 3 times faster. The process application is fairly limited (small thin tubes etc). Electro-hydraulic forming is really a hybrid of the hydro-punch and electro-dynamic press.

Electro-Dynamic Press:

Realising the importance of high speed blanking, a press has been built where the impulsive force is used to drive the punch (Figure 4.48 (a)). This press is likely to gain popularity for industrial blanking when compared with conventional high velocity machines. In this press, the energy control is infinite variable. Cycle speeds are of the order of 1 second and punch velocity is of the order of 19m/sec. and equipment is small in size. Applications have been tried to powder compaction.

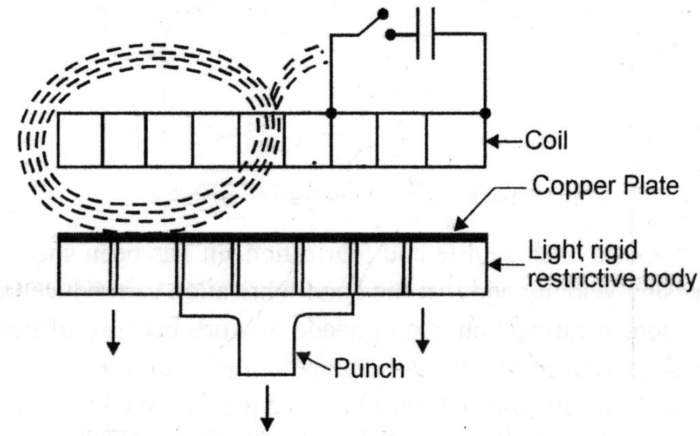

Figure 4.48(a) Principles of electro dynamic press

Applications

Metal to metal seal is a significant application and is useful for fabricating pressure vessels, joining of tubing and pipes, assembly of turbine shafts, assembly of missile nose cones, (Refer figure 4.48(b)) etc.

Such type of joint can be used for pressure vessels and also for household utensils (bottom of copper to the top of aluminium). Other applications may include ceramic assembly, forming of cups, embossing, etc.

In majority of formings made at high velocity, various claims have been made regarding the effects on forged components and die-materials e.g. better grain flow, superior mechanical properties, good fatigue life for the components, superior or equivalent die-lives and lower die-loads comparative to conventional processes. Fatigue-life of forged parts has been found to be improved. Contact loads and deformation

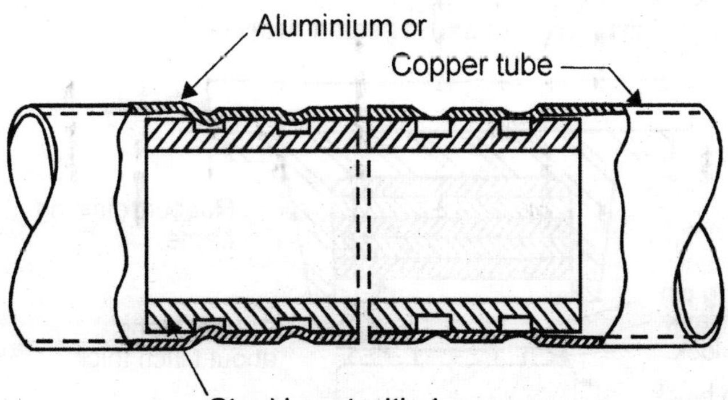

Figure 4.48(b) Pressure tied join

energy may in fact be reduced at large reductions due to the thermo-plastic softening by adiabatic heating. This depends on the shape of the die and the type of material (heat transfer, lubrication, strain-rate and rate of thermal softening), further the effects of adiabatic heating may be harmful, particularly if an unwanted phase change, grain-growth or incipient melting occurs.

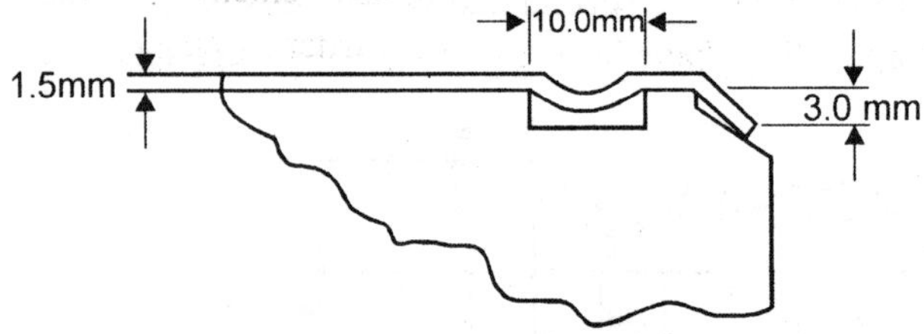

Figure 4.48(c) Design of groove

Great controversy exists over die-life and lubrication. It has been shown that the coefficient of friction falls with increasing velocity and that the good lubrication can reduce the energy required for a particular forging operation, if lubrication can be made to work because of the dynamic nature of the operation. The frictional conditions which will change during nature of the operation. The frictional conditions which will change during the high speed forging is indisputable, but exactly how, and what the predominant factors are, is not very well understood. Also, whether or not lubrication should even be used, is a matter of contention. This leads to the matter of die-design and life, which are intimately related.

Die-life, from one or two components to many thousands are quoted and claims are made for better component finish and tolerance. In most of the high speed deformation processes, dies are accurately guided that in slow speed. Die-life is obviously affected by die-design parameters (Drought angle, flash, venting of dies). The dies can show failure due to flash design, cracks caused by thermal fatigue, abrasive wear and plastic deformation. (Refer figure 4.48(c).

Because of the different nature of the deformation processes and the types of component suited to it, high velocity forming cannot be compared with conventional forming processes. Despite all the prognostication high velocity forming has a definite place in industry, but each potential application must be assessed on its merits after a through technical and economic investigations. It is not possible to generalise.

4.13.3 Rubber pad forming Guerin process and wheelon process

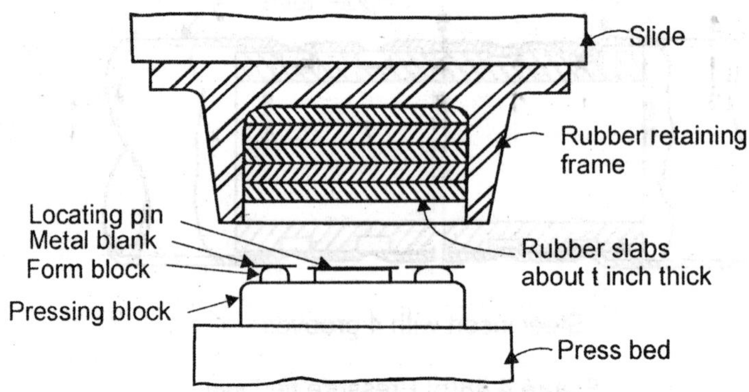

Figure 4.49 Important features in rubber forming or the Guerin process

The Guerin, or rubber forming, process requires only a punch, or the equivalent of half a die required by other forming methods. A flat rubber pad, surrounded by a strong metal frame to prevent lateral flow, serves as the other half of the die. This frame mounted on the slide closely fits around the pressing block on the bed of the press during the forming. In operation, forming blocks are placed on the pressing block, and a sheet-metal blank is properly located on top of each forming block . It is possible to place a number of loose forming blocks, which act as the punches, on the pressing block so that several parts can be formed simultaneously. If they contain holes, the blanks can be readily located by the use of short projecting finds at the top of the forming block. Sometimes use is made of a cover plate, which is placed over a portion of the blank. This clamps a flat portion of the part to the forming block in order to prevent tiping or moving of the blank or any bowing of the flat portion as a result of forces caused by the forming. Forming is accomplished by the rubber pad descending. The rubber, which is practically incompressible, is caused to flow around the form blocks forming the metal downward along the sides of the form blocks. (Refer figure 4.49)

Pressures exerted by the rubber on the blanks average form 1000 to 1300 psi. To ensure maximum life for the rubber pad, forming blocks should be as low in height as possible, and they should have generous forming radii. With this process it is possible to shear relatively thin metal sheets placed on steel blocks having sharp upper edges. Forming blocks may be made of hardwood, masonite, aluminum, or other materials. This process is suitable for small quantities because of the relatively low cost of the forming blocks. It may be used for forming straight and stetch flanges on flat or slightly curved parts. It cannot be used to form shrink flange free of wrinkles. It is also not suitable for deep drawing.

Wheelon Process:

In order to obtain higher pressures, 5000 to 10000 psi, against the work surface, the Wheelon process uses a rubber bag behind a somewhat thinner rubber pad. When the press slide has descended to its lowest position, hydraulic fluid is pumped into this bag forcing the rubber pad over the blank and form block. The Wheelon process, like the Guerin process, does not permit the forming of shrink flanges and deep drawing. Control over the metal edges to be formed during pressure build-up is not provided. Three

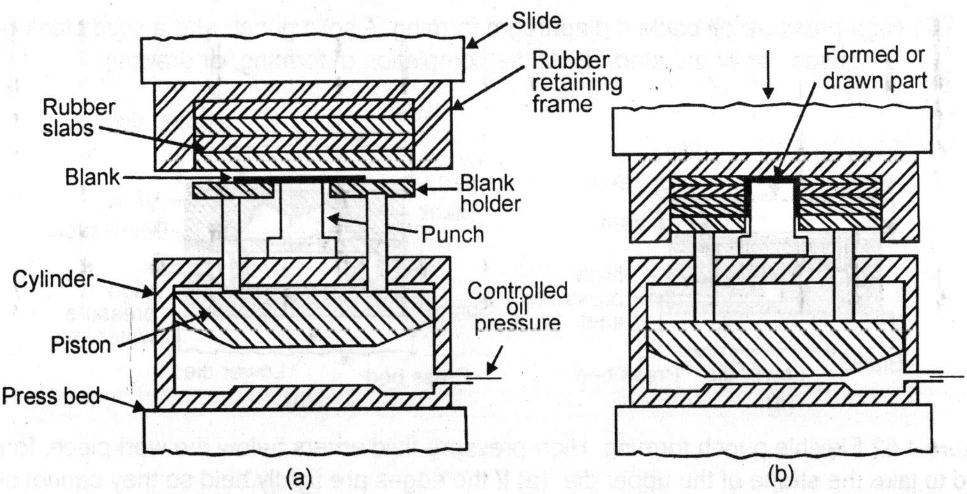

Figure 4.50 High-pressure rubber forming using a solid blank holder. Here the blank-holder pressure is controlled hydraulically. (a) At the start. (b) At the completion of forming, or drawing.

high-pressure flexible die-forming processes, which have been developed to form shrink flanges and to accomplish deep drawing, are shown in figures 4.50 to 4.53.

With these processes tooling costs are relatively low in comparison with conventional dies. These processes are slower than conventional drawing methods, but they are economically advantageous for sufficiently small quantities.

Bulging:

Bulging operations are those which force portions of walls of drawn shapes, tubing, or similar parts outward, as shown in figure 4.50. Hydraulic bulging is accomplished by introducing a liquid into the teiror of a hollow part while it is entirely enclosed in a split die. High liquid pressures force the walls outward to conform with the surrounding die cavity. Bulging is similar in many respects to the operations shown in figure 4.50. Bulging may also be accomplished with the use of a suitable rubber plug, which is compressed axially while inside the part. This rubber plug wears out rather quickly under the high pressures which are used.

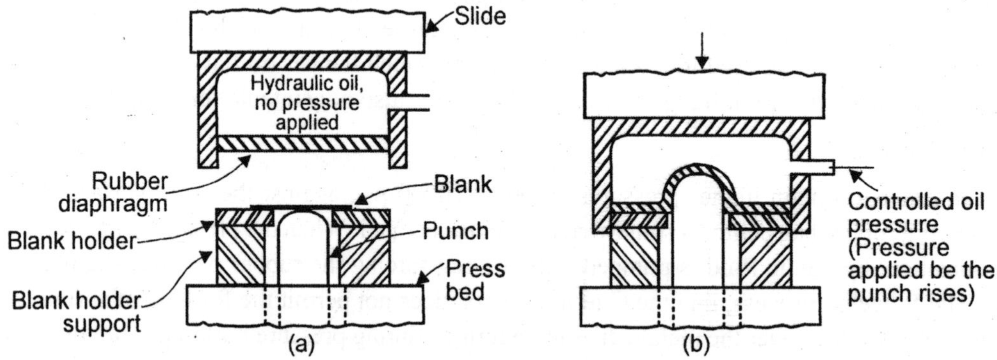

Figure 4.51 High-pressure oil-backed diaphragm forming. A solid punch and a solid blank holder are used. (a) At the start, (b) At the completion of forming, or drawing.

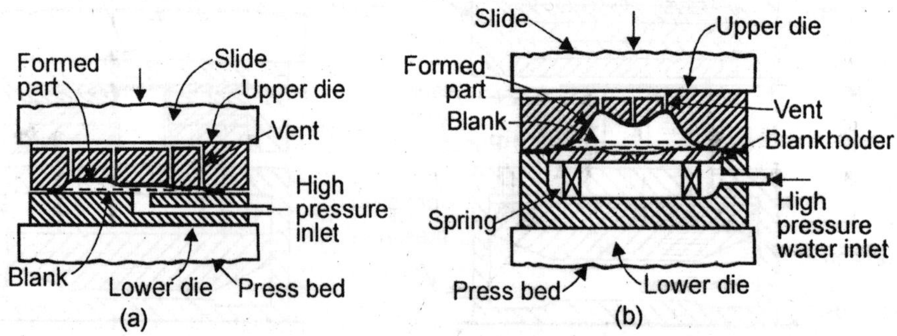

Figure 4.52 Flexible punch forming. High-pressure fluid enters below the workpiece, forcing it upward to take the shape of the upper die. (a) If the edges are tightly held so they cannot slip, this is called "stretching". (b) If a blank holder allows the blank edges to slip inward, this is called "drawing".

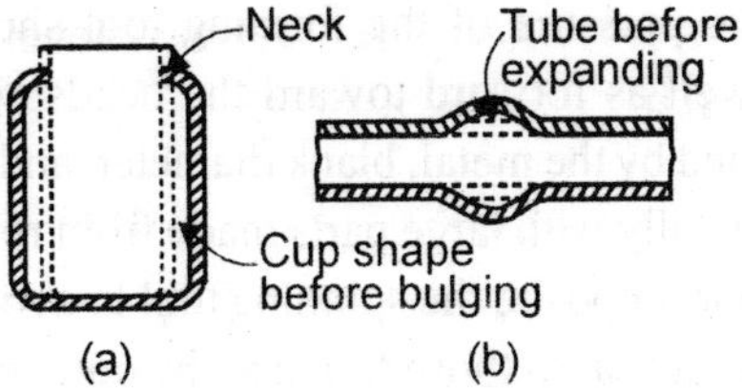

Figure 4.53 Examples of bulging. (a) Bulging a cup-shaped part made by drawing (b) Increase bulging or expanding of the tubing.

4.13.4 Metal spinning

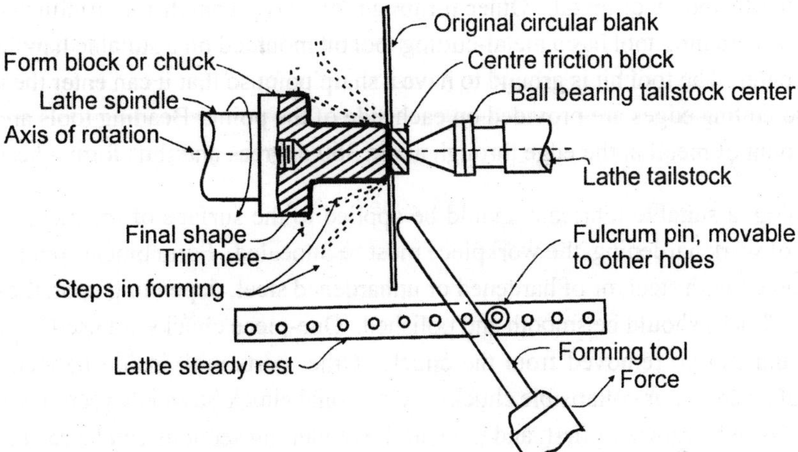

Figure 4.54 A typical spinning operation showing important features

Cup-shaped parts having rotational symmetry can be made by a process called "spinning." Stainless steels with thicknesses ranging from 0.012 to 0.125 inch are readily spun cold, and the diameters of parts which are spun cold range from a few inches up to about six feet. Above a certain thickness and up to reasonably thick plate, the metal is heated to forging temperature, and hot spinning equipment is used. Spinning is competitive with drawing. Due to its low tooling cost, spinning is normally used for producing smaller quantities. Since spinning requires considerably more time than drawing, drawing is preferred for larger quantities, because costs will be lower. As the size of the part and the complexity of the part shape increase, spinning becomes more competitive. Spinning is ordinarily done on a rigidly built lathe, similar to an engine lathe, with suitable rotational speeds and sufficient room to rotate the desired blank diameters. A circular blank is properly centered with a tool such as a hardwood stick as the blank rotates and is lightly held against the chuck by the pressure of a freely rotating pad on the lathe tailstock. As the blank is rotated rapidly about its axis of rotation, pressure is applied to the metal surface by means of bar type or roller-type tools, flowing the metal progressively over the chuck. Spinning is a highly specialized art. To avoid excessive thinning of the metal, the pressure of the forming tool should be directed some of the time backward toward the tailstock as well as forward toward the headstock. The spinning speeds used are comparatively high and are determined by the metal, blank diameter, and blank thickness. During "spinning on air" (not against the chuck), especially with large parts made from relatively thin metal, a hardwood bar is commonly used as a backup support opposite the spinning tool to avoid wrinkling at the outer edge. The forces involved at the outer edge are similar to those found at the outer edges of drawn parts. As in drawing the amount of cold working done to the metal is expressed as per cent reduction of the outer edge diameter. As its diameter is reduced, the metal becomes thicker, and this compensates somewhat for the thinning caused by the forward flow of the metal. (Refer figure 4.54)

Spinning tools include forming, trimming, and beading tools. A forming tool may consist of a well-rounded bar of hardwood or of metal. Other forming tools have a small roller mounted at the end of a suitable handle. A trimming tool has a metal-cutting tool bit mounted on a suitable handle so that it can be manipulated by hand. The tool bit is ground to have a sharp point so that it can enter the metal like a lathe parting tool; also cutting edges are provided on each side of the point. Beading tools are designed to roll over a small amount of metal at the edge through about 360 degrees and thus form a bead.

Before spinning, a suitable lubricant should be applied to the surface of the metal. When the metal reaches its limit of work hardening, the workpiece must be annealed. Form blocks, or chucks, are made of hardwood reinforced with steel, or of hardened or unhardened steel, depending upon the quantity of parts to be produced. Chucks should be smooth and polished. One-piece chucks are used for parts which have sufficient draft and can be removed from the chuck. Other parts with inside flangers or reverse draft require the use of sectional or collapsible chucks. A sectional chuck has a key section with parallel sides. This key section must be removed first, and then all the remaining sections can be readily removed. First, and then all the remaining sections can be readily removed. Chuck sections must fit together smoothly at joints, because slight irregularities will be reproduced on the spun parts.

(i) Applications of metal spinning

Some complicated shapes may require spinning using two or more different chucks. Certain operations on tubing, such as expanding or reducing, may be accomplished by spinning. Increasing the diameter of a portion of tubing is called "expanding", and decreasing the diameter of known as "reducing". The ends of cylindrical vessels, such as pressure tanks, are called "heads." Heads are usually dis-shaped to provide necessary strength, and often heads are spun either hot or cold depending upon the metal thickness. A hot-spinning machine is a combination of a vertical press and large vertical spinning lathe, which rotates the blank abut a vertical axis. Poer-actuated rollers do the forming, as the part is rotated.

4.13.5 Explosive Forming

(i) Introduction:

Explosive forming belongs to high energy rate forming (HERT) processes. This process has made rapid strides in the last two decades due to its special advantages over other forming processes. We know, the expression for kinetic energy is 1/2 mv2. to increase kinetic energy (K.E) either mass of the body or its velocity should be increased. Since the velocity is a quadrature term in the expression of kinetic energy, so a small increase in velocity will effect greater change in the kinetic energy. This advantage is taken use of by this process. The high rate of energy transfer is obtained by increasing the velocity instead of the mass. The process take place in a small interval of time and the large force for plastic deformation is supplied by inertia rather than by large frames of machines.

(ii) Classification of Explosive Forming:

The explosive forming of metals is classified mainly into the following categories:

a) Stand off operation or open type operation

b) Contact operation or closed type operation

c) Multiple type operation

We shall describe these one by one.

a) Stand off operation or open type operation:

The blank is secured over the female part of the die which is kept within a chamber. The chamber is filled usually with water for energy transfer (of the explosive to the blank) Though air, molten salt or oil could also be used. These are generally used in hot forming. Water is preferred due to its low compressibility and stock-impedance matching characteristics. The die is first evacuated completely. If air or any gas

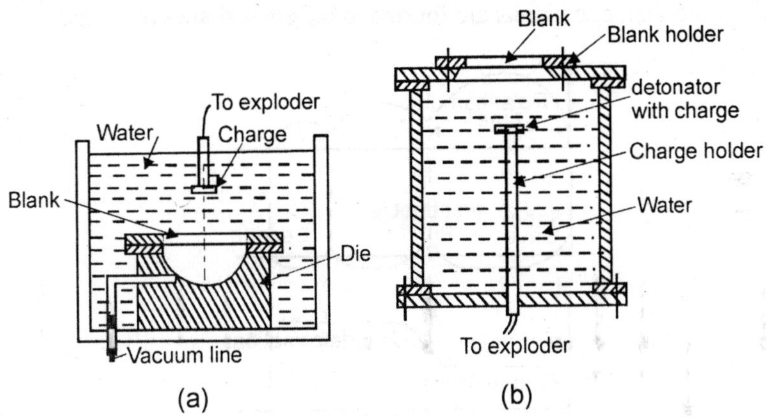

Figure 4.55 Stand OFF operation (or) Open type operation

remains in the die, it puts heavy resistance to the formation of the blank and due to air pockets proper shape becomes difficult to achieve. Moreover, the entrapped air gets compressed adiabatically very much thus raising its temperature and of the die as well. As a result the shape and size of the die may change. The explosive kept at a distance from the blank, connected to a detonator is exploded. The energy released in moment generates stock waves in water. The blank (or the workpiece) is deformed to desired shape in one or two milliseconds. As it appears from the above figures, only a portion of the total energy is utilised, which is greatest disadvantage of this arrangement. (Figure 4.55(a) and (b)).

b) Contact or closed type operation

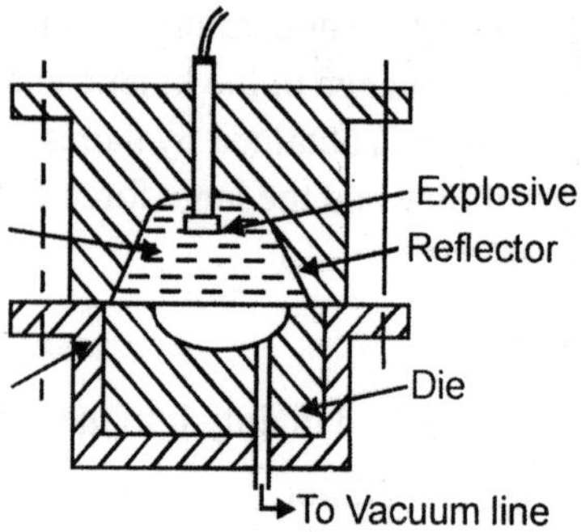

Figure 4.56 Contact (or) Closed type operation

The arrangement of this method is shown in figure 4.56. We observe that the explosive is detonated in closed chamber which is filled with water. As is obvious, all the energy is directed upon the blank thus its efficiency is greater than that of the previous type. In this method too the die is evacuated of the air or any gas. The design of the reflector is very important here. It must be robust and of such profile as to relect all the pressure wave on to the blank.

c) Multiple type forming

In this process more than one blanks are formed to different shapes at a time.

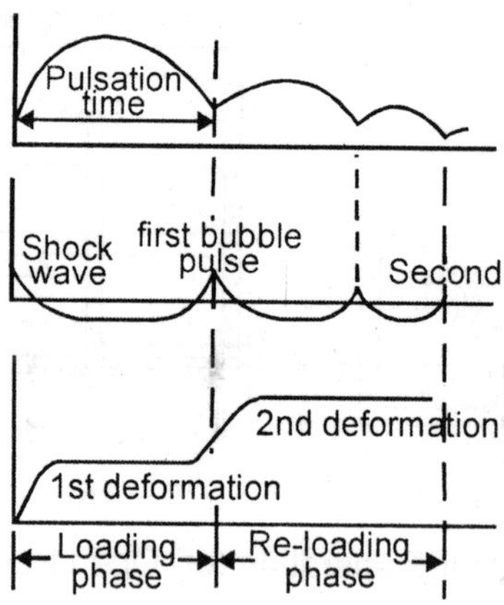

Figure 4.57 Mode of energy transfer

iii) Mode of Energy Transfer:

As the explosive is detonated, a mass of gaseous products, high in pressure and temperature is produced. These products impart high particle velocity (known as streaming velocity) to the particles beyond the gases. It comes to water particles layer by layer generating a shock wave or primary shock wave. This shock wave moves with a velocity of order of 1070 m. per. sec. and above. (Refer figure 4.57).

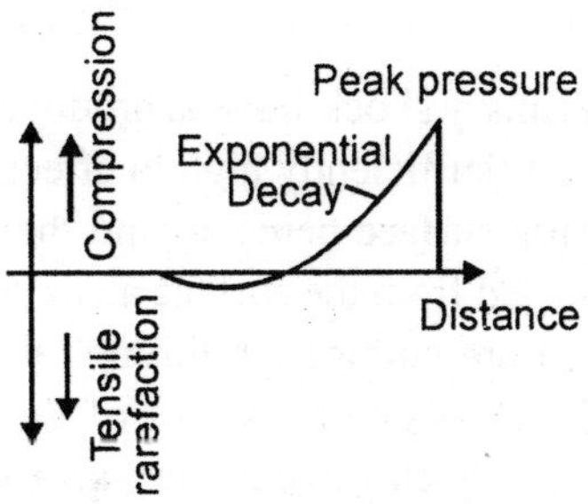

Figure 4.58(a)

Figure 4.58 shows the variation of pack pressure of the stock wave as it moves radially outward. The decrease in pressure, in fact is exponential in character. Figure 4.58 (b) shows the decrease in pressure with respect to time. It has been observed experimentally and related mathematically, as follows:-

$$P_{(t)} = P_k \cdot e^{\frac{-t}{t_1}}$$

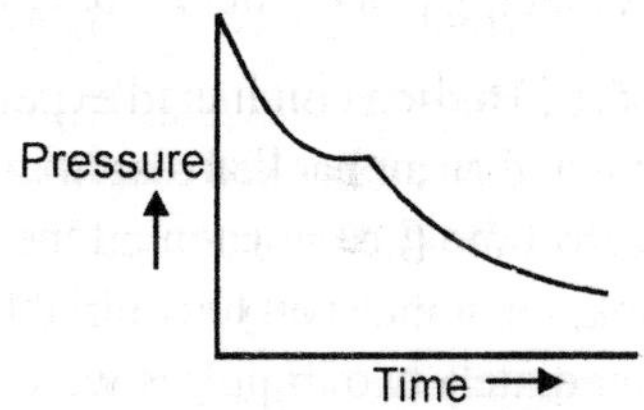

where P_t - Pressure at time, t

t_1 - time required for the maximum (peak) pressure to decay to 1/2.7 times the peak pressure.

P_k - peak pressure.

Figure 4.58(b)

Figure 4.58 shows the relative velocity of the stock wave and particle velocity for different peak pressure.

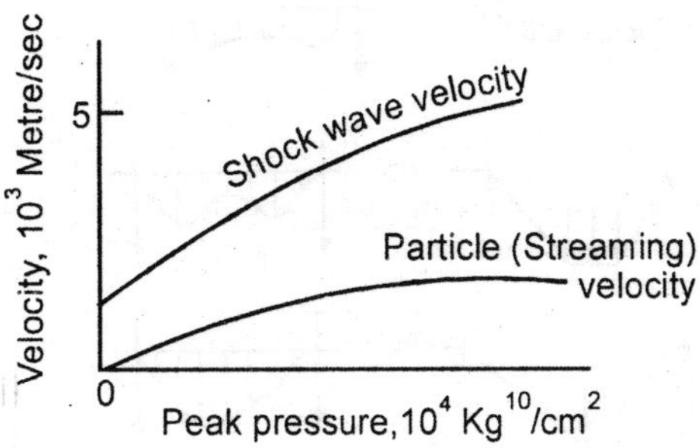

Figure 4.59

In addition to primary shock wave where is also secondary shock wave (also called bubble pulse). The gaseous products are termed as gas bubble. The high pressure of the gas bubble causes fast movement of the water particles of the water particles and with that the gas bubble also espands. The rate of expansion depend upon the difference between the gas pressure and the external hydrostatic head. The bubble gas on expanding due to its inertia and crosses the equilibrium position, the bubble then collapses due to higher hydrostatic pressure, hus generating fresh shock waves known as secondary shock wave (also known as bubble pulse). This wave is lesser in intensity than the primary wave. Such oscillations (in pressure) are finite in number and after these oscillations the bubble disappears. (Refer figure 4.59)

iv) Further explanation for deformation:

With the detonation of the charge, the primary shock wave is produced. It moves radially outwards and exerts pressure on the blank. This pressure is sufficiently high to effect deformation on the blank (Figure 4.60). The blank is shown as abcd. The top surface being ab and the bottom cd. The shock wave also behaves like light wave. A part of it is reflected from the surface ab as compressive wave the rest gets into the metal. A part is reflected from the bottom surface cd, through the metal as a low pressure tensile rarefaction wave, and goes back in the fluid medium. A part of the wave reaching the surface cd is transmitted to the low density air. Dur to the shock wave on the surfaces of the blank the particles of the blank move outwardly with sufficient velocity thus causing the deformation. Simultaneously due to movement of the tensile rarefaction wave cavitation of wate adjacent to the blank is produced but the expanding gas bubble produces a flow of water towards the blank which closes the cavitation as well as reloads the blank. The rushing water gathers sufficient kinetic energy and produces further deformation (secondary deformation) on the blank whose particles are in a state of accelerated outward motion due to first loading phase, the bubble pulses may produce a few additional reloading of lesser intensity.

G.E. Hudson conducted experiments to explain the deformation phenomenon in explosive forming. He subjected an air backed diaphragm, clamped at its periphery to an underwater explosion. He assumed that at zero time (i.e. as soon as the explosion occurs) a uniform velocity, normal to the initial plane of the blank, i.e. impulsively acquired (Figure 4.60) The radial velocity in the material remains zero at this stage. Immediately two impulses were propagated.

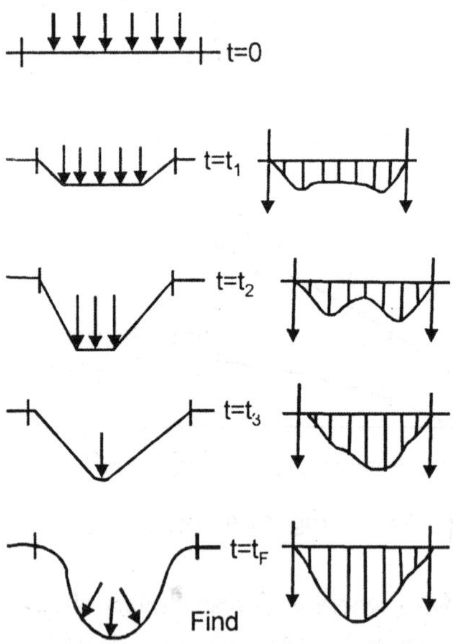

Figure 4.60

(i) A radial impulse travels inwards, leaving the interior in a state of plane plastic flow with an induced radial velocity.

(ii) A plastic bending impulse is generated near the edge because of the restraint and is propagated slowly in wards.

Each annular element is tilted and the central region gets gradually rounded up or forms to a shape of a cone.

v) Process parameters:

The following are the process parameters:-

– Stand off distance (i.e. the distance between the explosive and the blank)

– Charge weight and shape

– Reflector (shape and size)

– Entrapped air cell (within the die)

– Hydrostatic head

– Blank holding torque

The control of these parameters are very important for efficient operation. Their values are influenced by the material of the blank, thickness, size and shape of the blank and the degree of deformation to be inflicted upon the blank. Say for example, if the stand off distance is very large, sufficient force may not be developed for the forming. To make it effective more quantity of charge is required. If the entrapped air within the die is of appreciable volume, proper shape of blank will be almost impossible to obtain. The relector directs the shock wave toward the blank. So its size and shape are very important. Similarly other parameters are of paramount importance.

vi) Advantages and dísadvantages:

Advantages:

(i) Large components which are beyond the capabilities of the large conventional processess can be formed by explosive forming techniques economically.

(ii) Intricate shapes on materials of low form ability can be worked fairly easily.

(iii) Large machine frames are not required in this process.

(iv) No finishing operations are required on the process

(v) The process uses low-cost energy source.

(vi) There is flexibility in design.

(vii) The fabrication (of the set) or capital cost is less.

(viii) The production cost is not high.

(ix) With this process the product has high density.

(x) For a particular charge quantity or in other words, for a given explosive pressure, the ear of the blank (object or workpiece) is not limited as in the case of hydraulic piston press.

(xii) The process has high reliability and can give close tolerance.

Disadvantages

(i) There is obvious dangers while handling any type of explosives.

(ii) The noise from the operation is of too high intensity.

(iii) The overall rate of operation is slow.

(iv) It is mainly suited for small batch production.

vii) Applications:

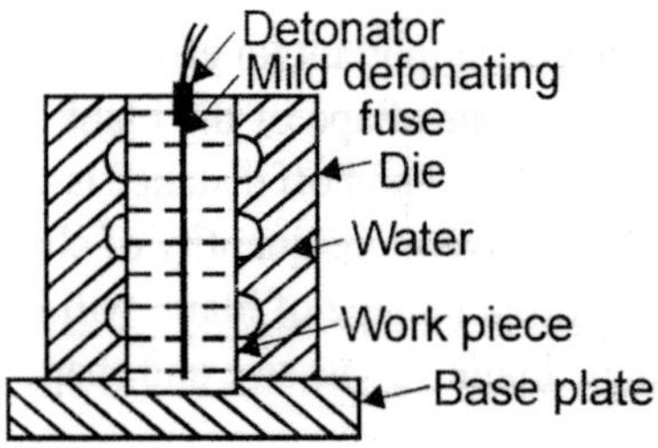

Figure 4.61

The explosive forming process is used for a variety of products such as:

Jet air-craft panel, Air duct-panel, Jet engine fuel filler, curved corrugated panel, Missiles domes, Tank car head, Corrugated cylinder (Figure 4.61). Hollow cones etc.,

4.13.6 Peen forming

This method is employed to set up a superficial state of surface compressive stress, causing the interior of the member to assume an opposite tensile stress. Because fatigue generally occurs from surface cyclically loaded in tension, the useful lives of such members are frequently extended by shot peening. This method is sometimes employed to achieve an ornamental effect.

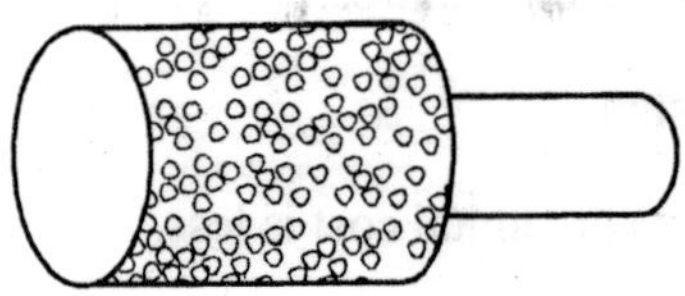

Figure 4.62

Shot peening is done by blasting or hurling a rain of small shot at high velocity against workpieces to cause slight indentations. Figure 4.62 shows shot peening action on a job.

4.13.7 Super plastic forming

Conventional metals and alloys typically exhibits tensile elongation in the range of 10 to 30%. By producing sheet materials with ultrafine-grain-size and performing the deformation at low strain rates and elevated temperature elongation can exceed 100% and may be as high as 2000 to 3000%. This superplastic behaviour can be used to form material into large, complex-shaped products with compound curves. Deep or complex shapes can be made as single-piece, single-operation pressings rather than multistop conventional pressings or multipiece assemblies.

At the elevated temperature required to promote superplasticity (about 900°C for titanium, between 450°C and 520°C for aluminium or generally above half of the melting point of the material on an absolute scale), the strength of material is sufficiently low so that many of the super plastic forming techniques are adaptations of processes used to form thermo plastics.

Limitations:

The major limitations to super plastic forming is the low forming rate that is required to maintain superplastic behaviour. Cycle time may range from 2 minutes to as much as 2 hours per part, compared tc the several seconds that is typical of conventional press work.

SUMMARY

❖ ***Mechanical working of metals (or) Metal forming:*** Metal forming is one of the manufacturing processes in which the metals are processed to have the desired size and shape by plastically deforming them in the solid state by the application of mechanical stresses imposed on them. The stresses induced during a metal process a greater than the yield strength but less than the fracture strength of the metal. Metal forming is also called as mechanical working of metals. The deformation metal forming may to bulk flow in three dimensionals, simple shearing, simple or compound bending or any combination of these and others.

❖ ***Cold working:*** Cold working is the process of plastic deformation of metal at room temperature or well below the recrystallization temperature of metals.

❖ ***Hot working:*** Hot working is defined as the plastic deformation of metal which are above their recrystallization temperature. It should be understood that the recrystallization temperature varies with different metals.

❖ ***Forging:*** Forging is a metal forming process. In this the product of desired shape and size is made by presssing and hammering.

❖ ***Special forging processes:*** Ring rolling, Orbital forging or rate forging, No-draft forging, High-energy rate forming, Cored forging, Wedge rolling, Incremental forging.

- ***Some hand forging operations:*** Bending, Drawing down or swaging, Upsetting, Punching, Setting down, Fullering.

- ***Forging defects:*** Surface cracks and cavities, Burnt metal, Pitting on of surface, Mismatched forging, Imcomplete component, Oversize component, Inclusion of dirt, slag and sand in the part.

- ***Examples of cold forging:*** The heads of small bolts, rivets, screws, pins, nails etc., small ball bearing.

- ***Forgeability:*** Forgeability is generally defined as the capability of a material to undergo deformation without cracking.

- ***Rolling:*** Rolling is a metal forming process in which the forming of metal is done by passing it through a fixed number of rotating rollers. The metal is passed into the rollers through dies.

- ***Tandem rolling mill:*** It has a set of two or more stands of rolls arranged in parallel so that a continuous pass can be made without changing the direction of feeding or stopping the rolling process.

- ***Roll bending:*** It is also a kind of roll forming operation through which metal sheets and plates are bent into cylindrical shapes. The roll bending machine caries three rolls: two being fixed and the third adjustable. Diameters of all the rolls are same. By adjusting the position of the adjustable roll the plates or sheets can be bent in different curvatures.

- ***Extrusion:*** Extrusion is a metal forming process in which a heated metal block called billet is reduced to the desired shape and size by confining it in a closed container and then forcing it through the opening in a die under high pressure.

- ***Some products obtained by extrusion:*** The extruded product may be solid or hollow objects examples are rods, aluminium, pipes, angles, automobile parts, metal frames, etc.

- ***Direct extrusion:*** In direct extrusion the direction of flow of extruded metal is the same as that of the ram movement.

- ***Indirect extrusion:*** In indirect extrusion the direction of flow of extruded metal is the opposite to that of the ram movement.

- ***Impact extrusion:*** It is cold extrusion process, the work metal in the form of blank is placed inside a die and punch forces suddenly on the blank in the die, due to this impact the metal flows around the punch in an upward direction. This is known as impact extrusion.

- ***Wire drawing:*** Wire drawing is a process in which the metal rod or wire with larger diameter is pulled through a die hole to produce a wire of smaller diameter.

- ***Deep drawing:*** If the depth is greater than the diameter it is called deep drawing or cupping. The speeds of drawing a considerably lower than those used in stamping and perforating.

❖ ***Hot spinning:*** Hot spinning is a process of plastic forming of sheet metals used for making cup shaped marts which are axi symmetrical. The process is carried out in a spinning, lathe which is similar to centre lathe.

❖ ***Cold drawing:*** Cold drawing is the process of pulling metal at room temperature through the opening of the die.

❖ ***Tube drawing:*** Tube drawing is the process of finishing hot rolled tubes by pulling them through a die opening with a stationary mandrel. Tube drawing is used for producing thin tubes with small diameters.

❖ ***Metal spinning:*** This is the process of producing circular shaped articles from very thin metal blanks by spinning action. The metal is worked at room temperature.

❖ ***Cutting operation:*** The workpiece is stressed beyond its ultimate strength. The stresses caused in the metal by the applied forces will be shearing stresses.

❖ ***Forming operation:*** The stresses are below the ultimate strength of the metal. In these operations, there is no cutting of the metal but only the contour of the workpiece is changed to get the desired product.

❖ ***Important forming operations:*** Bending, Drawing, Redrawing and Squeezing etc.

❖ ***General characteristics drawing:*** Production of long rod and wire, with round or various cross-section; similar cross section than extrusion; good surface finish; low to moderate die, equipment and labout costs; low to moderate operator skill.

❖ ***Trimming:*** Trimming is the process of removing the extra metal called flash that spread out near the parting lines of components produced in forging or die casting operations. The dies used for trimming are quite similar to the blanking dies.

❖ ***Nibbling:*** The nibbling operation which is used for only small quantities of components, is designed for cutting out flat parts from sheet metal. The flat parts range from simple to complex contours. This operation is generally substituted for blanking. The part is usually moved and guided by hand as the continuously operating punch cuts away at the edge of the desired contour.

❖ ***Uses of Progressive dies:*** Progressive dies are used when the production is of large numbers, stock material is not too thick so to avoid the problem of stock straightening.

❖ ***Compound dies:*** Several operations on the same strip may be performed in one stroke at one station with a compound die. Such combined operations are usually limited to relatively simple shapes, because they are somewhat slow and because the dies (with increasing complexity) rapidly become much more expensive than those for individual shearing operations.

❖ ***Combination die:*** Combination die is similar to that of compound die. The main difference from compound die in that a cutting operation is combined with a bending or drawing operation. In this die also, more than one operation may be performed at one station.

REVIEWS

1. Name some characteristics of sheet metal forming process.
2. Explain OBI press with neat sketch.
3. What are the types of drive system available in press working?
4. Explain the purpose of stops and strippers.
5. Explain with neat sketch progressive die.
6. Explain with neat sketch compound die.
7. Differentiate between punching and blanking operation.
8. Explain V and U bending process.
9. Explain Deep drawing process.
10. Explain Ironing process.
11. Explain electro hydraulic forming process.
12. Explain electro magnetic forming process.
13. Explain pad forming process with neat diagram.
14. Explain metal spinning process.
15. Explain explosive forming process.
16. Explain peen forming process.
17. Explain super plastic forming process.

CHAPTER - 5

FORMING AND SHAPING OF PLASTICS

5.1 INTRODUCTION

The term plastic includes a group of organic solids which can take any commercial shape when subjected to heat or pressure or both. These shapes can be made by using its important characteristics of flow by the application of heat/pressure, thus it can be casted/extruded. Today more than 30 chemically distinct families of plastics and their numerous compounds are available for use.

Plastics have been increasingly accepted because they offer unique combination of properties which fitted for many industrial developments because of weight, cost, chemical resistance, toughness, resistance to abrasive, strength, formability, machinability, appearance, etc. Most mechanical properties of plastics are usually inferior to those of metals. Plastics are having a large thermal expansion compared to metals and dimensional instability with appreciable creep and flow at all temperatures. The operating temperatures vary form 60°C to 315°C. However, the rigidity and fatigue strength of plastics are also below than that of metal. It can carry a good surface finish and a higher lustre can be added by buffing and colour impregmentation.

General Characteristics of Plastics:

1. Light weight, but possesses good strength and rigidity.
2. Easily moulded to desired shapes.
3. Less brittle than glass.
4. Resist corrosion and chemical action.
5. Low thermal conductivity.
6. Not affected by moisture.
7. Possesses good toughness.
8. Absorbs vibration and sound.
9. Good and attractive appearance.
10. Easily machinable.
11. Has high dielectric strength.
12. Not affected by weather.
13. Cheaper than other materials.

Advantages:

1. Plastic articles can be made rapidly with close dimensional tolerance.
2. Easy to fabricate than metals.
3. Low cost.
4. Vibration and sound damping application.
5. They can be made either transparent or in various color shades.

Limitations:

1. Many plastics are inflammable.
2. They are softer and ductile.
3. They possess low strength, low heat resistance, low dimensional stability.

5.2 TYPES OF PLASTICS

Majority of plastics are organic substances or resins containing carbon as central element together with oxygen, hydrogen, nitrogen, chlorine, etc. The properties of plastics depend on the atoms it contain and the way in which these atoms are arranged in molecules. These organic compounds are also known as polymers.

One major class of plastics called the *Thermoplastic* materials, composed of separate long molecules arranged side by side (synthesised in the shape of long threads). Thermoplastics become soft when heated and re-harden on cooling because the molecules are held together by secondary bond. The strength of these forces between separate particles of matter depend directly upon the masses and inversely upon a power of distances between the molecules. This explains why plastics of large molecules are strong and tough. Strong associative forces tend to make plastics brittle at low temperatures. Heat increases, atomic movement, separates the molecule, reduces the associative force and thus softens thermoplastics. Molecules with side branches cannot be packed closely and the material is not so strong but generally is soft and flexible.

A second major class of plastics called Thermosetting materials is composed of long molecules linked to each other in three dimensions by primary bonds. These bonds are not broken by heat until the compound is decomposed. Thermosetting plastics are therefore generally stronger, harder, more resistant to heat and solvent and lower in material cost. The secondary bonds of the thermoplastic materials allow deformation and flow under stress.

i) Thermo Plastics:

They when heated even at 40°C become soft and so can be remoulded without any change in structure. They can be ground up and remoulded, i.e. they can be re-shaped while in soft state. Usually, they become soft at about 65°C to 315°C and can flow to an appreciable extent underload. And that is why most of the plastics used today are thermoplastics. Today number of thermoplastics are available into the market such as nylons, polyetheline resins, polystyrene, polyropylne, polycarbonate, acrylic resins, vinyl plastics, etc. Amongst them nyloyns, acrylics, and vinyl plastics are commercially extensively used and therefore some of their properties are outlinead here in table 5.1.

Table 5.1

Thermoplastic type	Properties	Use
1. Polyethylene	Both flexible and rigid types are lightly break resistant, unaffected by food acids, and household solvents, resist moisture, flexible over a wide range of temperatures.	Widely used as a coating for paper and paperboard, film packaging, disposable bags, furniture components, etc.
2. Polyropylene	Good dimensional stability. Resistance to heat, chemicals, moisture, electricity, easy to process. Good impact resistance.	Appliance housings, refrigerator door liners, wall tiles, battery cases, toys insulators, ornaments, etc.

3. Polystyrene	Rigid, clear, transparent materials with good tensile strength, good combination of rigidity and toughness; good mouldability.	Wide applications in packaging, appliances; housewares; furniture; dentalplates; auto parts.
4. Polyvinyl Chloride (P.V.C.)	Chemical and abraison resistance, good weatherability, easily prigmented; good resistance to petroleum-based-solvents, acids, etc.	Sound records, blown bottles, rigid sheeting, upholstery, hose, flooring, rain-wear, insulation, etc.
5. Acrylic	Outstanding optical properties; inert to most chemicals; excellent mechanical properties for short term loading.	Aircraft and architectural domes, signs, machinery housings, optical components, goggles lenses, etc.
6. Flouracarbon (Teflon)	Low coefficient of friction, stable upto about 205°c; good chemical resistance and weatherability, excellent electrical insulation.	Non-lubricated bearings, chemical resistant pipe linings, seals, diaphram, valves, linings of cookware.
7. Nylon	High tensile strength, good impact strength and high abrasion resistance, good chemical and electrical propertises.	Gears, bushings, cans, tubings, rollers bearings, combs, and miscellaneous household items.

ii) Thermosetting Plastics:

They are made from chains which have been linked together known as cross-linked (three dimensional network of molecules). Usually they are hard and brittle and cannot be used after working once. The thermosetting materials undergo a chemical change when moulded and cannot be resoftened by heating to reshape them. The mould temperatures are usually considerable higher than those for the thermoplastics. The finished product is capable of withstanding much higher temperatures without deformations. Their properties and uses have been given in the table 5.2.

Table 5.2

Thermosetting	Properties	Use
1. Epoxy	Good flexibility, mechanical strength, adhesion, low shrinkage.	Laminated plastics, printed circuit boards, cast tooling, paints, and adhesives.
2. Phenolic	Strong, rigid, durable, and hard; resists heat upto 150°C; good weather resistance.	Laminates, foundry and grinding-wheel resins, gas and crude oil pipes, appliances housings, hot-melt adhesives.
3. Silicone	Exceptional heat resistance (utpto about 370°C); good moisture and chemical resistance.	Arc and thermal barriers, panel boards, terminal strips, electrical components.
4. Urethane	Rigid foam has good weatherability resistance to rot, vermin, and fungus; flexible and semi flexible types.	Arm rests, packaging for delicate instruments, rigid types used as inbulation, simulated wood beams, surfboards, and buoys.

Some other thermosetting materials may include Expoxides, Furan, Melamines, Polyesters, etc.

Elastomers:

Elastomers are also polymers which are lightly bound together and so exhibit the unique property of high elasticity, stretching five to ten times when tensile loading is applied and coming back to its original length or dimensions when the applied load is removed. These are non crystalline polymers at room temperature having long chain molecules in three dimensional network.

These elastomers can be classified into the following groups.

a) Thermosetting Elastomers:

These elastomers are special type of thermoset polymers having long chain molecular structure. In their green state they are tracky (shabby) and cold flow easily. To be useful they must be cross-linked. In case of rubbers, this cross-linking is known as Vulcanisation. Their stretching motion is limited to cross-links. The farther apart the cross-link points, the weaker the system and greater will be the elongation. This structure enables the elastomers to readily absorb and dissipate energy. This category includes the natural rubber; styrene butadiene, etc.

b) Thermoplastic Elastomer:

These elastomers are similar to the thermoset type in service properties, but these offer greater ease and speed of processing. Moulding or extrusion can be done on standard equipment within short cycle. The major thermoplastic elastomers are polyster-copolymers, polyurethane, etc.

5.3 MATERIALS FOR PROCESSING PLASTICS

In most of the cases, the plastic resins have got to be mixed with other substances before they are ready for processing and to make the final product. These mixing ingredients are vital for all class of polymers to impart special properties. These additives may include Fillers, pigments, stabilizers, plasticizers or lubricants.

Fillers are added to thermosetting plastics and also to some thermoplastics, and the amount to be added would depend on the purpose. Usually fibers (less than 50%) are added to impart strength, calcium carbonate adds the bulk, wood filler gives fair strength and good mouldability but tends to absorb moisture. Glass fibres also give strength while asbestos provide heat resistance and dimensional stability and addition of mica gives good electrical resistance. Sometimes, addition of carbon or silver makes the plastics conductive.

Generally these resins do not have good colours so the pigments are added to give the desired colour.

Stabilisers are added to initiate the reaction i.e., to allow polymerisation to begin and then stabilise the reaction of molecular chains. Hydrogen peroxide is a good stabiliser.

Solid or liquid plasticizer is to make the product more flexible and less brittle. These substances work as internal lubricants for improving the flowability inside the mould. Lubricants also help to prevent them from sticking together.

These plastics are usually available into three forms:

i) Powders, flakes, granules for moulding.

ii) Liquids for castings, impregnated laminates, adhesives, paints, mixed moulding compounds.

iii) Filaments, films, sheets, rods, tubes to be converted into finished products such as weaving into cloth, joining or cutting into useful wrappers.

5.4 BONDING OF THERMOPLASTICS

Many methods have been developed to join thermoplastics, ranging from the simple application of heat to the edges being joined, to the more exotic methods of induction, ultrasonic, and vibration welding.

i) Fusion Bonding:

In conventional fusion bonding, parts are placed in fixtures and held by chucks, gripping fingers, vacuum cups, or other mechanical devices. The fixtures then press the parts against a heating platen that is usually made of aluminum, aluminum-copper, or hot-rolled steel coated with Teflon or Teflon-impregnated fiber glass. The platen is heated to 500°F to soften the plastic to a predetermined depth. Holding fixtures are then opened and the heating platen removed, the fixtures are closed again until the melted material is cooled and fused.

ii) Solvent Welding:

Solvent welding is a widely used plastics-joining process. The solvent may be applied to the joint by dipping, spraying, and brushing. The most common method is to hold the solvent-soaked pad against the surfaces. The edges dissolve, producing tacky surfaces that are then brought together, moved slightly to obtain mixing, and held in contact, in correct relationship, to dry and harden. Drying time ranges from a few minutes to hours.

iii) Friction Welding:

Spin welding is limited to round parts that can be rotated while in contact to generate heat. The frictional heat melts the interfaces, and when the motion is stopped, the parts fuse together. Spinning speeds of 10 to 15 fps at the interface and pressures of 20 to 200 psi (138 to 1378 kPa) must be matched to the material to produce just the right amount of heat. Melting and fusion can occur in as little as 0.25 s. Maintaining pressure for 1 to 2 s after braking normally completes the cycle. Figure 5.1 is a sketch of the tooling used to spin an aerosol-bottle bottom in place.

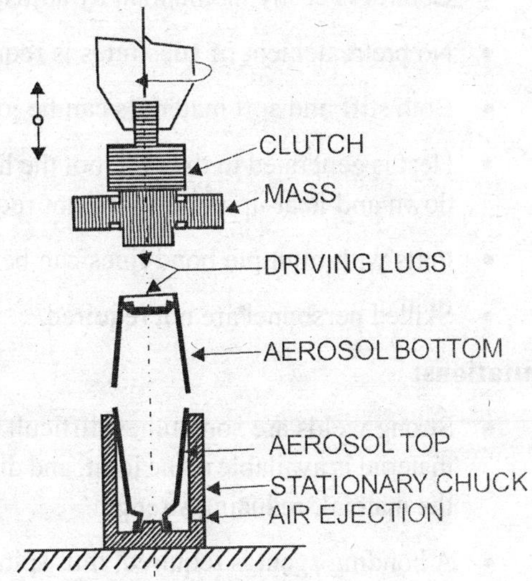

Figure 5.1 A spin weld made with an inertia tool running at a speed of 4000 rpm.

iv) Induction Welding:

Originally thermoplastics were induction welded by pressing two parts together around a metal insert. This combination was then subjected to an alternating magnetic field, which melted the surfaces of the plastics around the insert. Compression of the parts resulted in a fusion weld.

Improved methods employ impregnated plastics with either metallic particles or tiny, permanent magnets. Alternately, the metallic particles may be applied as a tape or in the form of a hot melt. In either case, the parts are passed through a magnetic field and the particles heat up to form the weld. Bonding of large parts is made easy by coextrusion of built-in bonding agents.

Advantages:

- Induction welding provides excellent polymer to polymer linkage. Molding irregularities at the substrate interfaces are filled during welding, thus, ensuring a strong bond. Induction welding can be used on a variety of materials and thicknesses, provided the materials are thermoplastic or the substrates are first coated with thermoplastic adherend.
- The configuration of the part is limited only by the ability to design a suitable coil. Several simple components can be joined to form complex shapes that would be expensive to mold. This approach is used in applications such as water pumps, plastic plumbing for appliances, heater ducts for automobiles, and complex plastic containers.
- Heat damage and distortion are reduced since they are localized to the bond. Parts need not be cooled but can be handled immediately.
- Bonds can be hidden inside the part. Hermetic seals are possible.
- Work coils need non touch the part and minimal compression is required to form the bond.
- Control is easily maintained by adjusting the power supply.
- No pretreatement of substrates is required, and bond agents have unlimited storage life.
- Both stiff and soft materials can be joined, and bonds can be flexible or rigid.
- Heat is generated in th bond, not the hardware, so the process is easily automated because cool-down and heat-up periods are not required.
- Parts with multiple bond lines can be welded in seconds.
- Skilled personnel are not required.

Limitations:

- Strong welds are sometimes difficult to make in reinforced thermoplastics because less plastic material is available at the joint, and direct application of heat may disrupt the fiber orientation of the mateial, reducing strength.
- A bonding agent is required; thus spin or vibration welding, which fuse at the bond line, may be more cost-effective.
- In bonds between metals and plastics, the substrate may overheat.
- The bonding agent is opaque, so transparent seals are not possible.

v) Ultrasonic Welding:

Ultrasonic welding requires the conversion of low-frequency electical power to high-frequency mechanical vibrations that are applied to thermoplastic components. When the plastic melts, the components are joined in a fusion bond. Joint designs for ultrasonic welding and staking are shown in Figure 5.2. Until recently, power supplies that could develop frequencies of 40 kHz at levels necessary to melt plastics were not available. Horn materials have also been improved.

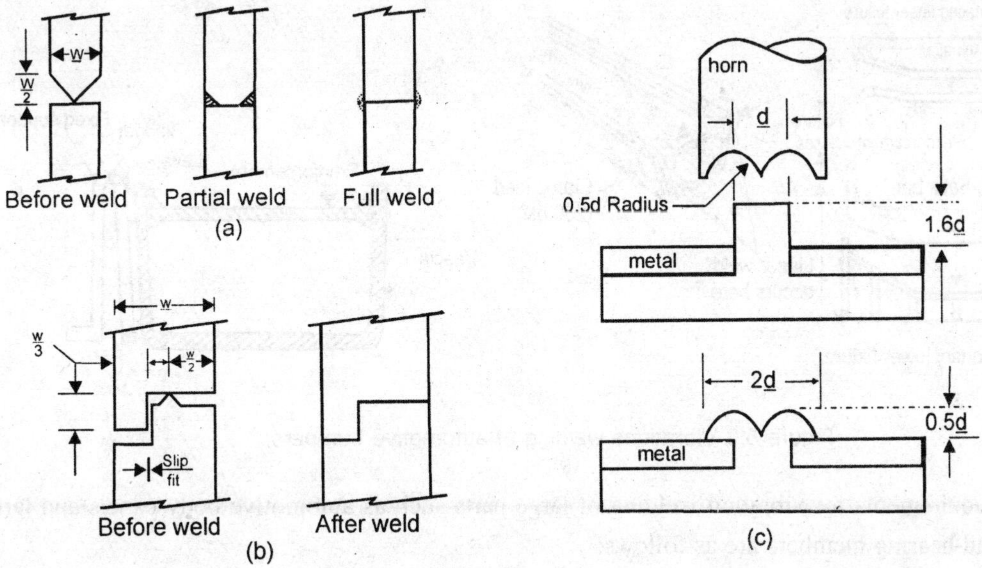

Figure 5.2 (a) and (b) Joint designs for ultrasonic welding. (c) Joint design for ultrasonic staking.

Advantages:

- High frequency, 40 kHz, equipment used at low amplitude of vibration is quite and does not require an enclosure.
- The smaller 40 kHz tools can fit into limited spaces or can be ganged together on one machine to join large complex parts.
- Energy transmitted into the part is localized. Thus the process can be used to fabricate delicate parts such as electrical switches and relays.
- Computer control of ultrasonic welders has eliminated cumbersome manual tuning. Also, automatic pressure adjustments during joining ensure quality regardless of variations in part height.
- Metal inserts such as interlooking threaded fasteners used to provide longer thread life can be ultrasonically driven into the plastic.

Limitations:

- The 20 kHz frequency that has generally been used can cause a great deal of mechanical vibration in the part, which is difficult to control. The tooling for 20 kHz is large and therefore limited to components with large weld areas.

vi) Vibration Welding:

In vibration welding, frictional heat is generated by clamping together two plastic parts under pressure and vibrating one part at a set amplitude and frequency. Common frequencies are 120 and 240 Hz, while vibration amplitudes can vary between 0.030 and 0.140 in.

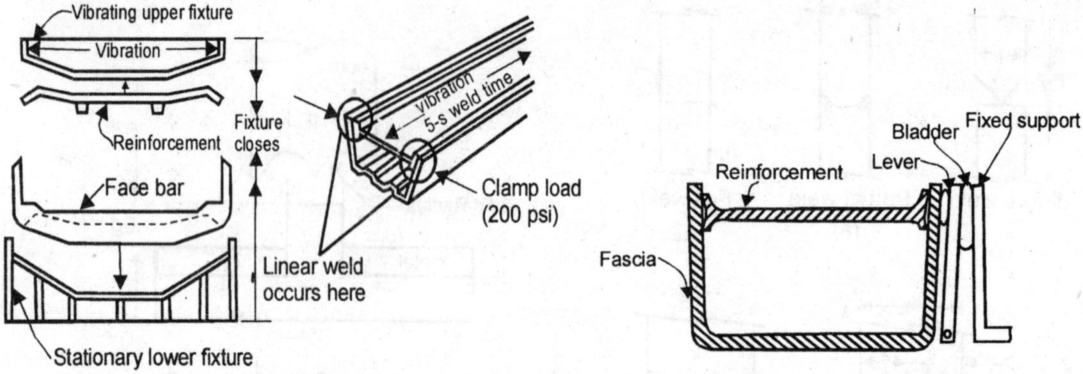

Figure 5.3 Vibrations welding of automotive bumpers.

The requirements for vibration welding of large parts such as automotive body panels and large structural load-bearing members are as follows:

i) The clearance between parts must be sufficient to allow movement of the halves.

ii) The fixtures must support the entire joint area.

iii) The parts must not flex during welding.

Figure 5.3 shows an example of how a reinforcing rib is attached to an automotive bumper by vibration welding. The upper vibrating structure supports and drives the bumper reinforcement. The lower fixture retains and applies clamping pressure to the faceplate. The faceplate nests in a urethane mold, and clamp pressure is applied to the side walls of the faceplate by a series of fingerlike levers that are pneumatically actuated.

5.5 THERMO PLASTICS AND THEIR APPLICATIONS

I) Polyethylene:

1) For making containers of chemicals
2) Insulation on wire and fabrics
3) Corrosion resistant coatings

II) Polystyrene:

1) Used to manufacturing toys, table ware, radio parts.
2) Packaging, chemical equipments, lenses etc.

III) Acrylics:

1) High transparency and good strength.
2) Dielectric property and colour ability.
3) Good wear resistance for making gears.

IV) Vinyl plasitics:

1) Rain coats, hoses and tubings. 2)Fuel tank, floor and wall covering 3) Abrasive resistant lining

V) Polyamide:

1) Used for making fibers, bearings, couplings, gears, combs etc.,

5.6 INJECTION MOULDING

In this reference, the oldest is the single stage plunger technique (Refer Figure 5.4(A)). When the plunger is drawn back, the charge material falls into a chamber from a hopper. Now the plunger drives the charge forward and force the heating cylinder where the charge becomes soft and fills the cavity of mould under pressure. The mould is a split die from where the product after solidification can be removed.

Later, this method was replaced by a single stage reciprocating screw method where the charge is prepared after throughly mixing (Figure 5.4 (B). It consists of a screw which turns and starts crushing the charge falling from a hopper. Charge fills into a heating chamber where mixing and crushing are being taken place by the screw. When sufficient material is prepared, the screw stops working and now it starts working as plunger which drives the charge forward into the die. The mould is of splitting type and so the product can be taken out easily.

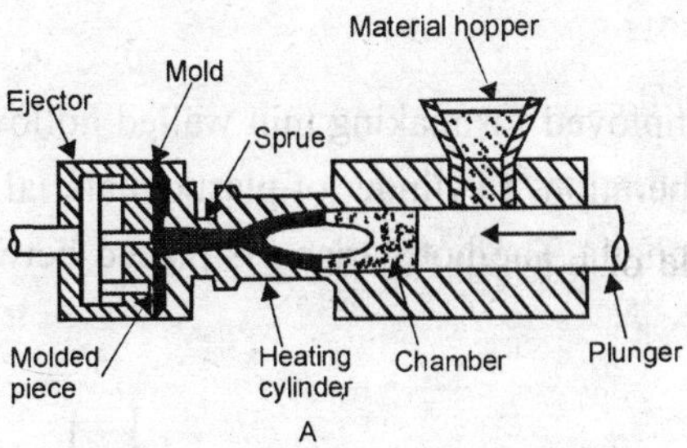

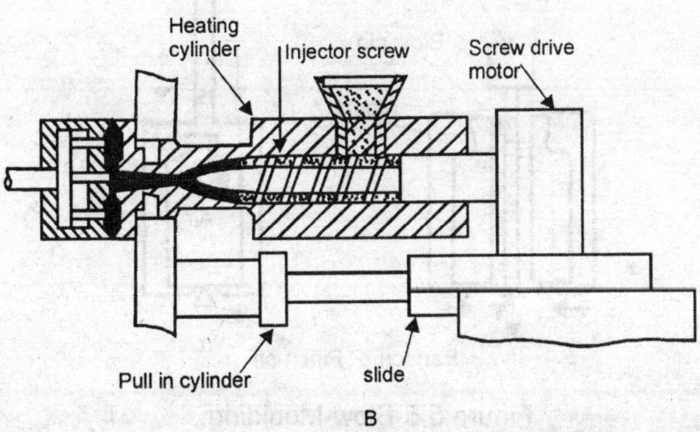

A - Conventional single stage plunger type, B- Single stage reciprocating screw type.

Figure 5.4 Injection moulding

Another method of injection moulding consists of two stages i.e., at the first stage the charge material is plasticised is one cylinder and a definite amount of charge is transferred by a screw plunger into a shot chamber, and in the second stage a simple plunger injects the softened charge into the die mould from shot chamber.

Basically, an injection moulding machine heats the material to soften state, injects the soft material ,into the die-mould which cools to harden a thermoplastic material. Operating temperatures vary between 150°c - 370°c and full pressure may vary between 35 - 350 MPa. The moulds are water cooled and the product solifies quickly and then ejected by knock out pins. It provides the highest rate of production thereby a low production cost. It can easily accomplish 300 to 400 shots per hour. (See figure 5.4(c))

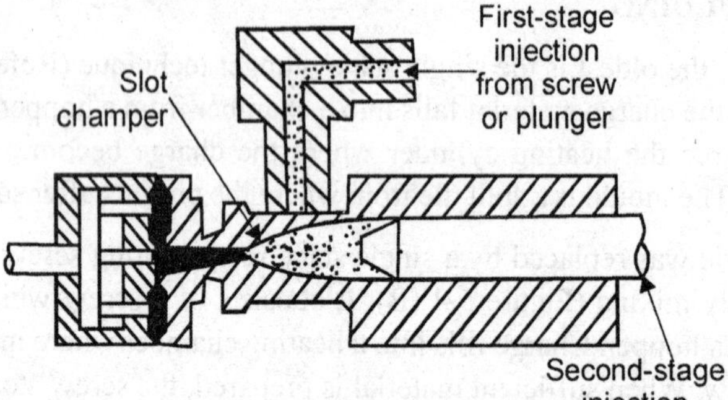

Figure 5.4 (c) Two stage plunger-or-screw-plasticiser type

5.7 BLOW MOULDING

This method is mostly employed for making thin walled hollow articles, especially bottles, from thermoplastic materials. In its operation, a cylinder of plastic material (known as parison) is extruded as rapidly as possible and cut to length. The hot parison is placed between the jaws of a split mould, as shown in figure 5.5.

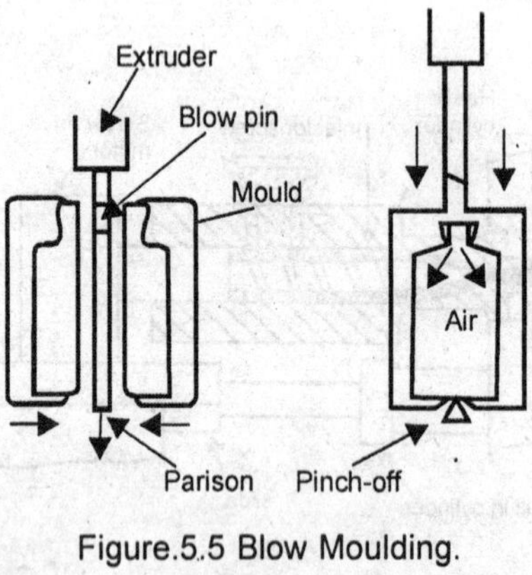

Figure.5.5 Blow Moulding.

When the mould is closed, it pinches off the parison and the product is completed by air pressure which forces the material against the mould surfaces. The mould should be adequately vented to eliminate poor surface finish. After the product is cooled sufficiently, the mould is opened and the product is removed.

5.8 ROTATIONAL MOULDING

Rotational moulding uses gravity inside a rotating mould to achieve a hollow form. Also called rotomoulding, it is an alternative to blow moulding for making large, hollow shapes. It is used principally for thermoplastic polymers, but applications for thermosets and elastomers are becoming more common. Rotomoulding tends to favor more complex external geometries, larger parts, and lower production quantitites than blow moulding. The process consists of the following steps: (1) A predetermined amount of polymer powder is loaded into the cavity of a split mould. (2) The mould is then heated and simultaneously rotated on two perpendicular axes, so that the powder impinges on all internal surfaces of the mould, gradually forming a fused layer of uniform thickness. (3) While still rotating, the mould is cooled so that the plastic skin solidifies. (4) The mould is opened, and the part is unloaded. Rotational speeds used in the process are relatively slow. It is gravity, not centrifugal force, that causes uniform coating of the mould surfaces.

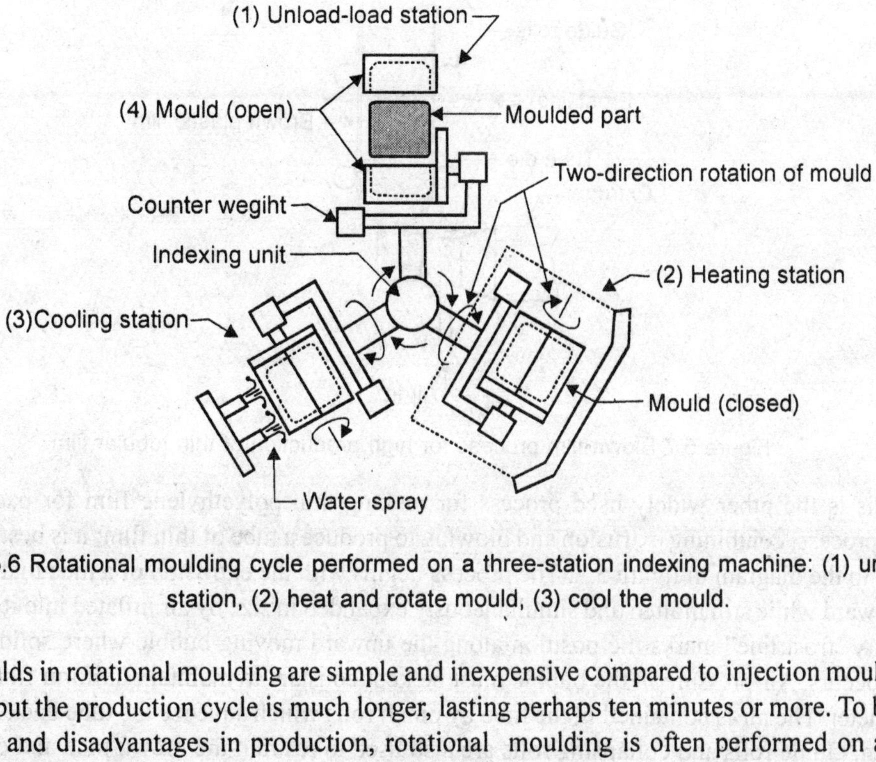

Figure 5.6 Rotational moulding cycle performed on a three-station indexing machine: (1) unload-load station, (2) heat and rotate mould, (3) cool the mould.

Moulds in rotational moulding are simple and inexpensive compared to injection moulding or blow moulding, but the production cycle is much longer, lasting perhaps ten minutes or more. To balance these advantages and disadvantages in production, rotational moulding is often performed on a multicavity indexing machine, such as the three-station machine shown in figure 5.6. The machine is designed so that three moulds are indexed in sequence through three workstations. Thus, all three moulds are working simultaneously. The first workstation is an unload-load station where the finished part is unloaded from the mould, and the powder for the next part is loaded into the cavity. The second station consists of a heating chamber where hot-air convection heats the mould while it is simultaneously rotated. Temperature inside the chamber are around 375°C (700°F), depending on the polymer and the item being moulded. The third station cools the mould, using forced cold air or water spray, to cool and solidify the plastic moulding inside.

A fascinating variety of articles are made by rotational moulding. The list includes hollow toys such as hobby horses and playing balls; boat and canoe hulls, sandboxes, small swimming pools; buoys and other flotation devices; truck body parts, automative dashboards, fuel tanks, luggage pieces, furniture, garbage cans, fashion mannequins, large industrial barrels, containers, and storage tanks, portable outhouses, and septic tanks. The most popular moulding material is polyethylene, especially HDPE. Other plastics include polypropylene, ABS, and high-impact polystyrene.

5.9 BLOWN-FILM EXTRUSION PROCESS

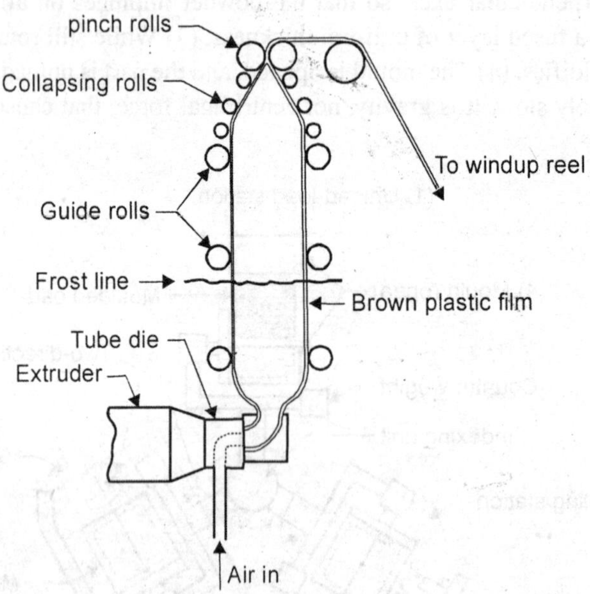

Figure 5.7 Blown-film process for high production of thin tubular film.

This is the other widely used process for making thin polyethylene film for packaging. It is a complex process, combining extrusion and blowing to produce a tube of thin film; it is best explained with reference to the diagram in figure 5.7. The process begins with the extrusion of a tube that is immediately drawn upward while still molten and simultaneously expanded in size by air inflated into it through the die mandrel. A "frost line" marks the position along the upward moving bubble where solidification of the polymer occurs. Air pressure in the bubble must be kept constant to maintain uniform film thickness and tube diameter. The air is contained in the tube by pinch rolls which squeeze the tube back together after it has cooled. Guide rolls and collapsing rolls are also used to restrain the blown tube and direct it into the pinch rolls. The flat tube is then collected onto a windup reel.

The effect of air inflation is to stretch the film in both directions as it cools from the molten state. This results in isotropic strength properties, which is an advantage over other processes in which the material is stretched primarily in one direction. Other advantages include the ease with which extrusion rate and air pressure can be changed to control stock width and gage. Comparing this process with slit-die extrusion, the blown-film method produces stronger film (so that a thinner film can be used to package a product), but thickness control and production rates are lower. The final blown film can be left in tubular form (e.g., for garbage bags), or it can be subsequently cut at the edges to provide two parallel thin films.

5.10 EXTRUSION MOULDING

The extrusion means the continuous flow of material through a die. The extrusion moulding is similar to injection moulding and is used for thermoplastic materials. This method is mostly used for making long tubes, rods, sheets, films, pipes, ropes, etc. In its operation, the thermoplastic material in the powered or granular form is fed into a hopper and forced by a spiral screw into the heating chamber, as shown in figure 5.8. The material becomes soft in the heating chamber and the rotating screw forces this material out through a die. The finished product coming out of the die, is cooled by blow air or spraying water and it is taken away by the belt conveyor.

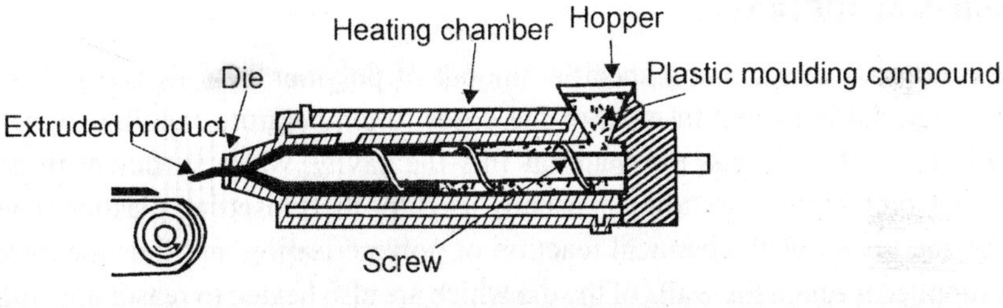

Figure 5.8. Plastic extrusion machine.

5.10.1 Extruded Plastics:

Extrusion is one of the most widely used plastic shaping processes. Several design recommendations are presented here for the conventional process.

- *Wall thickness:* Uniform wall thickness is desirable in an extruded cross section. Variations in wall thickness result in nonuniform plastic flow and uneven cooling that tend to wrap the extrudate.
- *Hollow sections:* Hollow sections complicate die design and plastic flow. It is desirable to use extruded cross section that are not hollow yet satisfy functional requirements.
- *Corners:* Sharp corners, inside and outside, should be avoided in the cross section, since they result in uneven flow during processing and stress concentrations in the final product.

5.11 TYPICAL INDUSTRIAL APPLICATIONS

1. The plastics are used for making fibers, bearings, couplings, gears and combs etc.
2. For producing transparent sheets, tubes and plates.
3. For making helmets and goggles.
4. For making rain coats and tubes.
5. For making industrial chemical containers.
6. For making insulation on wires and fabrics.
7. For making abrasive resistane lining.
8. For making fuel tanks.
9. For producing bullet proof safety glass.

5.12 JET MOULDING

The thermosetting plastics may be moulded by a process known as jet moulding. It is similar to injection moulding. The standard injection moulding machines used for thermoplastic material, may be converted for jet moulding, with few changes. The nozzle, which is the important part of the machine, must be both heated and cooled during the moulding cycle. In jet moulding, the resin is first slightly heated in the cylinder and then allowed to pass through the nozzle (which is preheated to a high temperature) to the mould. When the mould is full, the nozzle is rapidly cooled by water to prevent polymerisation of the remaining material.

5.13 COMPRESSION MOULDING

This process consists of placing a specific amount of polymer in a diecavity and squeezed by a punch. The plastic material is heated in most of the cases to temperature ranging form 125°C and this temperature softens the material, thus the material fills the cavity, when sufficient pressure is applied (3.5 MPa - 42 MPa). Compression moulding is usually used for thermosetting plastics. The mould is kept closed for enough time to complete chemical reaction or polymerisation and thus the material hardnens. Heat for curing is supplied through the walls of the die which are also heated to reasonably high temperature. Curing time can be reduced as much as 50% by preheating the charge.

There are three types of compression moulding which have been shown in the figure 5.9.

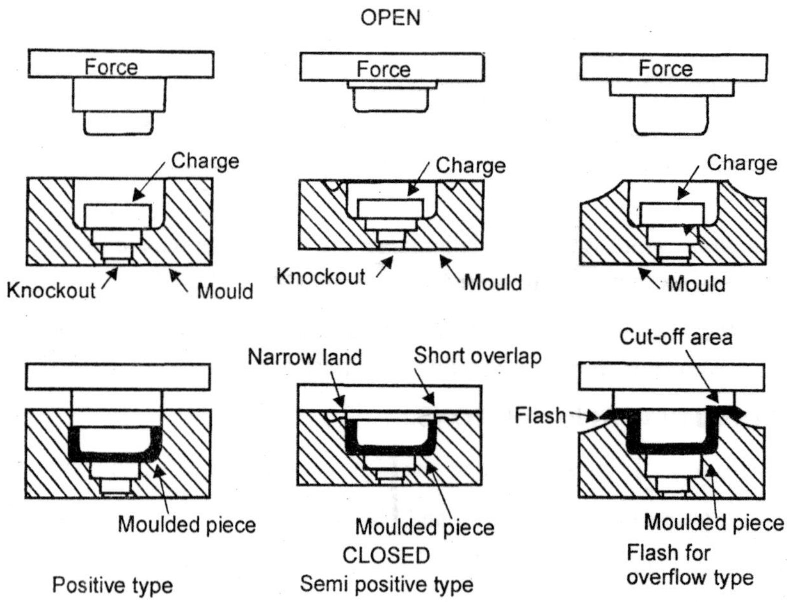

Figure 5.9. Three types of Compression moulds for plastics

i) Positive type:

Where full pressure is exerted to make the material fill out the mould, so the amount of charge material must be controlled closely to produce a part accurately.

ii) Semi-positive type:

Where full pressure is exerted at the final closing of the mould, but excess of material can escape. Obviously, close control on the charge is not needed. This process is usually used to produced the quality part in quantity production.

iii) Flash or Over-flow type:

Where flash cavity or area is provided in the die, so that the excessive material can flow out around the cavity in the form of thin flash. So it does not require any precise control over the charge. In this case, each piece of produce is to be trimmed. It is a cheaper process.

Cold moulding can also be carried out, however, in most of the cases, hot moulding is used. The charge material (refractory material mixed with binder) is compacted in a cold die and then baked in an oven until cured. This method of cold moulding is useful where close tolerance and good surface finish are not needed.

5.14 TRANSFER MOULDING

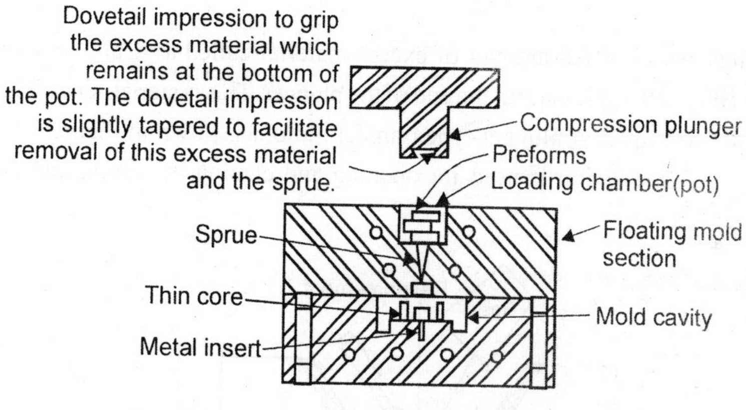

Figure 5.10 Sectional view through a sprue-type transfer mold.

This process is similar to compression molding, with the exception that in transfer molding, the molding material is loaded into a separate loading chamber. Under heat and pressure, the material becomes fluid and flows through a sprue from the loading chamber to all parts of the mold cavity or cavities. Additional heat then causes the thermosetting molding material to harden. Transfer molding is especially suitable for parts slender cores or metal inserts. If compression molded, these slender cores might be broken, and the inserts might be dislodged during the compressing of the solid material. Complex parts with thin sections also are made more easily by transfer molding. In comparison with compression molding, parts made by transfer molding ordinarily have better mechanical properties, more uniform densities, closer dimensional tolerances, and less flash at the parting line to clean off. In a sprue-type transfer mold, as shown in figure 5.10, the sprue will break near the molded part when the plunger returns upward at the end of the curing period.

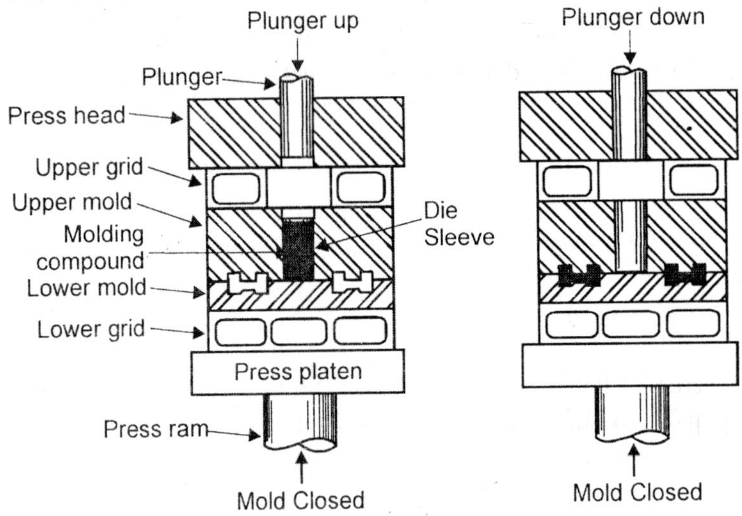

Figure 5.11. Sectional view through a positive-plunger-type transfer mold. (The Hydraulic Press Manufacturing Company).

The sprue, together with a small amount of excess material called a "Cull", may be removed by pushing it from a tapered dovetail impression at the end of the plunger. This material, since it is thermosetting, cannot be reclaimed and used again. A plunger-type transfer mold, as shown in Figure 5.11 must be used with a double-action press. One action is used for opening and closing the mold, and the other action operates the plunger.

5.15 CASTING

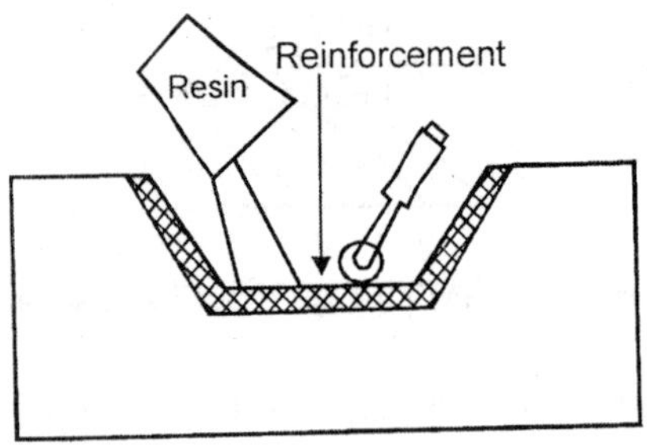

Figure 5.12 Hand Lay-up process

Casting, that is, pouring a liquid into a mould without pressure, can be done with several plastics. When there is not sufficient justification for making expensive dies, the casting process is generally used.

Open moulds into which the liquid resin, mixed with a catalyst or hardener, is poured can be used for most plastics.

Centrifugal castings, using the same methods discussed under the figure 1.26 can be used with several plastics to such as polyesters.

Shell moulding is somewhat similar to casting in that no pressure is used. In this process, powdered resin is placed in a heated mould until a layer fuses to the desired thickness. The unfused resin inside is taken out of the mould and the hollow moulding removed. Sometimes the mould is rotated in what is called rotational moulding.

Slush moulding is similar to shell moulding. It employs a thermoplastic resin slurry or "slush" which is poured into a preheated mould. The heat causes the slurry to set in a viscous layer of the desired wall thickness. Both shell and slush moulding produce thin-walled products such as toys and gloves.

Hand Lay-up is a simple method to form plastic materials into a required shape. Special shapes like boats, tanks and big-sized parts of plastics are produces ~~by~~ this method. First a mould of desirable shape is prepared. The mold is covered by fibre mats of desired thickness and plastic resin (such as epoxy) is draped over the mould and allowed to harden. Figure 5.12 shows the process.

5.16 LAMINATING

It is the process of bonding together a variety of materials by heat and pressure to form a single piece (known as laminates). The most laminated plastics consists of sheets of wood, asbestos, fabric and other similar materials. These materials are first impregnated or coated with thermosetting resin and then combined together under heat and pressure to form the commercial product. The laminated materials are hard, strong and impact resistant.

The following two processes are commonly used for producing laminated products.

i) High pressure process:

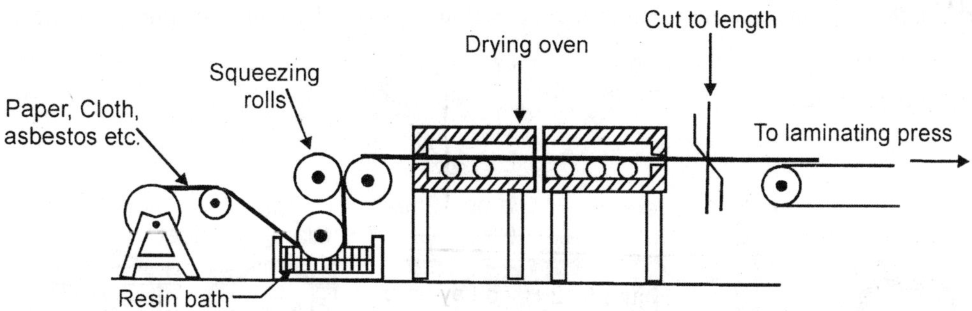

Figure. 5.13. Process for preparing sheets for lamination.

In this process, the resinoid material (e.g., phenol formaldehyde) is first dissolved by a solvent (e.g., alcohol) to convert it into a liquid varnish. The rolls of paper or fabric are passed through this liquid bath for impregnation, as shown in figure 5.13. It is then squeezed through a set of rollers and then dried in drying ovens at a temperature much lower (i.e., 150°C) than the curing temperature of the resin, to evaporate the solvent. In order to facilitate lamination, the sheets are cut into convenient sizes and piled one another (depending upon the desired thickness of the final sheet) between the polished metal plates at the top and bottom. These are then bonded together by a hydraulic press at about 180°C and under high pressure from 8 MPa to 14 MPa. Thus the laminated sheets are produced.

The tubes are made by rolling the impregnated material around a steel mandrel of a desired diameter and then placed in a tube mould for heating till they are cured.

The rods are produced by rolling and heating the impregnated material inside the cylindrical moulds, followed by grinding to size.

The paper base laminates are used in electrical products. The fabric base materials, which are stronger and harder, are better for stressed parts. The gears made of convass base are quiet in operation.

ii) Low pressure process:

It is also known as reinforced plastics moulding. In this process, a pressure upto 3 MPa is applied. The reinforcing materials used are glass fibres, asbestos, nylon, cotton, mat etc. The commonly used thermosetting resins are phenolics, furance, silicones, epoxies and polyesters. The pigments are added to obtain different colour shades. The fillers like clay (usually china clay), wood flour, silica sand, calcium carbonate and aluminium silicates are used to obtain better physical properties, better surface finish, reduced mould shrinkage and to lower cost.

Some of the processes used for low pressure laminates are as follows:

a) Contact Moulding:

It is the simplest of all the processes. In this process, the layers of reinforcing material are coated with catalysed resin and placed in layers on the prepared form. The number of layers depends upon the strength required in the finished product. In order to speed up the production of parts, both the resin and the reinforcing material are sprayed by means of guns, instead of being coated by hand.

5.17 THERMOFORMING

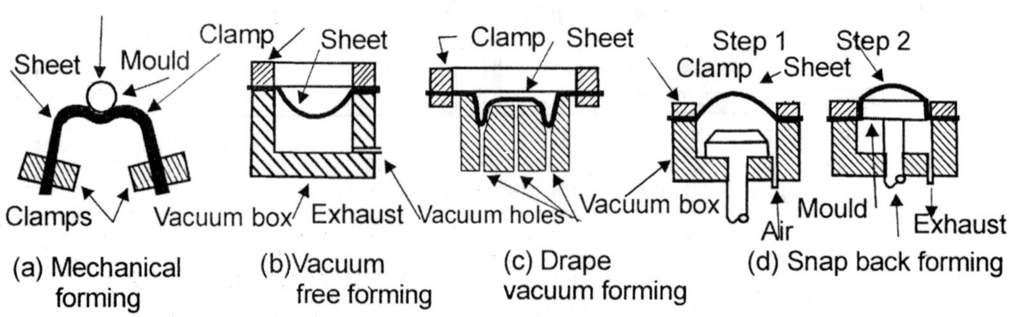

Figure 5.14. Thermoforming (methods of forming sheets)

Thermoforming sometimes called vacuum forming, is the shaping of hot sheets or strips of thermoplastic materials into a desired shape either by mechanical or pneumatic methods (Figure 5.14(a)). The sheets of plastic used in the thermoforming process are produced by either extrusion, calendering or pressing. Thickness of the sheets tat are processed range from 0.125 to 3.2 mm or even greater.

The basic process is quite simple. A piece of plastic sheet is clamed into a frame. The plastic is heated, usually with electric heaters, until it begins to sag. Vacuum, air or mechanical pressure is then applied through small holes in the mould and the plastic is rapidly pulled tightly against the mould creating close profile conformity. The frame is raised, the part is removed and then trimmed in a punch press. The mould may have several shapes, for the same part or for different parts.

In vacuum forming (Figure 5.14(b)), the sheet is clamped in a stationary frame and the heated sheet is vacuum-drawn into the mould.

A male mould is used in drape vacuum forming and a moveable frame or clamp drapes a heat-softened sheet over the male mould before vacuum is pulled (Figure 5.14(c)). A more complex shape can be developed by this method.

In the vacuum snap back process (Figure 5.14(d)), in the first step, vacuum is produced in the chamber and the sheet is drawn down into the cavity. The male plug descends and clamps against the chase which holds the sheet. In the second step, vacuum is applied to male plug and the material is drawn tot he mould. Deep drawn parts with generous tapers and corners with large radii are possible with this technique.

Among the items made by thermoforming process are small jelly containers used in restaurants, luggage, refrigerator inner panels lighting fixture, helmets and containers for packing.

5.17.1 Applications of Thermoforming:

Thermoforming is a secondary shaping process, the primary process being that which produces the sheet or film. Only thermoplastics can be thermoformed, since extruded sheets of thermosetting or elastomeric polymers have already been cross-linked and cannot be softened by reheating. Common thermoforming plastics are polystyrene, cellulose acetate and cellulose acetate butyrate, ABS, PVC, acrylic (polymethylmethacrylate), polyethene, and polypropylene.

Mass production thermoforming operations are performed in the packaging industry. The starting sheet or film is rapidly fed through a heating chamber and then mechanically formed into the desired shape. The operations are often designed to produce multiple parts with each stroke of the press using moulds with multiple punches and cavities. In some cases, the extrusion machine that produces the sheet or film is located directly upstream from the thermoforming process, thereby eliminating the need to reheat the plastic. And for best efficiency, the filling process to put the consumable food item into the container is placed immediately downstream from thermoforming.

Thin film packaging items that are mass produced by thermoforming include blister packs and skin packs. They offer an attractive way to display certain commodity products such as cosmetics, toiletries, small tools, and fasteners (nails, screws, etc.). Thermoforming applications include large parts that can be produced from thicker sheet stock. Examples include covers for business machines, boat hulls, shower stalls, diffusers for lights, advertising displays and signs, bathrubs, and certain toys. We had previously mentioned contoured skylights and internal door liners for refrigerators. These would be made, respectively, out of acrylic (because of its transparency) and ABS (because of its ease in forming and resistance to oils and fats found in refrigerators.

5.18 CALENDERING

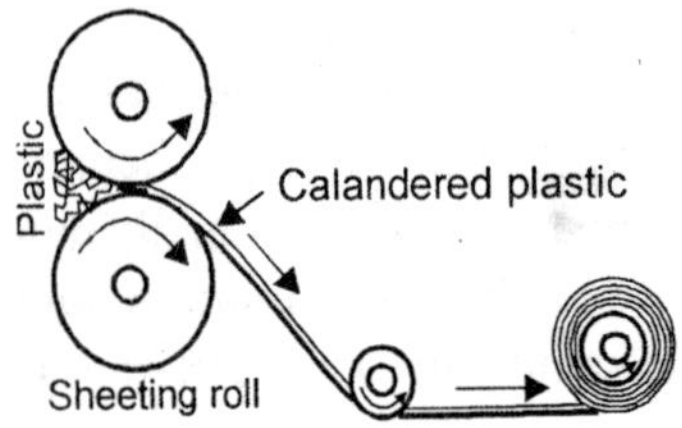

Figure 5.15 Forming sheet by calendering.

An important method of making film and sheet is known as calendering. In this process the plastic compound (composed of resin, filler, plasticizer and colour pigment) is passed between a series of heated rollers as illustrated in figure 5.15. It comes out from the rolls squeezed into film or sheet. Thickness is controlled by a combination of squeezing and altering the speed of the finishing rolls. The finished product is cooled by passing through water-cooled rolls. Vinyl floor tile, cellulose acetate sheeting and films are some of the applications.

SUMMARY

- ***The properties of plastics:*** depend on the atoms it contain and the way in which these atoms are arranged in molecules. These organic compounds are also known as polymers.

- ***One major class of plastics:*** called the *Thermoplastic* materials, composed of separate long molecules arranged side by side (synthesised in the shape of long threads). Thermoplastics become soft when heated and re-harden on cooling because the molecules are held together by secondary bond. The strength of these forces between separate particles of matter depend directly upon the masses and inversely upon a power of distances between the molecules.

- ***Thermo Plastics:*** heated even at 40°c become soft and so can be remoulded without any change in structure. They can be ground up and remoulded, i.e. they can be re-shaped while in soft state. Usually, they become soft at about 65°c to 315°c and can flow to an appreciable extent underload. And that is why most of the plastics used today are thermoplastics.

- ***Polyethylene:***

  Properties: Both flexible and rigid types are lightly break resistant, unaffected by food acids, and household solvents, resist moisture, flexible over a wide range of temperatures.

  Uses: Widely used as a coating for paper and paperboard, film packaging, disposable bags,furniture components, etc.

❖ ***Polyropylene:***

Properties: Good dimensional stability. Resistance to heat, chemicals, moisture, electricity, easy to process. Good impact resistance.

Uses: Appliance housings, refrigerator door liners, wall tiles, battery cases, toys insulators, ornaments, etc.

❖ ***Polystyrene:***

Properties: Rigid, clear, transparent materials with good tensile strength, good combination of rigidity and toughness; good mouldability.

Uses: Wide applications in packaging, appliances; housewares; furniture; dentalplates; auto parts.

❖ ***Polyvinyl Chloride (P.V.C.):***

Properties: Chemical and abraison resistance, good weatherability, easily prigmented; good resistance to petroleum-based-solvents, acids, etc.

Uses: Sound records, blown bottles, rigid sheeting, upholstery, hose, flooring, rain-wear, insulation, etc.

❖ ***Acrylic:***

Properties: Outstanding optical properties; inert to most chemicals; excellent mechanical properties for short term loading.

*Uses:*Aircraft and architectural domes,signs, machinery housings, optical components, goggles lenses, etc.

❖ ***Flouracarbon (Teflon):***

Properties: Low coefficient of friction, stable upto about 205°c; good chemical resistance and weatherability, excellent electrical insulation.

Uses: Non-lubricated bearings, chemical resistant pipe linings, seals, diaphram, valves, linings of cookware.

❖ ***Nylon::***

Properties: High tensile strength, good impact strength and high abrasion resistance, good chemical and electrical propertises.

*Uses:*Gears, bushings, cans, tubings, rollers bearings, combs, and miscellaneous household items.

❖ ***Thermosetting Plastics:*** are made from chains which have been linked together known as cross-linked (three dimensional network of molecules). Usually they are hard and brittle and cannot be used after working once. The thermosetting materials undergo a chemical change when moulded and cannot be resoftened by heating to reshape them. The mould temperatures are usually considerable higher than those for the thermoplastics.

- ***Elastomers:*** are also polymers which are lightly bound together and so exhibit the unique property of high elasticity, stretching five to ten times when tensile loading is applied and coming back to its original length or dimensions when the applied load is removed. These are non crystalline polymers at room temperature having long chain molecules in three dimensional network.

- ***Thermosetting Elastomers:*** are special type of thermoset polymers having long chain molecular structure. In their green state they are tracky (shabby) and cold flow easily. To be useful they must be cross-linked. In case of rubbers, this cross-linking is known as Vulcanisation.

- ***Thermoplastic Elastomers:*** are similar to the thermoset type in service properties, but these offer greater ease and speed of processing. Moulding or extrusion can be done on standard equipment within short cycle. The major thermoplastic elastomers are polyster-copolymers, polyurethane, etc.

- ***Stabilisers:*** are added to initiate the reaction i.e., to allow polymerisation to begin and then stabilise the reaction of molecular chains. Hydrogen peroxide is a good stabiliser.

- ***Solid or liquid plasticizer:*** is to make the product more flexible and less brittle. These substances work as internal lubricants for improving the flowability inside the mould. Lubricants also help to prevent them from sticking together.

- ***Fusion Bonding:*** in conventional fusion bonding, parts are placed in fixtures and held by chucks, gripping fingers, vacuum cups, or other mechanical devices. The fixtures then press the parts against a heating platen that is usually made of aluminum, aluminum-copper, or hot-rolled steel coated with Teflon or Teflon-impregnated fiber glass. The platen is heated to 500°F to soften the plastic to a predetermined depth. Holding fixtures are then opened and the heating platen removed, the fixtures are closed again until the melted material is cooled and fused.

- ***Solvent Welding:*** is a widely used plastics-joining process. The solvent may be applied to the joint by dipping, spraying, and brushing. The most common method is to hold the solvent-soaked pad against the surfaces. The edges dissolve, producing tacky surfaces that are then brought together, moved slightly to obtain mixing, and held in contact, in correct relationship, to dry and harden. Drying time ranges from a few minutes to hours.

- ***Friction Welding:*** spin welding is limited to round parts that can be rotated while in contact to generate heat. The frictional heat melts the interfaces, and when the motion is stopped, the parts fuse together. Spinning speeds of 10 to 15 fps at the interface and pressures of 20 to 200 psi (138 to 1378 kPa) must be matched to the material to produce just the right amount of heat. Melting and fusion can occur in as little as 0.25 s. Maintaining pressure for 1 to 2 s after braking normally completes the cycle.

- ***Induction Welding:*** Originally thermoplastics were induction welded by pressing two parts together around a metal insert. This combination was then subjected to an alternating magnetic field, which melted the surfaces of the plastics around the insert. Compression of the parts resulted in a fusion weld.

- ***Ultrasonic Welding:*** requires the conversion of low-frequency electical power to high-frequency mechanical vibrations that are applied to thermoplastic components. When the plastic melts, the components are joined in a fusion bond.

- ***Limitations:*** The 20 kHz frequency that has generally been used can cause a great deal of mechanical vibration in the part, which is difficult to control. The tooling for 20 kHz is large and therefore limited to components with large weld areas.

- ***Vibration Welding:*** frictional heat is generated by clamping together two plastic parts under pressure and vibrating one part at a set amplitude and frequency. Common frequencies are 120 and 240 Hz, while vibration amplitudes can vary between 0.030 and 0.140 in.

- ***Injection moulding:*** is the oldest is the single stage plunger technique (Refer Figure 5.4(A)). When the plunger is drawn back, the charge material falls into a chamber from a hopper. Now the plunger drives the charge forward and force the heating cylinder where the charge becomes soft and fills the cavity of mould under pressure. The mould is a split die from where the product after solidification can be removed.

- ***Blow moulding:***is mostly employed for making thin walled hollow articles, especially bottles, from thermoplastic materials. In its operation, a cylinder of plastic material (known as parison) is extruded as rapidly as possible and cut to length. The hot parison is placed between the jaws of a split mould.

- ***Rotational moulding:*** uses gravity inside a rotating mould to achieve a hollow form. Also called rotomoulding, it is an alternative to blow moulding for making large, hollow shapes. It is used principally for thermoplastic polymers, but applications for thermosets and elastomers are becoming more common. Rotomoulding tends to favor more complex external geometries, larger parts, and lower production quantitites than blow moulding.

- ***Blown-film extrusion process:*** is the other widely used process for making thin polyethylene film for packaging. It is a complex process, combining extrusion and blowing to produce a tube of thin film.

- ***Extrusion moulding:*** means the continuous flow of material through a die. The extrusion moulding is similar to injection moulding and is used for thermoplastic materials. This method is mostly used for making long tubes, rods, sheets, films, pipes, ropes, etc. In its operation, the thermoplastic material in the powered or granular form is fed into a hopper and forced by a spiral screw into the heating chamber. The material becomes soft in the heating chamber and the rotating screw forces this material out through a die. The finished product coming out of the die, is cooled by blow air or spraying water and it is taken away by the belt conveyor.

- ***Jet moulding:*** is thermosetting plastics may be moulded by a process known as jet moulding. It is similar to injection moulding. The standard injection moulding machines used for thermoplastic material, may be converted for jet moulding, with few changes. The nozzle, which is the important part of the machine, must be both heated and cooled during the moulding cycle. In jet moulding, the resin is first slightly heated in the cylinder and then allowed to pass through the nozzle (which is preheated to a high temperature) to the mould. When the mould is full, the nozzle is rapidly cooled by water to prevent polymerisation of the remaining material.

- ***Compression Moulding:*** process consists of placing a specific amount of polymer in a diecavity and squeezed by a punch. The plastic material is heated in most of the cases to temperature ranging form 125°C and this temperature softens the material, thus the material fills the cavity, when sufficient pressure is applied (3.5 MPa - 42 MPa). Compression moulding is usually used for thermosetting plastics. The mould is kept closed for enough time to complete chemical reaction or polymerisation and thus the material hardnens. Heat for curing is supplied through the walls of the die which are also heated to reasonably high temperature. Curing time can be reduced as much as 50% by preheating the charge.

- ***Positive type:*** is full pressure is exerted to make the material fill out the mould, so the amount of charge material must be controlled closely to produce a part accurately.

- ***Semi-positive type:*** is full pressure is exerted at the final closing of the mould, but excess of material can escape. Obviously, close control on the charge is not needed. This process is usually used to produced the quality part in quantity production.

- ***Flash or Over-flow type:*** is flash cavity or area is provided in the die, so that the excessive material can flow out around the cavity in the form of thin flash. So it does not require any precise control over the charge. In this case, each piece of produce is to be trimmed. It is a cheaper process.

- ***Transfer Moulding:*** process is similar to compression molding, with the exception that in transfer molding, the molding material is loaded into a separate loading chamber. Under heat and pressure, the material becomes fluid and flows through a sprue from the loading chamber to all parts of the mold cavity or cavities. Additional heat then causes the thermosetting molding material to harden. Transfer molding is especially suitable for parts slender cores or metal inserts. If compression molded, these slender cores might be broken, and the inserts might be dislodged during the compressing of the solid material.

- ***Casting:*** is pouring a liquid into a mould without pressure, can be done with several plastics. When there is not sufficient justification for making expensive dies, the casting process is generally used. Open moulds into which the liquid resin, mixed with a catalyst or hardener, is poured can be used for most plastics.

- ***Shell moulding:*** is somewhat similar to casting in that no pressure is used. In this process, powdered resin is placed in a heated mould until a layer fuses to the desired thickness. The unfused resin inside is taken out of the mould and the hollow moulding removed. Sometimes the mould is rotated in what is called rotational moulding.

- ***Laminating:*** is the process of bonding together a variety of materials by heat and pressure to form a single piece (known as laminates).

- ***The most laminated plastics:*** consists of sheets of wood, asbestos, fabric and other similar materials. These materials are first impregnated or coated with thermosetting resin and then combined together under heat and pressure to form the commercial product. The laminated materials are hard, strong and impact resistant.

- *High pressure process:* is the resinoid material (e.g., phenol formaldehyde) is first dissolved by a solvent (e.g., alcohol) to convert it into a liquid varnish. The rolls of paper or fabric are passed through this liquid bath for impregnation.

- *Low pressure process:* is also known as reinforced plastics moulding. In this process, a pressure upto 3 MPa is applied. The reinforcing materials used are glass fibres, asbestos, nylon, cotton, mat etc. The commonly used thermosetting resins are phenolics, furance, silicones, epoxies and polyesters.

- *Contact Moulding:* is the simplest of all the processes. In this process, the layers of reinforcing material are coated with catalysed resin and placed in layers on the prepared form. The number of layers depends upon the strength required in the finished product. In order to speed up the production of parts, both the resin and the reinforcing material are sprayed by means of guns, instead of being coated by hand.

- *Thermoforming:* sometimes called vacuum forming, is the shaping of hot sheets or strips of thermoplastic materials into a desired shape either by mechanical or pneumatic methods. The sheets of plastic used in the thermoforming process are produced by either extrusion, calendering or pressing. Thickness of the sheets tat are processed range from 0.125 to 3.2 mm or even greater.

- *Calendering:* is an important method of making film and sheet is known as calendering. In this process the plastic compound (composed of resin, filler, plasticizer and colour pigment) is passed between a series of heated rollers. It comes out from the rolls squeezed into film or sheet. Thickness is controlled by a combination of squeezing and altering the speed of the finishing rolls. The finished product is cooled by passing through water-cooled rolls. Vinyl floor tile, cellulose acetate sheeting and films are some of the applications.

REVIEWS

1. What are some of the reasons why the plastic shaping processes are important?
2. Identify the main categories of plastics shaping processes, as classified by the resulting product geometry.
3. Viscosity is an important property of a polymer melt in plastics shaping processes. Upon what two parameters does the viscosity of a given polymer depend?
4. Briefly describe the plastic extrusion process.
5. Why are the fabrication processes applied to plastics?
6. How does the fabrication of a thermoplastic polymer differ from the processing of a thermoplastic polymer?
7. What are some of the ways that plastic sheet, plate, and tubing can be cast?
8. Why do cast plastic resins typically have a lustrous appearance?
9. What types of polymers are most commonly blow moulded?

10. Why do blow moulding moulds typically contain a cooling system?
11. For what types of parts and production volumes would compression moulding be an appropriate process?
12. What are typical mould temperatures for compression moulding? What is the most common mould material?
13. What are some of the attractive features of the transfer moulding process?
14. In what ways in injection moulding of plastic similar to the die casting of metal?
15. Why is the cycle time for the injection moulding of thermosetting polymers significantly longer than that for the thermoplastics?
16. How are the individual components mixed in the reaction injection moulding process?
17. What are some of the typical production shapes that are produced by the extrusion of plastics?
18. What types of products are produced by rotational moulding?
19. What property of plastics is responsible for making snap-fit assembly a popular alternative for plastic products?
20. What are some of the attractive properties of plastics that favor their selection? What are some of the common limitations?
21. Why should adequate fillets be included between adjacent seactions of a mould? What is a major benefit of rounding exterior corners?
22. Why might threaded inserts be preferred over other means of producing threaded holes in a plastic components?
23. What are some of the ways in which metal inserts are held in place in a plastic part?
24. What types of products can be produced from elastomeric materials using the dipping process?
25. Describe the differences between the injection moulding of plastics and the injection moulding.
26. What are some of the forms in which reinforcement fibers can be used in composite materials?
27. What is a prepreg?
28. What are some of the various moulding processes that can be used to shape products from laminated sheets of woven fibers?
29. What type of reinforcing fibers can be incorporated in the spray moulding process? Injection moulding?
30. What is the major benefit of three-dimensional fiber reinforcement?

INDEX